Dynamic Programming

Richard Bellman

DOVER PUBLICATIONS, INC.
Mineola, New York

To Betty-Jo
whose decision processes defy analysis

Copyright

Bibliographical Note

This Dover edition, first published in 2003, is an unabridged reprint of the sixth (1972) printing of the work first published by Princeton University Press, Princeton, New Jersey, in 1957. A new Introduction has been written specially for this edition.

Library of Congress Cataloging-in-Publication Data

Bellman, Richard Ernest, 1920-
Dynamic programming / Richard Bellman.
p. cm.
Originally published: Princeton, N.J. : Princeton, University Press, 1957.
Includes bibliographical references and indexes.
ISBN 0-486-42809-5 (pbk.)
1. Dynamic programming. I. Title.

QA402.5 .B438 2003
519.7'03—dc21

2002072879

Manufactured in the United States of America
Dover Publications, Inc., 31 East 2nd Street, Mineola, N.Y. 11501

Introduction to the Dover Edition

IN THIS STUNNING BOOK, first published in 1957, Richard E. Bellman described, with grace and clarity, a potent method for solving optimization problems that entail sequences of decisions, and he applied this method to a wide range of subjects. The language he introduced to describe this method—*functional equation, principle of optimality, multistage decision problem, states, embedding, dynamic programming*—has now been adopted by computer science, control, economics, engineering, finance, operations research, mathematics, and statistics. This is surely one of the most influential mathematics books of the postwar era.

Typically, a good mathematics book—and this is a very good mathematics book—surveys a field whose fundamentals are deeply understood. By contrast, in this book Bellman described a research program that he was avidly pursuing. By writing it, he generously invited other researchers to jump in. Many did. It is no overstatement to say that *Dynamic Programming* launched a thousand professorships.

In the decades that followed its initial publication, the mathematics of dynamic programming has matured, its fields of application have evolved, and the subject has become better understood. A goodly number of introductory accounts of dynamic programming now exist, and beginning students may be better served by one of them.

Other than for its place in history, is there good reason to read this book? Yes! Much of it remains as fresh and vital to those of us who are familiar with the subject as Mozart's music is to those of us who listen to it repeatedly. Bellman's Preface is the best account I've found of the scope of dynamic programming. The analysis within each chapter is a refreshingly honest and lucid foray into its central issues. The problem set at the end of each chapter is a rich lode of applications and research topics. In concert, they remind us of the power of the methods, of what has been accomplished, and of what remains to do.

Bellman chose his language extraordinary well, and yet *functional equation* is less descriptive than *optimality equation*. He might agree that we should switch to the latter. In retrospect, Bellman would also cite Rufus Isaacs for his "tenet of transition" (Isaacs, "Games of Pursuit," P-257, The Rand Corporation, Santa Monica, 1951), which is quite like Bellman's principle of optimality.

ERIC V. DENARDO
Professor of Operations Research
Yale University

Preface

The purpose of this work is to provide an introduction to the mathematical theory of multi-stage decision processes. Since these constitute a somewhat formidable set of terms we have coined the term "dynamic programming" to describe the subject matter. Actually, as we shall see, the distinction involves more than nomenclature. Rather, it involves a certain conceptual framework which furnishes us a new and versatile mathematical tool for the treatment of many novel and interesting problems both in this new discipline and in various parts of classical analysis. Before expanding upon this theme, let us present a brief discussion of what is meant by a multi-stage decision process.

Let us suppose that we have a physical system S whose state at any time t is specified by a vector p. If we are in an optimistic frame of mind we can visualize the components of p to be quite definite quantities such as Cartesian coordinates, or position and momentum coordinates, or perhaps volume and temperature, or if we are considering an economic system, supply and demand, or stockpiles and production capacities. If our mood is pessimistic, the components of p may be supposed to be probability distributions for such quantities as position and momentum, or perhaps moments of a distribution.

In the course of time, this system is subject to changes of either deterministic or stochastic origin which, mathematically speaking, means that the variables describing the system undergo transformations. Assume now that in distinction to the above we have a process in which we have a choice of the transformations which may be applied to the system at any time. A process of this type we call a *decision process*, with a decision equivalent to a transformation. If we have to make a single decision, we call the process a single-stage process; if a sequence of decisions, than we use the term *multi-stage decision process*.

The distinction, of course, is not hard and fast. The choice of a point in three-dimensional space may be considered to be a single-stage process wherein we choose (x, y, z), or a multi-stage process where we choose first x, then y, and then z.

There are a number of multi-stage processes which are quite familiar to us. Perhaps the most common are those occurring in card games, such as the bidding system in contract bridge, or the raise-counter-raise system of poker with its delicate overtones of bluffing. On a larger scale, we continually in our economic life engage in multi-stage decision processes in connection with investment programs and insurance policies. In the scientific world, control processes and the design of experiments furnish other examples.

The point we wish to make is that in modern life, in economic, industrial, scientific and even political spheres, we are continually surrounded by multi-stage decision processes. Some of these we treat on the basis of experience, some we resolve by rule-of-thumb, and some are too complex for anything but an educated guess and a prayer.

Unfortunately for the peace of mind of the economist, industrialist, and engineer, the problems that have arisen in recent years in the economic, industrial, and engineering fields are too vast in portent and extent to be treated in the haphazard fashion that was permissible in a more leisurely bygone era. The price of tremendous expansion has become extreme precision.

These problems, although arising in a multitude of diverse fields, share a common property—they are exceedingly difficult. Whether they arise in the study of optimal inventory or stock control, or in an input-output analysis of a complex of interdependent industries, in the scheduling of patients through a medical clinic or the servicing of aircraft at an airfield, the study of logistics or investment policies, in the control of servo-mechanisms, or in sequential testing, they possess certain common thorny features which stretch the confines of conventional mathematical theory.

It follows that new methods must be devised to meet the challenge of these new problems, and to a mathematician nothing could be more pleasant. It is a characteristic of this species that its members are never so happy as when confronted by problems which cannot be solved—immediately. Although the day is long past when anyone seriously worried about the well of mathematical invention running dry, it is still nonetheless a source of great delight to see a vast untamed jungle of difficult and significant problems, such as those furnished by the theory of multi-stage decision processes, suddenly appear before us.

Having conjured up this preserve of problems, let us see what compass we shall use to chart our path into this new domain. The conventional approach we may label "enumerative." Each decision may be thought of as a choice of a certain number of variables which determine the transformation to be employed; each sequence of choices, or *policy* as we shall say, is a choice of a larger set of variables. By lumping all these choices together, we "reduce" the problem to a classical problem of determining the maximum of a given function. This function, which arises in the course of measuring some quantitative property of the system, serves the purpose of evaluating policies.

At this point it is very easy for the mathematician to lose interest and let the computing machine take over. To maximize a reasonably well-behaved function seems a simple enough task; we take partial derivatives and solve the resulting system of equations for the maximizing point.

There are, however, some details to consider. In the first place, the effective analytic solution of a large number of even simple equations as, for example, linear equations, is a difficult affair. Lowering our sights, even a computational solution usually has a number of difficulties of both gross and subtle nature. Consequently, the determination of this maximum is quite definitely not routine when the number of variables is large.

All this may be subsumed under the heading "the curse of dimensionality." Since this is a curse which has hung over the head of the physicist and astronomer for many a year, there is no need to feel discouraged about the possibility of obtaining significant results despite it.

However, this is not the sole difficulty. A further characteristic of these problems, as we shall see in the ensuing pages, is that calculus is not always sufficient for our purposes, as a consequence of the perverse fact that quite frequently the solution is a boundary point of the region of variation. This is a manifestation of the fact that many decision processes embody certain all-or-nothing characteristics. Very often then, we are reduced to determining the maximum of a function by a combination of analytic and "hunt and search" techniques.

Whatever the difficulties arising in the deterministic case which we have tacitly been assuming above, these difficulties are compounded in the stochastic case, where the outcome of a decision, or tranformation, is a random variable. Here any crude lumping or enumerative technique is surely doomed by the extraordinary manner in which the number of combinations of cases increases with the number of cases.

Assume, however, that we have circumvented all these difficulties and have attained a certain computational nirvana. Withal, the mathematician has not discharged his responsibilities. *The problem is not to be considered solved in the mathematical sense until the structure of the optimal policy is understood.*

Interestingly enough, this concept of the mathematical solution is identical with the proper concept of a solution in the physical, economic, or engineering sense. In order to make this point clear—and it is a most important point since in many ways it is the raison d'être for mathematical physics, mathematical economics, and many similar hybrid fields—let us make a brief excursion into the philosophy of mathematical models.

The goal of the scientist is to comprehend the phenomena of the universe he observes around him. To prove that he understands, he must be able to predict, and to predict, one requires quantitative measurements. A qualitative prediction such as the occurrence of an eclipse or an earthquake or a depression sometime in the near future does not have the same satisfying features as a similar prediction associated with a date and time, and perhaps backed up by the offer of a side wager.

To predict quantitatively one must have a mechanism for producing numbers, and this necessarily entails a mathematical model. It seems reasonable to suppose that the more realistic this mathematical model, the more accurate the prediction.

There is, however, a point of diminishing returns. The actual world is extremely complicated, and as a matter of fact the more that one studies it the more one is filled with wonder that we have even "order of magnitude" explanations of the complicated phenomena that occur, much less fairly consistent "laws of nature." If we attempt to include too many features of reality in our mathematical model, we find ourselves engulfed by complicated equations containing unknown parameters and unknown functions. The determination of these functions leads to even more complicated equations with even more unknown parameters and functions, and so on. Truly a tale that knows no end.

If, on the other hand, made timid by these prospects, we construct our model in too simple a fashion, we soon find that it does not predict to suit our tastes.

It follows that the Scientist, like the Pilgrim, must wend a straight and narrow path between the Pitfalls of Oversimplification and the Morass of Overcomplication.

Knowing that no mathematical model can yield a complete description of reality, we must resign ourselves to the task of using a succession of models of greater and greater complexity in our efforts to understand. If we observe similar structural features possessed by the solutions of a sequence of models, then we may feel that we have an approximation to what is called a "law of nature."

It follows that from a teleological point of view the particular numerical solution of any particular set of equations is of far less importance than the understanding of the nature of the solution, which is to say the influence of the physical properties of the system upon the form of the solution.

Now let us see how this idea guides us to a new formulation of these decision processes, and indeed of some other processes of analysis which are not usually conceived of as decision processes. In the conventional formulation, we consider the entire multi-stage decision process as essentially one stage, at the expense of vastly increasing the dimension of the problem. Thus, if we have an N-stage process where M decisions are to be made at each stage, the classical approach envisages an MN-dimensional single-stage process. The fundamental problem that confronts us is: How can we avoid this multiplication of dimension which stifles analysis and greatly impedes computation?

In order to answer this, let us turn to the previously enunciated principle that it is the *structure* of the policy which is essential. What does this mean precisely? It means that we wish to know the characteristics of the system which determine the decision to be made at any particular stage of the process. Put another way, in place of determining the optimal sequence of decisions from some *fixed* state of the system, we wish to determine the optimal decision to be made at *any* state of the system. Only if we know the latter, do we understand the intrinsic structure of the solution.

The mathematical advantage of this formulation lies first of all in the fact that it reduces the dimension of the process to its proper level, namely the dimension of the decision which confronts one at any particular stage. This makes the problem analytically more tractable and computationally vastly simpler. Secondly, as we shall see, it furnishes us with a type of approximation which has a unique mathematical property, that of monotonicity of convergence, and is well suited to applications, namely, "approximation in policy space".

The conceptual advantage of thinking in terms of policies is very great. It affords us a means of thinking about and treating problems which cannot be profitably discussed in any other terms. If we were to hazard a guess as to which direction of research would achieve the greatest success in the future of multi-dimensional processes, we would unhesitatingly choose this one.

The theme of this volume will be the application of this concept of a solution to a number of processes of varied type which we shall discuss below.

The title is also derived in this way. The problems we treat are programming problems, to use a terminology now popular. The adjective "dynamic," however, indicates that we are interested in processes in which time plays a significant role, and in which the order of operations may be crucial. However, an essential feature of our approach will be the reinterpretation of many static processes as dynamic processes in which time can be artificially introduced.

Let us now turn to a discussion of the contents.

In the first chapter we consider a multi-stage allocation process of deterministic type which is a prototype of a general class of problems encountered in various phases of logistics, in multi-stage investment processes, in the study of optimal purchasing policies, and in the treatment of many other economic processes. From the mathematical point of view, the problem is related to multi-dimensional maximization problems, and ultimately, as will be indicated below, to the calculus of variations.

We shall first discuss the process in the conventional manner and observe the dimensional difficulties which arise from the discussion of even very simple processes. Then we shall introduce the fundamental technique of the theory, the conversion of the original maximization problem into the problem of determining the solution of a functional equation.

The functional equations which arise in this way are of a novel type, completely different from any of the functional equations encountered in classical analysis. The particular one we shall employ for purposes of discussion in this chapter is

$$(1) \qquad f(x) = \operatorname*{Max}_{0 \leq y \leq x} [g(y) + h(x - y) + f(ay + b(x - y))].$$

where g and h are known functions and a and b are known constants, satisfying the condition $0 \leq a, b < 1$.

After establishing an existence and uniqueness theorem, we shall derive some simple properties of the optimal policy which can be deduced from simple functional properties of g and h. In particular, we shall present the explicit solution of some equations where g and h have various special forms.

The advantage of obtaining these solutions lies in the fact that they can be utilized to obtain approximations to the solutions of more complicated equations, and, what is more important, approximations to the associated optimal policies. The subject of approximation leads us to the concept of approximation in policy space, of importance and utility in both theoretical and practical discussion, and to the discussion of the question of the stability of f under changes in g and h.

In the second chapter we consider a multi-stage decision process of stochastic type in the guise of a gold-mining venture with a delicate gold-mining machine. Here we encounter the equation

$$(2) \qquad f(x, y) = \operatorname{Max} \begin{bmatrix} A: & p_1 [r_1 x + f((1 - r_1) x, y)] \\ B: & p_2 [r_2 y + f(x, (1 - r_2) y)] \end{bmatrix}$$

In addition to pursuing an investigation similar to that given in Chapter I, we actually obtain a solution to this equation, and some of its generalizations. The solution has a particularly simple and intuitive form, and introduces the useful idea of "decision regions."

We show, however, that some other generalizations do not have as simple a structure, and, indeed, pose as yet unresolved problems. An attempt to obtain approximate solutions to these problems for a particular region of parameter space will lead us to the continuous versions treated in Chapter VIII.

Chapter III is devoted to a synthesis of these processes which seem so different at first glance. In this chapter we analyze the common features of the two processes treated in the preceding chapters, and then proceed to formulate general versions of these processes. In this way we obtain the functional equation

$$(3) \qquad f(p) = \operatorname*{Max}_{q} [g(p, q) + h(p, q) f(T(p, q))],$$

which includes both of the preceding, and a number of equations of still more general type.

Also in this chapter we explicitly state the "principle of optimality" whose mathematical transliteration in the case of any specific process yields the functional equation governing the process. The concept of "approximation in policy space" is also discussed in more detail.

In the following chapter, Chapter IV, a number of existence and uniqueness theorems are established for several frequently occurring classes of equations having the above form. Our proofs hinge upon a simple lemma which enables us to compare two solutions of the equation in (3). Although these equations are highly non-linear, in many ways they constitute a natural generalization of linear equations. For this reason alone, aside from their applications, they merit study.

In Chapter V, we discuss a functional equation derived from a problem of much economic interest at the current time, the "optimal inventory" problem. Here we show that the various techniques we have discussed in the preceding chapters yield the solutions of some interesting particular cases. In particular, we show that the method of successive approximations is an efficient analytic tool for the discovery of properties of the solution and the policy, rather than merely a humdrum means of obtaining existence and uniqueness theorems. There are many different versions of the optimal inventory problem and we restrict ourselves to a discussion of the mathematical model first proposed by Arrow, Harris, and Marschak, and treated also by Dvoretzky, Kiefer, and Wolfowitz.

A particular equation of the type we shall consider is

$$(4)\quad f(x) = \operatorname*{Min}_{y \geq x} [g(y - x) + a\{\int_y^\infty p(s - y)\, dG(s) + f(0) \int_y^\infty dG(s) + \int_0^y f(y - s)\, dG(s)\}]$$

We then turn to a study of what we call "bottleneck processes." These we define as processes where a number of interdependent activities are to be combined for one common purpose, with the level of this principal activity dependent upon the *minimum* level of activity of the components.

Two chapters are devoted to these problems, the first, Chapter VI, of theoretical nature, and the second, Chapter VII, given over to the actual details of the complete solution of one particular process.

The problems that we encounter are particular cases of the general problem, apparently not treated before in any mathematical detail, of determining the maximum over z of the inner product $(x(T), a)$, where x and z are connected by means of the vector-matrix equation

$$(5)\qquad dx/dt = Ax + Bz,\ x(0) = c,$$

and where there is a constraint of the form $Cz + Dx \leq f$. Here x, z, c and f are vectors and A, B, C and D are matrices. The linearity of the operators and functionals constitutes the principal difficulty.

We might observe parenthetically that it is often thought that linearizing a problem facilitates its solution. On occasion, however, particularly in variational problems, it frequently complicates affairs to an enormous degree, since this linearization renders classical variational techniques largely inapplicable. In return, however, the computational solution of particular cases may often be obtained by routine procedures.

In Chapter VIII, we return to the gold-mining process, and consider a continuous version. There are many problems, some of a quite recondite nature, associated with the formulation of continuous stochastic decision processes. In the processes at hand, we are fortunate in being able to sidestep these difficulties. In the continuous version, combining the classical variational approach with the techniques employed in previous chapters, we are able to solve completely the continuous versions of a number of problems that were resolutely intractable in the discrete case.

We now turn to the calculus of variations in Chapter IX, and show that various characteristic problems may be viewed as dynamic programming processes of continuous and deterministic type.

In geometric terms, the classical formulation is equivalent to considering an extremal curve as a locus of points, while the dynamic programming formulation conceives of the extremal as the envelope of tangents.

Taking this latter point of view, we are able to obtain a new formulation of some parts of the classical theory. In particular, we show how to obtain partial differential equations, in terms of suitably introduced state variables, for the principal eigen-value of the differential equation

$$(6) \qquad u'' + \lambda^2 \varphi(t) u = 0, u(0) = u(1) = 0.$$

Furthermore, we provide a new computational approach to variational problems with constraints.

In Chapter X, we consider dynamic programming processes involving two decision-makers, essentially opposed to each other in their interests. This leads to the discussion of multi-stage games, and, in particular, to the very interesting class of games called "games of survival." With the aid of some heuristic reasoning, we are able to obtain a new rationale for non-zero sum games, as a by-product.

The functional equations encountered in this domain have the general form

$$(7) \qquad f(p, p') = \underset{G}{\text{Max}} \ \underset{G'}{\text{Min}} \left[\int\int [g(p, p', q, q') + h(p, p', q, q') f[T_1(p, p', q, q'), T_2(p, p', q, q')]] \, dG(q) \, dG'(q') \right].$$

They may be treated by means of the same general methods used in Chapter IV to discuss the equation in (3) above.

In the final chapter, we consider a class of continuous decision processes which lead to non-linear differential equations of the form

$$(8) \quad \frac{dx_i}{dt} = \underset{q}{\text{Max}} \left[\sum_{j=1}^{N} a_{ij}(t; q) x_j + b_i(q) \right], x_i(0) = c_i, i = 1, 2, \ldots, N,$$

together with the corresponding equations derived from the discrete process.

These equations possess amusing connections with some classical non-linear equations, as we indicate.

In addition to a number of exercises inserted for pedagogical purposes, we have included a cross-section of problems designed to indicate the scope of the application of the methods of dynamic programming.

There may be some who will frown upon some of the less than profound subjects which are occasionally discussed in the exercises, and used to illustrate various types of processes. We are prepared to defend ourselves against the charges of lèse majesté in a number of ways, but we prefer the two following. In the first place, interesting mathematics is where you find it, sometimes in a puzzle concerning the bridges of Koenigsberg, sometimes in a problem concerning the coloring of maps, or perhaps the seating of schoolgirls, perhaps in the determining of winning play in games of chance, perhaps in an unexpected regularity in the distribution of primes. In the second place, all thought is abstract, and mathematical thought especially so. Consequently, whether we introduce our mathematical entities under the respectable sobriquets of A and B, or by the more charming Alice and Betty, or whether we speak of stochastic processes, or the art of gaming, it is the mathematical analysis that counts. Any mathematical study, such as this, must be judged, ultimately upon its intrinsic content, and not by the density of high-sounding pseudo-abstractions with which a text may so easily be salted.

This completes our synopsis of the volume. Since the processes we consider, the functional equations which arise, and the techniques we employ are in the main novel and therefore unfamiliar, we have restricted ourselves to a moderate mathematical level in order to emphasize the principles involved, untrammeled by purely analytic details. Consistent with this purpose we have not penetrated too deeply into any one domain of application of the theory from either the mathematical, economic, or physical side.

In every chapter we have attempted to avoid any discussion of deeper results requiring either more advanced training on the part of the reader or more high-powered analytic argumentation. Occasionally, as in Chapter VI and Chapter IX, we have not hesitated to waive rigorous discussion and proceed in a frankly heuristic manner.

In a contemplated second volume on a higher mathematical level, we propose to rectify some of these omissions, and present a number of topics of a more advanced character which we have either not mentioned at all here, mentioned in passing, or sketched in bold outline. It will be apparent from the text how much remains to be done.

In this connection it is worth indicating a huge, important, and relatively undeveloped area into which this entire volume represents merely a small excursion. This is the general study of the computational solution of multi-dimensional variational problems. Specifically we may pose the general problem as follows: Given a process with an associated variational problem, how do we utilize the special features of the process to construct a computational algorithm for solving the variational problem?

Dynamic programming is designed to treat multi-stage processes possessing certain invariant aspects. The theory of linear programming is designed to treat processes possessing certain features of linearity, and the elegant "simplex method" of G. Dantzig to a large extent solves the problem for these processes. For certain classes of scheduling processes, there are a variety of iterative and relaxation methods. In particular, let us note the methods of Hitchcock, Koopmans, and Flood for the Hitchcock-Koopmans transportation problem, and the "flooding technique" of A. Boldyreff for railway nets. Furthermore, there is the recent theory of non-linear programming of H. Kuhn and A. W. Tucker and E. Beale. The study of computational techniques is, however, in its infancy.

Let us now discuss briefly some pedagogical aspects of the book. We have taken as our audience all those interested in variational problems, including mathematicians, statisticians, economists, engineers, operations analysts, systems engineers, and so forth. Since the interests of various members of this audience overlap to only a slight degree, some parts of the book will be of greater interest to one group than another.

As a mathematics text the volume is suitable for a course on the advanced calculus level, either within the mathematics department proper, or in conjunction with engineering or economics departments, in connection with courses in applied mathematics or operations research.

For first courses, or first readings, we suggest the following programs:

Mathematician:	Chapters I, II, III, IV, IX, X
Economist:	Chapters I, II, III, V, IX
Statistician:	Chapters I, II, III, IX, X, XI
Engineer:	Chapters I, II, III, IX
Operations Analyst:	Chapters I, II, III, V, IX, X

Finally, before ending this prologue, it is a pleasure to acknowledge my indebtedness to a number of sources: First, to the von Neumann theory of games as developed by J. von Neumann, O. Morgenstern, and others, a theory which shows how to treat by mathematical analysis vast classes of problems formerly far out of the reach of the mathematician—and relegated, therefore, to the limbo of imponderables—and, simultaneously, to the Wald theory of sequential analysis, as developed by A. Wald, D. Blackwell, A. Girshick, J. Wolfowitz, and others, a theory which shows the vast economy of effort that may be effected by the proper consideration of multi-stage testing processes; second, to a number of colleagues and friends who have discussed various aspects of the theory with me and contributed to its clarification and growth.

Many of the results in this volume were obtained in collaboration with fellow mathematicians. The formulation of games of survival was obtained in conjunction with J. P. LaSalle; the results on the optimal inventory equation were obtained together with I. Glicksberg and O.

Gross; the results on the continuous gold-mining process in Chapter VIII and the results in Chapter VII concerning specific bottleneck processes were obtained together with S. Lehman; a number of results obtained with H. Osborn on the connection between characteristics and Euler equations, and on the convergence of discrete gold-mining processes to the continuous versions will not appear in this volume. Nor shall we include a study of the actual computational solution of many of the processes discussed below, in which we have been engaging in conjunction with S. Dreyfus.

I should particularly like to thank I. Glicksberg, O. Gross and A. Boldyreff who read the final manuscript through with great care and made a number of useful suggestions and corrections, and S. Karlin and H. N. Shapiro who have done much valuable work in this field and from whose many stimulating conversations I have greatly benefited.

Finally, I should like to record a special debt of gratitude to O. Helmer and E. W. Paxson who early appreciated the importance of multi-stage processes and who, in addition to furnishing a number of fascinating problems arising naturally in various important applications, constantly encouraged me in my researches.

A special note should be made here of the fact that most of the mathematicians cited above are either colleagues at The RAND Corporation, or are consultants. Our work has been conducted under a broad research program for the United States Air Force.

Santa Monica, California RICHARD BELLMAN

Contents

CHAPTER II

A STOCHASTIC MULTI-STAGE DECISION PROCESS

SECTION

CHAPTER III

THE STRUCTURE OF DYNAMIC PROGRAMMING PROCESSES

CHAPTER IV

EXISTENCE AND UNIQUENESS THEOREMS

SECTION

CHAPTER V

THE OPTIMAL INVENTORY EQUATION

CHAPTER VI

BOTTLENECK PROBLEMS IN MULTI-STAGE PRODUCTION PROCESSES

CHAPTER VII

BOTTLENECK PROBLEMS: EXAMPLES

CHAPTER VIII

A CONTINUOUS STOCHASTIC DECISION PROCESS

SECTION

CHAPTER X

MULTI-STAGE GAMES

CHAPTER XI

MARKOVIAN DECISION PROCESSES

SECTION

DYNAMIC PROGRAMMING

CHAPTER I

A Multi-Stage Allocation Process

§ 1. Introduction

In this chapter we wish to introduce the reader to a representative class of problems lying within the domain of dynamic programming and to the basic approach we shall employ throughout the subsequent pages.

To begin the discussion we shall consider a multi-stage allocation process of rather simple structure which possesses many of the elements common to a variety of processes that occur in mathematical analysis, in such fields as ordinary calculus and the calculus of variations, and in such applied fields as mathematical economics, and in the study of the control of engineering systems.

We shall first formulate the problem in classical terms in order to illustrate some of the difficulties of this straightforward approach. To circumvent these difficulties, we shall then introduce the fundamental approach used throughout the remainder of the book, an approach based upon the idea of imbedding any particular problem within a family of similar problems. This will permit us to replace the original multi-dimensional maximization problem by the problem of solving a system of recurrence relations involving functions of much smaller dimension.

As an approximation to the solution of this system of functional equations we are lead to a single functional equation, the equation

$$f(x) = \operatorname*{Max}_{0 \leq y \leq x} [g(y) + h(x - y) + f(ay + b(x - y))]. \tag{1}$$

This equation will be discussed in some detail as far as existence and uniqueness of the solution, properties of the solution, and particular solutions are concerned.

Turning to processes of more complicated type, encompassing a greater range of applications, we shall first discuss time-dependent processes and then derive some multi-dimensional analogues of (1), arising from multi-stage processes requiring a number of decisions at each stage. These multi-dimensional equations give rise to some difficult, and as yet unresolved, questions in computational analysis.

In the concluding portion of the chapter we consider some stochastic versions of these allocation processes. As we shall see, the same analytic methods suffice for the treatment of both stochastic and deterministic processes.

§ 2. A multi-stage allocation process

Let us now proceed to describe a multi-stage allocation process of simple but important type.

Assume that we have a quantity x which we divide into two non-negative parts, y and $x - y$, obtaining from the first quantity y a return of $g(y)$ and from the second a return of $h(x - y)$.[1] If we wish to perform this division in such a way as to maximize the total return we are led to the analytic problem of determining the maximum of the function

$$R_1(x, y) = g(y) + h(x - y) \tag{1}$$

for all y in the interval $[0, x]$. Let us assume that g and h are continuous functions of x for all finite $x \geq 0$ so that this maximum will always exist.

Consider now a two-stage process. Suppose that as a price for obtaining the return $g(y)$, the original quantity y is reduced to ay, where a is a constant between 0 and 1, $0 \leq a < 1$, and similarly $x - y$ is reduced to $b(x - y)$, $0 \leq b < 1$, as the cost of obtaining $h(x - y)$. With the remaining total, $ay + b(x - y)$, the process is now repeated. We set

$$ay + b(x - y) = x_1 = y_1 + (x_1 - y_1), \tag{2}$$

for $0 \leq y_1 \leq x_1$, and obtain as a result of this new allocation the return $g(y_1) + h(x_1 - y_1)$ at the second stage. The total return for the two-stage process is then

$$R_2(x, y, y_1) = g(y) + h(x - y) + g(y_1) + h(x_1 - y_1) \tag{3}$$

and the maximum return is obtained by maximizing this function of y and y_1 over the two-dimensional region determined by the inequalities

$$\text{(4)} \qquad \text{a. } 0 \leq y \leq x$$

$$\text{b. } 0 \leq y_1 \leq x_1$$

Let us turn our attention now to the N-stage process where we repeat

[1] The units of the return are, in this case, different from the units of x. Thus, for example, x may be in dollars, and $g(y)$ may be man-hours of service from machines purchased with the y dollars. In other cases, occurring in multi-stage investment problems, or multi-stage production problems, this will not be so, in that the units of the return will be the same as that of the resources, or a mixture of both situations will occur. We are considering the simplest case here.

the above operation of allocation N times in succession. The total return from the N-stage process will then be

$$(5) \qquad R_N(x, y, y_1, \ldots, y_{N-1}) = g(y) + h(x - y) + g(y_1) + h(x_1 - y_1) + \ldots + g(y_{N-1}) + h(x_{N-1} - y_{N-1}),$$

where the quantities available for subsequent allocation at the end of the first, second, ..., $(N-1)$st stage are given by

$$(6) \qquad \begin{aligned} x_1 &= ay + b(x - y), \; 0 \leq y \leq x, \\ x_2 &= ay_1 + b(x_1 - y_1), \; 0 \leq y_1 \leq x_1, \\ &\cdot \\ &\cdot \\ &\cdot \\ x_{N-1} &= ay_{N-2} + b(x_{N-2} - y_{N-2}), \\ &\qquad 0 \leq y_{N-2} \leq x_{N-2}, \quad 0 \leq y_{N-1} \leq x_{N-1} \end{aligned}$$

The maximum return will be obtained by maximizing the function R_N over the N-dimensional region in the space of the variables $y, y_1, \ldots, y_{N-1}$, described by the relations in (6).

§ 3. Discussion

In setting out to solve this problem, the temptation is, quite naturally, to use calculus. If the absolute maximum occurs inside the region, which is to say if all the y_i satisfy the strict inequalities $0 < y_i < x_i$, and if the functions $g(x)$ and $h(x)$ possess derivatives, we obtain for the determination of the maximizing y_i the system of equations,

$$(1) \qquad \begin{aligned} & g'(y_{N-1}) - h'(x_{N-1} - y_{N-1}) = 0 \\ & g'(y_{N-2}) - h'(x_{N-2} - y_{N-2}) + (a - b)\, h'(x_{N-1} - y_{N-1}) = 0 \\ & \cdot \\ & \cdot \\ & \cdot \\ & g'(y) + h'(x - y) + (a - b)\, h'(x_1 - y_1) + \ldots = 0, \end{aligned}$$

upon taking partial derivatives. However, in the absence of this knowledge, since we are interested not in *local* maxima, but in the *absolute* maximum, we must also test the boundary values $y_i = 0$ and x_i, and all combinations of boundary values and internal maxima. Furthermore, if the solution of the equations in (1) is not unique, we must run through a set of conditions sufficient to ensure our having a maximum and not a minimum or a mere local maximum. It is evident that for problems of large dimension, which is to say for processes involving a large number of stages, a systematic procedure for carrying out this program is urgently required to keep the problem from getting out of hand.

Suppose that we abdicate as an analyst in the face of this apparently formidable task and adopt a defeatist attitude. Turning to the succor of modern computing machines, let us renounce all analytic tools. Consider, as a specific example, the problem posed by a 10-stage process. Then, if we wish to go about the determination of the maximum in a rudimentary fashion by computing the value of the function $R_{10} = R_{10}(y, y_1, \ldots, y_9)$ at suitably chosen lattice points, we may proceed to divide all the intervals of interest, $0 \leq y \leq x$, $0 \leq y_1 \leq x_1$, ..., $0 \leq y_9 \leq x_9$, into, say, ten parts, and compute the value of R_{10} at each of the 10^{10} points obtained in this manner. 10^{10} is, however, a number that commands respect. Even the fastest machine available today or in the near future, will still require an appreciable time to determine the solution in this manner.

To give some idea of the magnitude of 10^{10}, note that if the machine took one second for the calculation of R_{10} at a lattice point, storage and comparison with other values, the computation of 10^{10} values would require 2.77 million hours; if one millisecond, then 2.77 thousand hours; if one micro-second, then 2.77 hours. This last seems fairly reasonable. Observe, however, that if we consider a 20-stage process, we must multiply any such value by 10^{10}, i.e., $10^{20} = 10^{10} \cdot 10^{10}$.

Needless to say, there are various ingenious techniques that can be employed to cut this time down. Nonetheless, the method sketched above is still an unwieldy and inelegant method of attack.

Furthermore, it should be realized that if we are sufficiently interested in the solution of the above decision process to engage in computations, we will, in general, wish to compute the answer not only for one particular value of x, but for a range of values, not only for one set of values of a and b but for a set of values, and not only for one set of functions g and h, but for a class of functions. In other words, we will perform a *sensitivity analysis* or *stability analysis* of the solution. Any such sensitivity analysis attempted by the above methods will run into fairly large computing times.

One of the aspects of the situation viewed in these terms which is really disheartening is that this problem is, after all, only the consequence of a very, almost absurdly, simple version of an applied problem. It is clear that any modification of the problem in the direction of realism, say subdivision of x into more than two parts, which is to say an increase in the number of activities we can engage in, or an increase in the types of resources, will increase the computing time at an exponential rate.

Furthermore, as we have pointed out in the Preface, we must realize that the essential purpose in formulating many of these mathematical models of the universe, economic, physical, biologic, or otherwise, is not so much to calculate numbers, which are in many cases of dubious value because of the lack of knowledge of some of the basic constants and functions involved, but rather to determine the *structure* of the solution. Concepts are, in many processes, more important than constants.

The two, however, in general go hand-in-hand. If we have a thorough understanding of the process, we have means, through approximation techniques of various sorts, of determining the constants we require. Furthermore, in the processes occurring in applications, of such enormous complexity that trial and error computation is fruitless, it is only by having an initial toe-hold on the solution that we can hope to use computing machines effectively.

Going back to the idea of the intrinsic structure of a solution, we may ask what it is that we really wish to know if we are studying a process of this type. Naturally, we would like to obtain the point $(y, y_1, \ldots, y_N)$ at which the maximum occurs, and any solution must furnish this. But from the point of view of a person carrying out the process, all that is really required at any particular stage of the process is the value of y in terms of x, the resources available, and N, the number of stages ahead; that is to say, the allocation to be made when the quantity available is x and the number of stages of the process remaining is N. Viewed as a multi-stage process, at each stage a *one-dimensional* choice is made, a choice of y in the interval $[0, x]$. It follows [2] that there should be a formulation of the problem which preserves this dimensionality and saves us from becoming bogged down in the complexities of multidimensional analysis.

§ 4. Functional equation approach

Taking this as our goal, namely the preservation of one-dimensionality, let us proceed as follows. We first observe that the maximum total return over an N-stage process depends only upon N and the initial quantity x. Let us then define the function,

(1) $f_N(x) =$ the maximum return obtained from an N-stage process starting with an initial quantity x, for $N = 1, 2, \ldots,$ and $x \geq 0$.

[2] As an application of the useful principle of wishful thinking.

We have

(2) $$f_N(x) = \underset{\{y, y_i\}}{\text{Max}}\, R_N(x, y, \ldots, y_{N-1}), \; N = 2, 3, \ldots$$

with

(3) $$f_1(x) = \underset{0 \leq y \leq x}{\text{Max}} \left[g(y) + h(x-y)\right].$$

Our first objective is to obtain an equation for $f_2(x)$ in terms of $f_1(x)$. Considering the two-stage process, we see that the total return will be the return from the first stage plus the return from the second stage, at which stage we have an amount $ay + b(x-y)$ left to allocate. It is clear that whatever the value of y chosen initially, this remaining amount, $ay + b(x-y)$, must be used in the best possible manner for the remaining stage, if we wish to obtain a two-stage allocation which maximizes.

This observation, simple as it is, is the key to all of our subsequent mathematical analysis. It is worthwhile for the reader to pause here a moment and make sure that he really agrees with this observation, which has the deceptive simplicity of a half-truth.

It follows that as a result of an initial allocation of y we will obtain a total return of $f_1(ay + b(x-y))$ from the second stage of our two stage process, if y_1 is chosen optimally. Consequently, for the total return from the two stage process resulting from the initial allocation of y, we have the expression

(4) $$R_2(x, y, y_1) = g(y) + h(x-y) + f_1(ay + b(x-y)).$$

Since y is to be chosen to yield the maximum of this expression, we derive the recurrence relation

(5) $$f_2(x) = \underset{0 \leq y \leq x}{\text{Max}} \left[g(y) + h(x-y) + f_1(ay + b(x-y))\right],$$

connecting the functions $f_1(x)$ and $f_2(x)$. Using precisely the same argumentation for the N-stage process, we obtain the basic functional equation

(6) $$f_N(x) = \underset{0 \leq y \leq x}{\text{Max}} \left[g(y) + h(x-y) + f_{N-1}(ay + b(x-y))\right]$$

for $N \geq 2$, with $f_1(x)$ defined as in (3) above.

Starting with $f_1(x)$, as determined by (3), we use (6) to compute $f_2(x)$, which, in turn, repeating the process, yields $f_3(x)$, and so on. At each step of the computation, we obtain, not only $f_k(x)$, but also $y_k(x)$, the optimal allocation to be made at the beginning of a k-stage process, starting with an amount x.

The solution, then, consists of a tabulation of the sequence of functions $\{y_k(x)\}$ and $\{f_k(x)\}$ for $x \geq 0$, $k = 1, 2, \ldots$.

Given the sequence of functions $\{y_k(x)\}$, the solution of a specific problem, involving a given N and a given x has the form

$$(7) \qquad \begin{aligned} \bar{y} &= y_N(x), \\ \bar{y}_1 &= y_{N-1}(a\bar{y} + b(x - \bar{y})), \\ \bar{y}_2 &= y_{N-2}(a\bar{y}_1 + b(x_1 - \bar{y}_1)), \\ &\cdot \\ &\cdot \\ &\cdot \\ \bar{y}_{N-1} &= y'_1(a\bar{y}_{N-2} + b(x_{N-2} - \bar{y}_{N-2})), \end{aligned}$$

where $(\bar{y}, \bar{y}_1, \ldots, \bar{y}_{N-1})$ is a set of allocations which maximizes the total N-stage return.

A digital computer may be programmed to print out the sequence of values $\bar{y}, \bar{y}_1, \ldots, \bar{y}_{N-1}$, in addition to tabulating the sequences $\{f_k(x)\}$ and $\{y_k(x)\}$.

§ 5. Discussion

The important fact to observe is that we have attempted to solve a maximization problem involving a particular value of x and a particular value of N by first solving the general problem involving an arbitrary value of x and an arbitrary value of N. In other words, as we promised in the first section, we have imbedded the original problem within a family of similar problems. We shall exploit this basic method of mathematical analysis throughout the book.

What are the advantages of this approach? In the first place, we have reduced a single N-dimensional problem to a sequence of N one-dimensional problems. The computational advantages of this formulation are obvious, and we shall proceed in the next sections to show that there are analytic advantages as well, as might be suspected. As we shall see, we will be able to obtain explicit solutions for large classes of functions g and h, which can be used for approximation purposes. This point will be discussed again below. Furthermore, we will be able to determine many important structural features of the solution even in those cases where we cannot solve completely. The utilization of structural properties of the solution and the reduction in dimension combine to furnish computing techniques which greatly reduce the time required to solve the original problem. We shall return to this point in connection with some multi-dimensional versions.

§ 6. A multi-dimensional maximization problem

Before proceeding to a more detailed theory of the processes described above, let us digress for a moment and briefly present two further applications of the general method.

For the first application, consider the problem of determining the maximum of the function

$$(1) \qquad F(x_1, x_2, \ldots, x_N) = \sum_{i=1}^{N} g_i(x_i),$$

over the region defined by

$$(2) \qquad \begin{aligned} &(a)\ x_1 + x_2 + \ldots + x_N = c, \\ &(b)\ x_i \geq 0. \end{aligned}$$

Each function $g_i(x)$ is assumed to be continuous for all $x \geq 0$.

Since the maximum of F depends only upon c and N, let us define the sequence of functions

$$(3) \qquad f_N(c) = \operatorname*{Max}_{\{x_i\}} F(x_1, x_2, \ldots, x_N),$$

for $c \geq 0$ and $N = 1, 2, \ldots$.

Then, arguing as above, we have the recurrence relation

$$(4) \qquad f_N(c) = \operatorname*{Max}_{0 \leq x \leq c} [g_N(x) + f_{N-1}(c - x)],$$

for $N = 2, 3, \ldots$, with

$$(5) \qquad f_1(c) = g_1(c).$$

§ 7. A "smoothing" problem

As the second application, let us consider the problem of determining the sequence $\{x_k\}$ which minimizes the function

$$(1) \qquad F(x_1, x_2, \ldots, x_N) = \sum_{k=1}^{N} g_k(x_k - r_k) + \sum_{k=1}^{N} h_k(x_k - x_{k-1}).$$

Here $\{r_k\}$ is a given sequence, $x_0 = c$ a given constant, and we assume that the functions $g_k(x)$ and $h_k(x)$ are continuous for all finite x, and that $g_k(x), h_k(x) \to \infty$ as $|x| \to \infty$.

The genesis of this problem, explaining its name, will be discussed in the exercises.

Let us define the sequence $\{f_R(c)\}$, $R = 1, 2, \ldots, N$, by the property that $f_R(c)$ is the minimum over all $x_R, x_{R+1}, \ldots, x_N$ of the function

$$(2) \qquad F_R = \sum_{k=R}^{N} g_k(x_k - r_k) + \sum_{k=R}^{N} h_k(x_k - x_{k-1}),$$

where $x_{R-1} = c$.

We have

$$(3) \qquad f_N(c) = \operatorname*{Min}_{x} [g_N(x - r_N) + h_N(x - c)],$$

and

$$(4) \qquad f_R(c) = \operatorname*{Min}_{x} [g_R(x - r_R) + h_R(x - c) + f_{R+1}(x)],$$

for $R = 1, 2, \ldots, N - 1$.

§ 8. Infinite stage approximation

Let us now return to the allocation process. The treatment we present here serves as a prototype for the discussion of a number of multi-stage processes, of diverse origin, but similar analytic structure.

If N is large, it is reasonable to consider as an approximation to the N-stage process, the infinite stage process defined by the requirement that the process continue indefinitely. Although an unbounded process is always a physical fiction,[3] as a mathematical process it has many attractive features. One immediate advantage of this approximation lies in the fact that in place of the sequence of equations given by (4.6), we now have the single equation

$$(1) \qquad f(x) = \operatorname*{Max}_{0 \le y \le x} [g(y) + h(x - y) + f(ay + b(x - y))]$$

satisfied by $f(x)$, the total return of the process, with a single allocation function $y = y(x)$, determined by the equation.

To balance this, we encounter many of the usual difficulties associated with infinite processes. It is, first of all, no longer clear that a maximum exists rather than a supremum. This is to say, there may be no allocation policy which actually yields the total return $f(x)$. Furthermore, if we wish to employ (1) in an unrestricted fashion to determine properties of the infinite process, we must show that it possesses *no* extraneous solutions. In other words, we must establish existence and uniqueness theorems if this equation is to serve a useful purpose.

[3] We shall occasionally use the word "physical" to describe the "real" world. It should be interpreted to mean economic, biological, engineering, etc., depending upon the background and interests of the reader.

§ 9. Existence and uniqueness theorems

The result we obtain in this section is actually a special case of a more general result we shall derive in a later chapter. Repetition, however, no matter how dismaying as a social or literary attribute, is no great mathematical sin, and it is important to present the simpler case first, enabling the basic ideas to appear unimpeded by technicalities of lesser import.

Let us now demonstrate

THEOREM 1. *Let us assume that*

(1) a. *$g(x)$ and $h(x)$ are continuous functions of x for $x \geq 0$, $g(0) = h(0) = 0$.*

b. *If* $m(x) = \underset{0 \leq y \leq x}{\text{Max}} \ \text{Max}\,(|g(y)|, |h(y)|)$, *and*

$c = \text{Max}\,(a, b)$, *then* $\sum_{n=0}^{\infty} m(c^n x) < \infty$ *for all* $x \geq 0$.

c. $0 \leq a < 1, 0 \leq b < 1$.

Under these assumptions, there is a unique solution to (8.1) *which is continuous at $x = 0$, and has the value 0 at this point; moreover, this function is continuous.*

Before proceeding to the proof, let us digress for a moment and consider the important special case where g and h are both non-negative. The sequence $\{f_N(x)\}$ as given by (4.6) is a monotone increasing sequence, with boundedness a consequence of condition (1b), as we shall show below in a moment. Consequently, for all $x \geq 0$, $f_N(x)$ converges to a function $f(x)$ as $N \to \infty$.

Let us show that this function satisfies the equation

$$f(x) = \sup_{0 \leq y \leq x} [g(y) + h(x-y) + f(ay + b(x-y))]. \tag{2}$$

To simplify our notation, let us set

$$T(f, y) = g(y) + h(x-y) + f(ay + b(x-y)). \tag{3}$$

The basic recurrence relation is then

$$f_{N+1}(x) = \underset{0 \leq y \leq x}{\text{Max}} \ T(f_N, y). \tag{4}$$

From (4) we obtain as a consequence of the monotonicity in N,

$$f(x) \geq \max_{0 \leq y \leq x} T(f_N, y). \tag{5}$$

For any y in the interval $[0, x]$, this means that the inequality

$$f(x) \geq T(f_N, y) \tag{6}$$

holds. Letting $N \to \infty$, this yields

$$f(x) \geq T(f, y) \tag{7}$$

for all y in $[0, x]$, which, in turn, leads to the result

$$f(x) \geq \sup_{0 \leq y \leq x} T(f, y). \tag{8}$$

We cannot write $\max_{0 \leq y \leq x}$ since we have no guarantee that the limit function $f(x)$ is actually continuous as a function of x. On the other hand, from (4) we also obtain

$$f_{N+1}(x) \leq \sup_{0 \leq y \leq x} T(f, y), \tag{9}$$

for all N, and thus

$$f(x) \leq \sup_{0 \leq y \leq x} T(f, y). \tag{10}$$

Comparing (8) and (10), we obtain (2).

One of the defects of this proof based solely upon monotonicity is that it does not yield the continuity of the limit function, a result which implies the existence of an optimal policy. This optimal policy is a function $y(x)$ which yields the maximum in

$$f(x) = \max_{0 \leq y \leq x} T(f, y), \tag{11}$$

when the maximum exists.

The existence of an optimal policy for the infinite process is directly of no particular importance computationally, or as far as applications are concerned. It is, however, of great importance in connection with the determination of the structure of optimal policies for the infinite process. Thus, indirectly, the question of the existence of continuous solutions is significant as far as numerical results are concerned, since the solution of the infinite process can be used as an approximation to the solution of the finite.

In order to establish the existence and uniqueness of a continuous solution of (11), we shall employ a technique that is applicable to a large class of equations of this type, the method of *successive approximations.* We shall, however, encounter monotonicity arguments again in later chapters.

Turning to the recurrence relations in (4), let us begin with the observation that $f_1(x)$ is continuous for all $x \geq 0$ by virtue of the assumptions made concerning $g(x)$ and $h(x)$. It follows, inductively, that each element of the sequence $\{f_N(x)\}$ is continuous. It is worth pointing out, however, that the location of the maximizing y need not depend continuously upon x. In other words, the policy is not necessarily a continuous function of x. An example of this is given in § 15.

Let $y_N(x)$ be a value of y which yields the maximum in (4). It is a matter of indifference as to which value of y we choose, if there is more than one value producing the maximum. Then we have

$$\begin{aligned} f_{N+1}(x) &= T(f_N, y_N), \\ f_{N+2}(x) &= T(f_{N+1}, y_{N+1}), \end{aligned} \tag{12}$$

and, as a consequence of the maximum property of the y_N, the further inequalities

$$\begin{aligned} f_{N+1}(x) &= T(f_N, y_N) \geq T(f_N, y_{N+1}) \\ f_{N+2}(x) &= T(f_{N+1}, y_{N+1}) \geq T(f_{N+1}, y_N). \end{aligned} \tag{13}$$

These, in turn, yield

$$\begin{aligned} f_{N+1}(x) - f_{N+2}(x) &\geq T(f_N, y_{N+1}) - T(f_{N+1}, y_{N+1}) \\ &\leq T(f_N, y_N) - T(f_{N+1}, y_N) \end{aligned} \tag{14}$$

The two inequalities combined yield the important estimate

$$\begin{aligned} |f_{N+1}(x) - f_{N+2}(x)| \leq \text{Max}\,[\,&|T(f_N, y_{N+1}) - T(f_{N+1}, y_{N+1})|, \\ &|T(f_N, y_N) - T(f_{N+1}, y_N)|\,]. \end{aligned} \tag{15}$$

Turning to the definition of $T(f, y)$ given in (3), we see that

$$\begin{aligned} &|T(f_N, y_N) - T(f_{N+1}, y_N)| \\ &= |f_N(ay_N + b(x - y_N)) - f_{N+1}(ay_N + b(x - y_N))| \end{aligned} \tag{16}$$

Let us now define

$$u_N(x) = \underset{0 \leq z \leq x}{\text{Max}} |f_N(z) - f_{N+1}(z)|, \; N = 1, 2, \ldots \tag{17}$$

Since $ay + b(x - y) \leq cx$ for all y in $[0, x]$, the relation in (16) yields

$$u_{N+1}(x) \leq u_N(cx). \tag{18}$$

It remains to estimate $u_1(x)$. We have, referring to the equations for $f_1(x)$ and $f_2(x)$, the relation

$$\begin{aligned} |f_1(x) - f_2(x)| \leq \operatorname{Max} [\, |f_1(ay_1 + b(x - y_1))|, \\ |f_1(ay_2 + b(x - y_2))|\,] \leq m(cx), \end{aligned} \tag{19}$$

using the definition of $m(x)$ given in (1b).

Hence we see that $u_1(x) \leq m(cx)$, and thus, using (18), that $u_N(x) \leq m(c^{N+1}x)$. By virtue of our assumption concerning $m(x)$ it follows that $\sum_{N=1}^{\infty} u_N(x)$ converges for all x, and what is important, uniformly in any finite interval. The limit function $f(x) = \lim_{N \to \infty} f_N(x)$, in consequence, exists and is continuous for all x. Furthermore, the uniformity of convergence ensures that $f(x)$ is a solution of (8.1).

It remains to establish the uniqueness of the solution. Let $F(x)$ be any other solution which exists for all x and is continuous at $x = 0$, with $F(0) = 0$.

In the equation

$$f(x) = \operatorname*{Max}_{0 \leq y \leq x} T(f, y), \tag{20}$$

let $y = y(x)$ be a value of y which yields the maximum, and let $w = w(x)$ play the similar role in

$$F(x) = \operatorname*{Max}_{0 \leq y \leq x} T(F, y). \tag{21}$$

Then, as above, we obtain the two inequalities,

$$\begin{aligned} f(x) &= T(f, y) \geq T(f, w) \\ F(x) &= T(F, w) \geq T(F, y), \end{aligned} \tag{22}$$

and, as before, this leads to

$$\begin{aligned} |f(x) - F(x)| &\leq \operatorname{Max} [\, |T(f, y) - T(F, y)|, |T(f, w) - T(F, w)|\,]. \\ &\leq \operatorname{Max} [\, |f(ay + b(x - y)) - F(ay + b(x - y))|, \\ &\qquad |f(aw + b(x - w)) - F(aw + b(x - w))|\,]. \end{aligned} \tag{23}$$

Let us now define

$$u(x) = \operatorname*{Sup}_{0 \leq z \leq x} |f(z) - F(z)|. \tag{24}$$

Since $f(x)$ is continuous for all $x \geq 0$ and $F(x)$ is, by assumption, continuous at $x = 0$, we see that $u(x)$ is continuous at $x = 0$ and has the value 0 there.

From (23) we obtain

$$u(x) \leq u(cx), \tag{25}$$

whence, by iteration,

$$u(x) \leq u(c^N x), \tag{26}$$

for all $N \geq 1$. Since $u(x)$ is continuous at $x = 0$ and $u(0) = 0$, upon letting $N \to \infty$ we obtain $u(x) \leq 0$, and thus that $f(x) = F(x)$. This completes the proof of the existence and uniqueness of a solution of the functional equation associated with the infinite process.

§ 10. Successive approximations

In considering the equation

$$f(x) = \operatorname*{Max}_{0 \leq y \leq x} T(f, y), \tag{1}$$

we have shown that a particular sequence of successive approximations converged to the unique solution which is continuous at $x = 0$ and zero there. It is important for both analytic and computational purposes to know that actually any sequence whose initial function satisfies certain simple requirements converges.

The methods we have used above may also be employed to prove the following

THEOREM 2. Let $f_0(x)$ *satisfy the following conditions:*

(1) a. $f_0(x)$ *is continuous for* $x \geq 0$.

b. $f_0(0) = 0$.

Then, if the conditions of Theorem 1 are fulfilled, the sequence defined by

$$f_{N+1}(x) = \operatorname*{Max}_{0 \leq y \leq x} T(f_N, y), N = 0, 1, \ldots, \tag{2}$$

converges to the solution $f(x)$ *obtained above, uniformly in any finite interval.*

§ 11. Approximation in policy space

We have employed above the classical technique of successive approximations in order to obtain a solution to the nonlinear functional equation

$$f(x) = \operatorname*{Max}_{0 \leq y \leq x} T(f, y). \tag{1}$$

We now wish to exploit a certain duality which is present in these decision processes to show that we can choose the initial approximation in such a way that we can always ensure this approximation being monotone. This means that we have uniformly better convergence with each iteration.

Let us begin by introducing some terminology. We shall call a sequence of allocations; i.e., a sequence of admissible choices of y, a *policy*, and a policy which yields $f(x)$ an *optimal policy*.

The duality that exists in the theory of dynamic programming arises from the interconnection between the functions $f(x)$ which measure the maximum return and the policies which yield these maximum returns. Actually a policy is a function, since a policy is a determination of y as a function of x. It is worthwhile nonetheless to preserve this terminology since it possesses certain advantages derived from intuition. If the policy is not unique, y will not be a single-valued function of x.

It follows from the functional equation that a knowledge of $f(x)$ yields $y(x)$, and conversely any $y(x)$ determines $f(x)$, iteratively by means of the functional equation

$$f(x) = T(f, y(x)). \tag{2}$$

Thus, for example, if the optimal policy consisted of the choice $y = 0$ continually, $f(x)$ would satisfy the functional equation

$$f(x) = h(x) + f(bx), \tag{3}$$

which would yield the result

$$f(x) = \sum_{n=0}^{\infty} h(b^n x). \tag{4}$$

As we have mentioned above, the purpose of our investigation is not so much to determine $f(x)$, which is really a by-product, but more importantly, to determine the structure of the optimal policy, which is to say to determine $y(x)$.

This leads to an important and useful idea. Just as we can approximate in the space of the functions $f(x)$, so we can approximate in the space of policies, $y(x)$. Furthermore, in many ways, this is a more natural and simpler form of approximation. The advantage of this type of approximation analytically is that it always leads to monotone approximations. From the standpoint of applications, it is by far the more natural approximation since it is usually the one part of the problem about which a certain amount is known as a result of experience.

Let $y_0(x)$ be an initial guess for an optimal policy and let $f_0(x)$ be the return function derived from this policy function, which is to say that $f_0(x)$ satisfies the functional equation

$$f_0(x) = T(f_0, y_0(x)), \tag{5}$$

an equation which we solve iteratively. To improve $y_0(x)$, we determine $y_1(x)$ as a function of x which maximizes $T(f_0, y)$ for $0 \leq y \leq x$. Assume for the moment that $y_1(x)$ is itself continuous in x, (which need not necessarily be the case), and that the return function $f_1(x)$ computed using this policy is also continuous. This will always be the case, as we point out again below, under the assumptions we have made. We now continue in this way, generating a sequence of policies, $\{y_N(x)\}$, and a sequence of return function, $\{f_N(x)\}$.

It is easy to show, utilizing the methods described in the foregoing sections, that under the assumptions we have made the sequence $\{f_N(x)\}$ is monotone increasing. A rigorous proof of the existence and convergence of the sequences $\{y_N(x)\}$ and $\{f_N(x)\}$ described above seems difficult to obtain. Consequently, we compromise for the following.

THEOREM 3. *Let $f_0(x)$ be the result of an initial approximation in policy space, that is,*

$$f_0(x) = T(f_0, y_0(x)), \tag{6}$$

where $y_0(x)$ is any continuous function of x satisfying the conditions

$$0 \leq y_0(x) \leq x. \tag{7}$$

Under the assumptions of Theorem 1, the sequence defined by

$$f_{N+1}(x) = \operatorname*{Max}_{0 \leq y \leq x} T(f_N, y), \; N = 0, 1, 2, \ldots, \tag{8}$$

converges uniformly to the solution $f(x)$ obtained, and this convergence is monotone.

PROOF. Let us demonstrate the monotonicity, which is the essential feature, first. We have

$$f_1(x) = \operatorname*{Max}_{0 \leq y \leq x} T(f_0, y). \tag{9}$$

Comparing the definition of f_0 given in (5) with the definition of f_1 above, we see that $f_1 \geq f_0$ for all values of x. From this it follows inductively that $f_{N+1} \geq f_N$ for all values of $x \geq 0$.

It remains to prove the continuity of the function $f_0(x)$ for $x \geq 0$.

The conditions upon g and h which we have imposed above show that the formal series for $f_0(x)$

$$f_0(x) = g(y_0) + h(x - y_0) + \ldots, \tag{10}$$

obtained iteratively, converges uniformly in any finite interval and represents a continuous function of x for $x \geq 0$, if $y_0(x)$ is a continuous function of x.

§ 12. Properties of the solution—I: Convexity

Let us now show that we can derive certain structural properties of the optimal policy from various simple structural properties of the functions g and h. The structure of the optimal policy $y(x)$ and that of the return function $f(x)$ turn out to be intimately entwined.

Our first result in this direction is

THEOREM 4. *If, in addition to the assumptions in Theorem 1, we impose the conditions that* g *and* h *be convex functions of* x*, then* $f(x)$ *will be a convex function, and for each value of* x*,* y *will equal* 0 *or* x*.*

PROOF. The proof will be inductive. Since

$$f_1(x) = \max_{0 \leq y \leq x} (g(y) + h(x - y)) \tag{1}$$

and $g(y) + h(x - y)$ is convex as a function of y for $0 \leq y \leq x$, it follows that

$$f_1(x) = \operatorname{Max}(g(x), h(x)), \tag{2}$$

since the maximum of a convex function must occur at one of the end-points. As the maximum of two convex functions, $f_1(x)$ is convex.

Since $g(y) + h(x - y) + f_1(ay + b(x - y))$ is a convex function of y for y in $[0, x]$ it follows by repetition of the above argument that

$$f_2(x) = \operatorname{Max}(g(x) + f_1(ax), h(x) + f_1(bx)), \tag{3}$$

is a convex function of x. We see then, inductively that $f_N(x)$ is convex, and thus that the limit function $f(x)$ is convex.

Turning to the equation $f(x) = \max\limits_{0 \leq y \leq x} T(f, y)$, the convexity of f reduces this to the simpler equation

$$f(x) = \operatorname{Max}(g(x) + f(ax), h(x) + f(bx)), \tag{4}$$

showing that $y = 0$ or x for each value of x. This equations is, surprisingly, still a difficult equation to solve in general. We shall consider a particular case of it below.

§ 13. Properties of the solution—II: Concavity

Let us now demonstrate that an analogous result holds for the case where g and h are both strictly concave functions of x for $x \geq 0$.

THEOREM 5. *If, in addition, to the assumptions in Theorem 1, we impose the conditions that g and h be strictly concave functions of x, then $f(x)$ will be a strictly concave function of x.*

In this case, the optimal policy will be unique.

PROOF. Let us consider the one-stage case first, and perform some simple calculations which will show us why the result should be true, before proceeding to a rigorous proof using a different and more general technique.

We have

$$f_1(x) = \max_{0 \leq y \leq x} [g(y) + h(x-y)]. \tag{1}$$

Since g and h are strictly concave functions, the function $g(y) + h(x-y)$ is a strictly concave function of y. There is, in consequence, a single maximum, which may, nonetheless, occur at an end point $y = 0$ or $y = x$. Let us suppose for the moment that it occurs at an interior point, and that g and h possess second derivatives. Then,

$$f_1(x) = g(y) + h(x-y) \tag{2}$$

where y is determined as a function of x by means of the relation

$$g'(y) = h'(x-y). \tag{3}$$

Differentiation of (2) yields

$$f_1'(x) = (g'(y) - h'(x-y))\, dy/dx + h'(x-y) = h'(x-y), \tag{4}$$

and thus

$$f_1''(x) = h''(x-y)(1 - dy/dx). \tag{5}$$

Differentiating the relation in (3), we obtain

$$g''(y)\, dy/dx = h''(x-y)(1 - dy/dx), \tag{6}$$

which yields

$$dy/dx = h''(x-y)/(g''(y) + h''(x-y)). \tag{7}$$

This shows that $1 > dy/dx > 0$, and thus, returning to (5), that $f_1''(x) < 0$.

If the maximum is not actually inside, we can force it to be by various modifications of the functions g and h which prevent the maximum from ever being at $y = 0$ or $y = x$; e.g., by addition of a term $\varepsilon \log y (x - y)$, where ε is a small positive quantity. We can then proceed inductively and establish the same result for all the members of the sequence $f_N(x)$. This is, however, a rather clumsy method which does not extend without pain to multi-dimensional problems. We shall therefore use a more elegant and simple method.

LEMMA 1. *If $G(x, y)$ is a concave function*[4] *of x and y for $x, y \geq 0$, then $f(x)$ as defined by*

$$f(x) = \operatorname*{Max}_{0 \leq y \leq x} G(x, y) \tag{8}$$

is a concave function of x for $x \geq 0$.

PROOF. We have, for $0 \leq \lambda \leq 1$,

$$f(\lambda x + (1 - \lambda) z) = \operatorname*{Max}_{0 \leq y \leq \lambda x + (1-\lambda) z} G(\lambda x + (1 - \lambda) z, y). \tag{9}$$

We may replace y by the quantity $y = \lambda y_1 + (1 - \lambda) y_2$ where y_1 and y_2 range independently over the intervals $0 \leq y_1 \leq x$, $0 \leq y_2 \leq z$. Then

$$f(\lambda x + (1 - \lambda) z) = \operatorname*{Max}_{\substack{0 \leq y_1 \leq x \\ 0 \leq y_2 \leq z}} G(\lambda x + (1 - \lambda) z, \lambda y_1 + (1 - \lambda) y_2). \tag{10}$$

Since $G(x, y)$ is concave in x and y, we have

$$G(\lambda x + (1 - \lambda) z, \lambda y_1 + (1 - \lambda) y_2) \geq \lambda G(x, y_1) + (1 - \lambda) G(z, y_2) \tag{11}$$

Hence

$$\begin{aligned} f(\lambda x + (1 - \lambda) z) &\geq \operatorname*{Max}_{\substack{0 \leq y_1 \leq x \\ 0 \leq y_2 \leq z}} [\lambda G(x, y_1) + (1 - \lambda) G(z, y_2)] \\ &\geq \lambda \operatorname*{Max}_{0 \leq y_1 \leq x} G(x, y_1) + (1 - \lambda) \operatorname*{Max}_{0 \leq y_2 \leq z} G(z, y_2) \\ &\geq \lambda f(x) + (1 - \lambda) f(z). \end{aligned} \tag{12}$$

Let us now apply this lemma to prove Theorem 5. It is easily verified that $g(y) + h(x - y)$ is a concave function of x and y if g and h are concave functions. This shows immediately that $f_1(x)$ is concave. Similarly, since $f_1(ay + b(x - y))$ is a concave function of x and y, $f_2(x)$

[4] Concavity in both x and y means the $G(\lambda x_1 + (1 - \lambda) x_2, \lambda y_1 + (1 - \lambda) y_2) \geq \lambda G(x_1, y_1) + (1 - \lambda) G(x_2, y_2)$, for $0 \leq \lambda \leq 1$.

as defined by the basic recurrence relation is a concave function. We thus proceed inductively and show that each function in the sequence, $\{f_N(x)\}$, is a strictly concave function, and hence that the limit function is concave. That it is strictly concave follows from the strict concavity of g and h, using Lemma 1 upon the functional equation for $f(x)$.

Once we have established the strict concavity of $f(x)$, the uniqueness of the maximizing y and thus of the optimal policy follows immediately. This completes the proof of Theorem 5.

§ 14. Properties of the solution—III: Concavity

Let us now show that the assumption of concavity enables us to tell quite a bit more about the nature of the solution.

THEOREM 6. *Let us assume that*

(1) a. $g(x)$ *and* $h(x)$ *are both strictly concave for* $x \geq 0$, *monotone increasing with continuous derivatives and that* $g(0) = h(0) = 0$.

b. $g'(0)/(1-a) > h'(0)/(1-b)$, $h'(0) > g'(\infty)$, $b > a$.

Then the optimal policy has the following form:

(2) a. $y = x$ *for* $0 \leq x \leq \bar{x}$, *where* $\bar{x}$ *is the root of* $h'(0) = g'(x) + (b-a)g'(ax) + (b-a)ag'(a^2x) + \dots$

b. $y = y(x)$ *for* $x \geq \bar{x}$ *where* $y(x)$ *is a function satisfying the inequalities* $0 < y(x) < x$, *and* $y(x)$ *is the solution of*

$$g'(y) - h'(x-y) + (a-b)f'(ay + b(x-y)) = 0. \tag{3}$$

Remark. We have given the solution for only one of the possible combinations of inequalities connecting $g'(0)$, $h'(0)$, b and a. It will be easily seen from the procedure below, that corresponding results hold for the other cases. Furthermore, the number of cases can be halved by the observation that the interchange of y and $x-y$ results in an interchange of a and b.

PROOF. Let us employ the method of successive approximations. Set

$$f_1(x) = \operatorname*{Max}_{0 \leq y \leq x} [g(y) + h(x-y)]. \tag{4}$$

Since, by assumption, $g'(0) > h'(0)$, for small x, we have $g'(y) - h'(x-y) > 0$, for y in the interval $[0, x]$. Hence $g(y) + h(x-y)$ is monotone increasing in $0 \leq y \leq x$ and the maximum occurs at $y = x$. As x increases, the equation $g'(y) - h'(x-y) = 0$ will ultimately have a root at $y = x$, and then as x increases further a root inside the interval $[0, x]$. The critical value of x is given as the solution of $g'(x) - h'(0) = 0$. This equation has precisely one solution, which we call x_1. For $x \geq x_1$ let $y_1 = y_1(x)$ be the unique solution of $g'(y) = h'(x-y)$. The uniqueness of solution is a consequence of the concavity assumptions concerning g and h, and the existence of a solution is a consequence of the continuity of g' and h'.

Thus we have

$$
\begin{aligned}
(5) \qquad f_1(x) &= g(x), && 0 \leq x \leq x_1, \\
&= g(y_1) + h(x - y_1), && x \geq x_1.
\end{aligned}
$$

and

$$
\begin{aligned}
(6) \quad f_1'(x) &= g'(x), \qquad 0 \leq x < x_1 \\
&= [g'(y_1) - h'(x - y_1)]\, dy_1/dx + h'(x - y_1) = h'(x - y_1), \\
&\quad \text{for } x > x_1.
\end{aligned}
$$

Since $y_1(x_1) = x_1$, we see that $f_1'(x)$ is continuous at $x = x_1$, and hence, for all values of $x \geq 0$. Furthermore $f_1(x)$ is a concave function of x; cf. the analysis of § 11.

Now let us turn to the second approximation

$$(7) \quad f_2(x) = \operatorname*{Max}_{0 \leq y \leq x} [g(y) + h(x-y) + f_1(ay + b(x-y))].$$

The critical function is now $D(y) = g'(y) - h'(x-y) + f_1'(ay + b(x-y))(a-b)$. Since $g'(0) - h'(0) + f_1'(0)(a-b) = g'(0) - h'(0) + g'(0)(a-b) > h'(0)[\{(1-a)(1+a-b)/(1-b)\} - 1] > 0$, we see that $D(y)$ is again positive for all y in $[0, x]$ for small x. Hence the maximum occurs in (7) at $y = x$ for small x. As x increases, there will be a first value of x where $D(x) = 0$. This value, x_2, is determined by the equation $g'(x) = h'(0) + (b-a) f_1'(ax)$. Comparing the two equations

$$
\begin{aligned}
(8) \qquad g'(x) &= h'(0) \\
g'(x) &= h'(0) + (b-a) f_1'(ax),
\end{aligned}
$$

we see that $0 < x_2 < x_1$.

Hence the equation for x_2 has the simple form

$$(9) \qquad g'(x) = h'(0) + (b-a)\, g'(ax).$$

Thus $y = x$ for $0 \leq x \leq x_2$ in (7) and $y = y_2(x)$ for $x \geq x_2$, where $y_2(x)$ is the unique solution of

$$(10) \qquad g'(y) = h'(x-y) + (b-a) f_1'(ay + b(x-y)).$$

Furthermore

$$(11) \quad \begin{aligned} f_2'(x) &= g'(x), \qquad 0 \leq x \leq x_2 \\ &= h'(x - y_2) + bf_1'(ay_2 + b(x - y_2)), \qquad x \geq x_2, \end{aligned}$$

and $f_2'(x)$ is continuous at $x = x_2$.

Comparing (10) with the equation $g'(y) = h'(x - y)$ defining y_1, we see that $y_2(x) < y_1(x)$. In order to carry out the induction and obtain the corresponding results for all members of the sequence $\{f_n\}$, defined recurrently by the relation

$$f_{n+1} = \underset{0 \leq y \leq x}{\text{Max}} \left[g(y) + h(x - y) + f_n(ay + b(x - y))\right],$$

we require the essential inequality $f_2'(x) \geq f_1'(x)$. There are three intervals $[0, x_2]$, $[x_2, x_1]$, $[x_1, \infty]$, to examine, each one requiring a separate argument. Using (10) and (11) we have

$$(12) \qquad f_2'(x) = \frac{bg'(y_2) - ah'(x - y_2)}{b - a}$$

for $x \geq x_2$. Combining (6) and the equation for y_1 we have

$$(13) \qquad f_1'(x) = \frac{bg'(y_1) - ah'(x - y_1)}{b - a}.$$

The function $[bg'(y) - ah'(x - y)]/(b - a)$ is monotone decreasing in y for $0 \leq y \leq x$. Since $y_2 < y_1$ we see that $f_2'(x) > f_1'(x)$. This completes the proof for the interval $[x_1, \infty]$. The interval $[0, x_2]$ yields equality. The remaining interval is $[x_2, x_1]$. In this interval, we have

$$(14) \qquad \begin{aligned} f_1'(x) &= g'(x) \\ f_2'(x) &= \frac{bg'(y_2) - ah'(x - y_2)}{b - a}. \end{aligned}$$

Hence in this interval, since $0 \leq y_2 \leq x$,

$$(15) \qquad f_2'(x) \geq \frac{bg'(x) - ah'(0)}{b - a} > g'(x),$$

since $g'(x) \geq h'(0)$ is a consequence of $g'(y) \geq h'(x - y)$ for $0 \leq y \leq x$ and $0 \leq x \leq x_1$. This completes the proof that $f_2'(x) \geq f_1'(x)$.

We now have all the ingredients of an inductive proof which shows that

$$(16) \qquad \begin{aligned} &\text{a. } x_1 > x_2 > \ldots x_n > \ldots > 0 \\ &\text{b. } f_1'(x) \leq f_2'(x) \leq \ldots f_n'(x) \leq \ldots \\ &\text{c. } y_1(x) > y_2(x) > \ldots \end{aligned}$$

Since $f_n(x)$ converges to $f(x)$, $f_n'(x)$ to $f'(x)$, $y_n(x)$ to $y(x)$ and x_n to $\bar{x}$, we see that the solution has the indicated form.

§ 15. An "ornery" example

Having imposed successively the conditions that g and h be both convex or both concave, let us now show by means of an example that the solution can be exceedingly complicated if we allow more general functions possessing points of inflection.

Let us consider the equation

$$(1) \qquad f(x) = \max_{0 \leq y \leq x} \left[e^{-10/y} + e^{-15/(x-y)} + f(.8y + .9(x-y)) \right].$$

The function $e^{-c/x}$ is used since it is one of the simplest possessing a point of inflection. Determining $f(x)$ by means of the method of successive approximations, we obtain a well-behaved curve

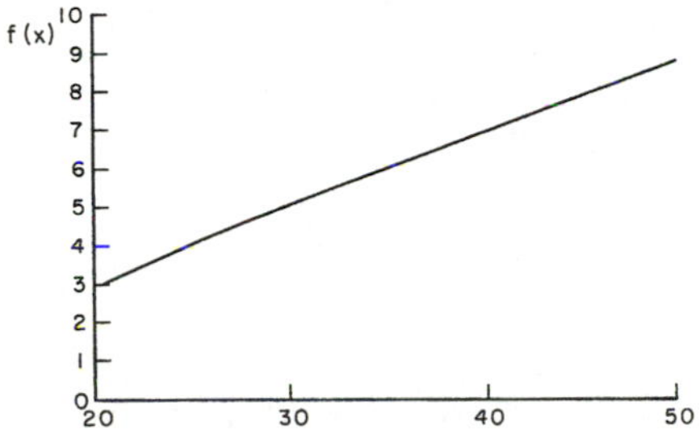

Figure 1

Note, however, the strange behavior of $y(x)$!

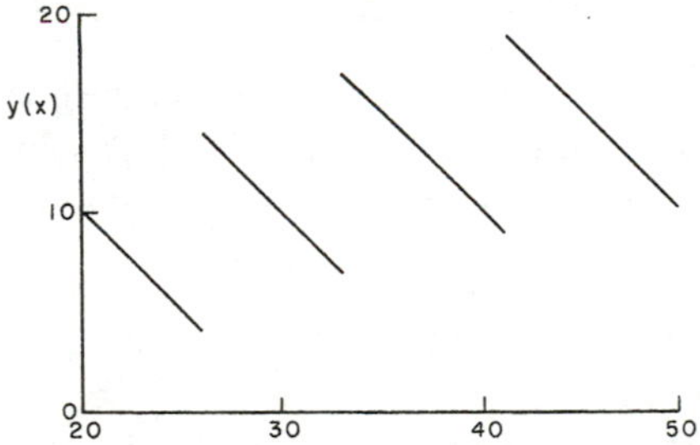

Figure 2

As soon as we allow changes of sign on the part of $g''(x)$ and $h''(x)$, we seem to encounter functional equations which defy precise analysis.

§ 16. A particular example—I

Figures 1 and 2 show the difficulties that can be encountered in the pursuit of general solutions. Let us then consider some simpler equations which can be used for approximation purposes.

THEOREM 7. *The continuous solution of*

$$f(x) = \text{Max}\,[cx^d + f(ax),\, ex^g + f(bx)],\, f(0) = 0, \tag{1}$$

subject to

$$\begin{aligned} &\text{a. } 0 < a, b < 1;\ c, d, e, g > 0, \\ &\text{b. } 0 < d < g, \end{aligned} \tag{2}$$

is given by

$$\begin{aligned} f(x) &= \frac{cx^d}{1 - a^d},\ 0 \leq x \leq \bar{x}, \\ f(x) &= ex^g + f(bx),\ x \geq \bar{x}, \end{aligned} \tag{3}$$

where

$$\bar{x} = \left[\frac{c/(1 - a^d)}{e/(1 - b^d)}\right]^{1/(g-d)} \tag{4}$$

Since $0 < b < 1$, $f(x)$ *may be found explicitly in the intervals*

$[\bar{x}, \bar{x}/b\,], \ldots [\bar{x}/b^n, \bar{x}/b^{n+1}] \ldots,$ *for* $n = 0, 1, 2, \ldots$

PROOF. Let us represent by A the operation of choosing $cx^d + f(ax)$, and by B the operation of choosing $ex^g + f(bx)$. A solution corresponding to an optimal sequence of choices, S may then be represented symbolically by

$$S = A^{a_1} B^{b_1} A^{a_2} B^{b_2} \ldots, \tag{5}$$

where a_i and b_i are positive integers or zero, and A^{a_i} means the choice A repeated a_i times, with B^{b_i} having a similar meaning.

Let us assume for the moment that the solution does have the indicated form and show how to calculate $\bar{x}$. At the point $\bar{x}$ either an A or B

decision is optimal, while below $\bar{x}$ only an A decision is optimal. Consequently, symbolically, $\bar{x}$ is the point where

$$BA^{\infty} = A^{\infty}. \tag{6}$$

To compute A^{∞} we write

$$\begin{aligned} f(x) &= cx^d + f(ax) = cx^d + c\,(ax)^d + c\,(a^2x)^d + \dots \\ &= cx^d/(1 - a^d). \end{aligned} \tag{7}$$

Similarly BA^{∞} yields

$$f(x) = ex^g + cb^dx^d/(1 - a^d). \tag{8}$$

Equating the two expressions, we find that $\bar{x}$ has the stated value.

It remains to prove that the solution has the desired form. Let us begin by showing that A is always used when x is small. To do this it is sufficient to show that $f(x) = cx^d/(1 - a^d)$ is a solution for small x, and then to invoke the uniqueness theorem.[5] We must assure ourselves that

$$\frac{cx^d}{1 - a^d} = \operatorname{Max}\left[\frac{cx^d}{1 - a^d},\; ex^g + \frac{cb^d\,x^d}{1 - a^d}\right] \tag{9}$$

for small x. This, however, is clear if $g > d > 0$ and $0 < b < 1$. We now proceed inductively. Let z be the smallest value of x for which a B-choice is optimal. At this point $BA^{\infty} = A^{\infty}$. This means that $z = \bar{x}$. Let us now consider the interval $x > \bar{x}$, and begin by asking for the point p where AB and BA are equally effective as a set of first two choices.

We have, using an obvious notation,

$$\begin{aligned} f_{AB}(x) &= cx^d + ea^gx^g + f(abx) \\ f_{BA}(x) &= ex^g + cb^dx^d + f(abx). \end{aligned} \tag{10}$$

Hence the required point p is given by

$$p = [c\,(1 - b^d)/e\,(1 - a^g)]^{1/(g-d)}. \tag{11}$$

Since $g > d$, we see that $p < \bar{x}$.

It follows then from the fact that $f_{AB}(x) < f_{BA}(x)$ for $x > p$ that for $x > \bar{x}$, AB plus an optimal continuation is inferior to BA plus an optimal continuation. From this we see that A cannot be used for $x > \bar{x}$

[5] Strictly speaking, we haven't established this uniqueness theorem yet. However, it is easy to see that the method used to establish Theorem 1 works equally well in this case.

unless followed by A^∞, which we know is also impossible. This completes the proof.

§ 17. A particular example—II

Another interesting case is that where g and h are quadratic in x. We leave as an exercise the following result:

THEOREM 8. *Let* $c, d > 0$ *and* $0 < b \leq a < 1$. *Let*

$$(1)\quad f(x) = \max_{0 \leq y \leq x} [cy - y^2 + d(x-y) - (x-y)^2 + f(ay + b(x-y))],$$
$$f(0) = 0.$$

Then, in the interval[6] $0 \leq x \leq \mathrm{Min}(c/2, d/2)$, $f(x)$ *has the following form, which depends on the sign of* $c/(1-a) - d/(1-b)$:

Case I: $c/(1-a) = d/(1-b)$.

$$(2)\qquad f(x) = \frac{(c-d)\alpha + d}{1 - b + (b-a)\alpha}x - \frac{\alpha^2 + (1-\alpha)^2}{1 - [(a-b)\alpha + b]^2}x^2,$$

where

$$(3)\qquad \alpha = \left\{1 + \frac{1}{2}\left(\frac{a^2 - b^2}{1 - ab}\right) + \sqrt{1 + \frac{1}{4}\left(\frac{a^2 - b^2}{1 - ab}\right)^2}\right\}^{-1}$$

Case II: $c/(1-a) < d/(1-b)$.

$$(4)\qquad f(x) = \left(\frac{d}{1-b}\right)x - \left(\frac{1}{1-b^2}\right)x^2,$$

for $0 \leq x \leq \mathrm{Min}(\lambda, c/2, d/2)$, *where*

$$(5)\qquad \lambda = \frac{(1+b)[d(1-a) - c(1-b)]}{2(1-ab)}.$$

When $\lambda < \mathrm{Min}(c/2, d/2)$ *use of* (1) *as a recursion formula enables one to obtain* $f(x)$ *over the entire interval of interest.*

Case III: $c/(1-a) > d/(1-b)$.

$$(6)\qquad f(x) = \left(\frac{c}{1-a}\right)x - \left(\frac{1}{1-a^2}\right)x^2$$

for $0 \leq x \leq \mathrm{Min}(\mu, c/2, d/2)$ *where*

$$(7)\qquad \mu = (1+a)[c(1-b) - d(1-a) / 2(1-ab).$$

[6] This is the maximum interval over which the g and h functions are both increasing.

§ 18. Approximation and stability

It is, of course, interesting to have the explicit solutions of as many equations as possible available. However, the true importance of the explicit solutions of simple equations lies in the use of these solutions as approximate solutions to more obdurate equations, and in furnishing clues to the nature of optimal policies for more complicated processes.

In the above sections we have derived explicit solutions for the case where g and h have monomial forms cx^d, and for the case where they are quadratic. Note that approximation to $g(x)$ by means of cx^d is equivalent to an approximation to $\log g(e^x)$ by means of $\log c + dx$, a straight line, which is readily accomplished.

Observe that as x changes, we may change our approximating curves so as to obtain better fits if we wish closer approximations. Furthermore, let us point out that in general the approximation is most useful as an approximation in policy space rather than in function space.

In order to use approximation techniques, we require an estimate for the difference between the solutions [7] of the two equations

$$(1) \qquad f(x) = \max_{0 \le y \le x} [u(x, y) + f(ay + b(x - y))], \qquad f(0) = 0,$$
$$F(x) = \max_{0 \le y \le x} [v(x, y) + F(ay + b(x - y))], \qquad F(0) = 0,$$

in terms of the difference between $u(x, y)$ and $v(x, y)$. This is a *stability* theorem in the classical sense.

Let us prove

THEOREM 9. *Let $f(x)$ and $F(x)$ be the continuous solutions of the above equations under the assumptions that $u(x, y)$ and $v(x, y)$ are continuous in x and y for all $x, y \ge 0$, with $0 < a, b < 1$, and that $\sum_{n=0}^{\infty} m(c^n z) < \infty$ where $m(z) = \max_{0 \le x \le z} [\max_{0 \le y \le x} \max \{ |u(x, y)|, |v(x, y)| \}]$.*

If

$$(2) \qquad \max_{0 \le x \le z} \{\max_{0 \le y \le x} |u(x, y) - v(x, y)|\} = D(z),$$

and $\sum_{n=0}^{\infty} D(c^n z) < \infty$, $c = \max(a, b)$, *then*

$$(3) \qquad |f(x) - F(x)| \le \sum_{n=0}^{\infty} D(c^n x).$$

[7] The existence and uniqueness of these solutions is assured by the natural modification of the proof of Theorem 1. When we speak of *the* solution, we shall mean the continuous solution, or, generally, the solution furnished by the existence theorem.

PROOF. Define

$$f_1(x) = \operatorname*{Max}_{0 \le y \le x} u(x, y) \tag{4}$$

$$f_{N+1}(x) = \operatorname*{Max}_{0 \le y \le x} [u(x, y) + f_N(ay + b(x - y))]$$

$$F_1(x) = \operatorname*{Max}_{0 \le y \le x} v(x, y)$$

$$F_{N+1}(x) = \operatorname*{Max}_{0 \le y \le x} [v(x, y) + F_N(ay + b(x - y))].$$

We know, using the methods given previously that $f_N(x)$ converges to $f(x)$, and $F_N(x)$ converges to $F(x)$ as $N \to \infty$.

Let us estimate the difference between f_1 and F_1. Clearly,

$$|f_1(x) - F_1(x)| \le \operatorname*{Max}_{0 \le y \le x} |u(x, y) - v(x, y)| \le D(x). \tag{5}$$

Proceeding, as in § 7, we have

$$|f_{N+1}(x) - F_{N+1}(x)| \le \operatorname*{Max}_{0 \le y \le x} |f_N(ay + b(x - y)) - F_N(ay + b(x - y))| + \operatorname*{Max}_{0 \le y \le x} |u(x, y) - v(x, y)| \tag{6}$$

It now follows inductively that

$$|f_{N+1}(x) - F_{N+1}(x)| \le \sum_{n=0}^{N} D(c^n x). \tag{7}$$

Letting $N \to \infty$, we obtain (3).

§ 19. Time-dependent processes

We have tacitly assumed in the foregoing pages that the processes under consideration were time-independent in that the total return depended only upon the initial quantity x and the duration of the process N, and not upon the time at which the process were initiated. Let us now see how we can handle situations in which this is not the case.

Let us assume that as a result of the division of x into y and $(x - y)$ at the kth stage, we receive a return $g_k(x,y)$ and are left with a quantity $a_k(x, y)$. It is required to determine the allocation policy which maximizes the total N-stage return.

We shall assume that $g_k(x, y)$ is continuous in x and y for $x \ge 0$ and $0 \le y \le x$ and that $a_k(x, y)$ is likewise continuous in this region and satisfies the inequality $0 \leq a_k(x, y) \leq ax$, $a < 1$, for $k = 1, 2, \ldots$

Define

(1) $f_{k,N}(x) =$ total N-stage return obtained starting with a quantity x at stage k and employing an optimal policy.

We have

$$f_{k,1}(x) = \max_{0 \leq y \leq x} g_k(x, y), \tag{2}$$

and for $N \geq 2$, arguing as in the preceding pages,

$$f_{k,N}(x) = \max_{0 \leq y \leq x} (g_k(x, y) + f_{k+1, N-1}(a_k(x, y))). \tag{3}$$

Since the double subscript is distressing both analytically, esthetically, and above all, computationally, let us see whether or not we can restore the single subscript relation. Having made up our mind that we are interested in an N-stage process starting at stage 1, let us define

(4) $f_k(x) =$ total return obtained starting with a quantity x at stage k and ending at stage N, employing an optimal policy, $k = 1, 2, \ldots, N$.

Then

(5) $$f_N(x) = \max_{0 \leq y \leq x} g_N(x, y)$$

$$f_k(x) = \max_{0 \leq y \leq x} [g_k(x, y) + f_{k+1}(a_k(x, y))], \ k = 1, 2, \ldots, N-1.$$

This simplification is essential if we are interested in computational solutions, since the difference between the effort involved in the tabulation of functions of one variable and functions of two variables is enormous, while that between the tabulation of functions of two variables and functions of three variables may be the difference between a feasible and unfeasible approach.

The case of unbounded processes, i.e., $N = \infty$, yields the set of functional equations

$$f_k(x) = \max_{0 \leq y \leq x} [g_k(x, y) + f_{k+1}(a_k(x, y))]. \tag{6}$$

It is not difficult to obtain the analogues of Theorem 1 for these systems.

§ 20. Multi-activity processes

The process we have been using for expository purposes is the simplest of its category since we allow only one type of resource, and require only one allocation at each stage. Let us now discuss the formulation of more general and more realistic processes.

Let there be M different kinds of resources, in quantities $x_1, x_2, \ldots, x_M$ respectively. At each stage, a quantity x_{ij} of the ith resource is utilized to produce an additional quantity of the jth resource. Hence we have the equations, relating the resources at the $(k+1)^{\text{st}}$ stage to the resources at the kth stage,

$$(1) \quad x_i(k+1) = x_i(k) - \sum_{j=1}^{M} x_{ij}(k) + g_i(x_{1i}(k), x_{2i}(k), \ldots, x_{Mi}(k)),$$

for $i = 1, 2, \ldots, M$, where

$$(2) \qquad \text{(a)} \quad x_{ij}(k) \geq 0,$$

$$\text{(b)} \quad \sum_{j=1}^{M} x_{ij}(k) \leq x_i(k),$$

and the production functions, g_i, are assumed known, together with the initial quantities, $x_i(0) = c_i$.

The $x_{ij}(k)$ are to be chosen so as to maximize some pre-assigned function

$$(3) \qquad R_N = F(x_1(N), x_2(N), \ldots, x_M(N)),$$

of the final resources.

In many cases, as we shall see in Chapter 6, there are other constraints in addition to those of (2).

If we set

$$(4) \qquad f_N(c_1, c_2, \ldots, c_M) = \operatorname*{Max}_{\{x_{ij}\}} R_N,$$

we obtain, as before, the recurrence relations

$$(5) \quad f_N(c_1, c_2, \ldots, c_M) = \operatorname*{Max}_{\{y_{ij}\}} f_{N-1}(c_1 - \sum_{j=1}^{M} y_{1j} + g_1(y_{11}, y_{21}, \ldots, y_{m1}), \ldots)$$

for $N \geq 2$, where the y_{ij} are restricted by the relations

$$(6) \qquad \text{(a)} \quad y_{ij} \geq 0$$

$$\text{(b)} \quad \sum_{j=1}^{M} y_{ij} \leq c_i, \quad i = 1, 2, \ldots, M,$$

and

$$(7) \qquad f_1(c_1, c_2, \ldots, c_M) = F(c_1, c_2, \ldots, c_M).$$

Existence and uniqueness theorems covering the unbounded versions of these general processes will be given in Chapter IV, in conjunction with a better notation. We shall encounter a particular example of this equation further along in connection with the bottleneck processes of Chapter VI. In the present chapter we shall discuss briefly some of the difficult computational problems raised in maximizing over a multi-dimensional domain.

§ 21. Multi-dimensional structure theorems

It is not difficult to extend the results we obtained in the one-dimensional case concerning convexity and concavity of the solutions of the functional equation of (8.1) to the multi-dimensional equations of § 20.

Let $G(x)$ be a scalar function of a vector variable x. It is said to be *convex* if

$$G(\lambda x + (1-\lambda) y) \leq \lambda G(x) + (1-\lambda) G(y) \tag{1}$$

for all λ in the range $0 \leq \lambda \leq 1$. The function is *concave* if the inequality goes the other way.

The multi-dimensional analogue of Lemma 1, proved in § 13, is valid and the proof is precisely the same. Using the lemma, we can establish the result below.

Before stating the result, let us introduce a more convenient notation. Let x denote the vector whose components are x_i, and $y^{(i)}$ denote the vector whose components are y_{ij}, for $1 \leq i, j \leq M$. Then, in terms of the process described above, we have

(2)
$$\text{(a)}\quad x = \sum_i y^{(i)},$$
$$\text{(b)}\quad y^{(i)} \geq 0,$$

where the notation $y \geq 0$ signifies that all components of y are non-negative. Let $D(x, y)$ denote the domain defined by (2).

THEOREM 10. *If $r(x, y)$ and $a(x, y)$ are continuous concave functions of x and y for all $x, y \geq 0$, and $r(x, y)$, $a(x, y)$ are monotone increasing in the components of x, then the functions $\{f_N(x)\}$ defined by the equations*

(2)
$$f_1(x) = \operatorname*{Max}_{D(x,y)} r(x, y),$$
$$f_{N+1}(x) = \operatorname*{Max}_{D(x,y)} [r(x, y) + f_N(a(x, y))]$$

are all concave functions of x for $x \geq 0$.

This implies a unique optimal policy for each N, if $r(x, y)$ is strictly concave.

The importance of this result resides in the following. If we have an N-stage process where k decisions must be made at each stage, the functional equation approach reduces the Nk-dimensional maximization problem to a set of N k-dimensional problems. Although this is an essential reduction, the k-dimensional maximization problems themselves possess thorny features.

If, however, the function of k variables we are maximizing is *strictly concave*, we know that it possesses a unique relative maximum which is the absolute maximum. Given this additional information that the function under investigation has a *unique* relative maximum, we should be able to determine a search procedure for the location of this maximum which is far more efficient than the search procedure we would employ for a general function.

§ 22. Locating the unique maximum of a concave function

The determination of optimal search procedures [8] for the location of the maximum of a concave function or, conversely, for the minimum of a convex function, is an extremely important and difficult problem which has not been solved to date. The solution has, however, been obtained in the one-dimensional case for the more general situation where the function is unimodal, which is to say possesses a single relative maximum.

Let us pose the problem in the following terms. The function $y = f(x)$ is a strictly unimodal function defined on the interval $[0, L_n]$. We wish to determine the maximum L_n with the property that we can *always* locate the maximum of $y = f(x)$ on a sub-interval of unit length by calculating at most n values of the function $f(x)$. Since the maximum may not exist, it is safer to begin by setting

$$F_n = \text{Sup } L_n \tag{1}$$

We then have the following result

THEOREM 11. *F_n is the nth Fibonacci number*; i.e., $F_0 = F_1 = 1$ *and*

$$F_n = F_{n-1} + F_{n-2} \tag{2}$$

for $n \geq 2$.

PROOF. The definition of F_0 is a matter of convention, on the other hand the value of F_1 is determined by the process.

[8] It is actually not easy to specify precisely what we mean by an optimal search procedure. It clearly depends upon the type of equipment we have, the type of operations we permit, the "cost" of these operations, and so on. Consequently, there are a variety of problems of the above type which may be posed. The subject has not been explored to any extent.

Let us now proceed inductively. Fix n and calculate the values $y_1 = f(x_1)$, $y_2 = f(x_2)$ where $0 < x_1 < x_2 < L_n$. If $y_1 > y_2$, the maximum occurs on $(0, x_2)$ since $f(x)$ is strictly unimodal. If $y_2 > y_1$, the maximum is on (x_1, L_n). If $y_1 = y_2$, choose either of the above intervals, even though we know the maximum occurs on (x_1, x_2). Thus, at each stage after the first computation we are left with a subinterval and the value of $f(x)$ at some interior point x. Since values at the ends of an interval furnish no information per se, we restrict our attention to the interior points.

For $n = 2$, $L_n = 2 - \varepsilon$, $x_1 = 1 - \varepsilon$, $x_2 = 1$, for arbitrarily small $\varepsilon > 0$. From the preceding argument it follows that $F_2 = 2 = F_1 + F_0$.

Consider the case where $n > 2$ and assume that $F_k = F_{k-1} + F_{k-2}$ for $k = 2, \ldots, n - 1$. Let us begin by showing that

$$F_n \leq F_{n-1} + F_{n-2}. \tag{3}$$

For if we calculate $f(x)$ at x_1 and x_2 on $(0, L_n)$ we have

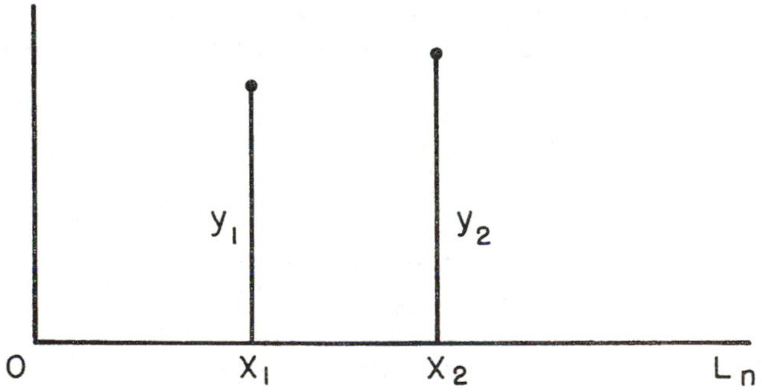

Figure 3

If $y_1 > y_2$, we obtain the new picture

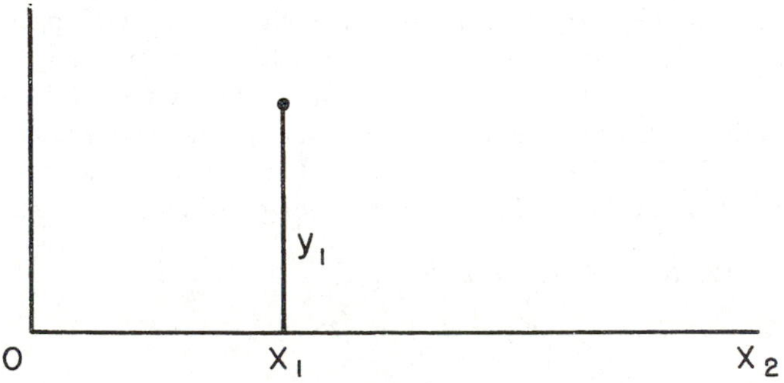

Figure 4

In this case $x_2 < F_{n-1}$ since we have only $(n - 2)$ additional choices with x_1 a first choice, for the case $k = n - 1$. Moreover, $x_1 < F_{n-1}$, since the maximum could occur on $[0, x_1]$, with two choices of x already used.

Similarly if $y_2 > y_1$, we have $L_n - x_1 < F_{n-1}$

Thus in all cases $L_n < F_{n-1} + F_{n-2}$, which yields (3). Now chose L_n, x_1, x_2 arbitrarily close to their respective upper bounds $F_{n-1} + F_{n-2}$, F_{n-1} and F_{n-2} respectively. Then $F_n = F_{n-1} + F_{n-2}$. This yields the proof of Theorem 11. Furthermore, it yields the optimal policy, since each x_i is either discarded or is the optimal first choice for the remaining subinterval.

The sequence $\{F_n\}$ has as its first few terms

$$1, 1, 2, 3, 5, 8, 13, 21, 34, 55, \ldots, \tag{4}$$

with $F_{20} > 10{,}000$. Hence the maximum of a strictly unimodal function can always be located within 10^{-4} of the original interval length with at most 20 calculations of the value of the function.

It is easy to obtain an explicit representation for F_n, namely

$$F_n = \frac{(r_2 - 1)}{(r_2 - r_1)} r_1^n + \frac{(1 - r_1)}{(r_2 - r_1)} r_2^n, \tag{5}$$

where

$$r_1 = \frac{1 + \sqrt{5}}{2} \cong 1.61 \tag{6}$$

$$r_2 = \frac{1 - \sqrt{5}}{2} \cong -.61$$

From this we see that $F_{n+1}/F_n \to r_1 \cong 1.61$ as $n \to \infty$. Thus, for large n, a uniform approximate procedure is to choose the two first values at distances L/r_1 from either end, where L is the length of the interval. This is a useful technique for machine computation.

Consider now the related problem where the unimodal function is defined only for discrete values of x. Let K_n be the maximum number of points such that the maximum of the function can always be identified in n computations. The same type of proof as above establishes.

THEOREM 12. $K_0 = 1$, $K_1 = 1$, $K_2 = 2$, $K_3 = 4$, *and*

$$K_n = 1 + F_n, \; n \geq 3. \tag{7}$$

§ 23. Continuity and memory

Let us suppose that we have a function of two variables, $f(x, y)$, depending continuously on x and y for $x \geq 0$ and $0 \leq y \leq x$. Define the function

$$g(x) = \max_{0 \leq y \leq x} f(x, y). \tag{1}$$

It is clear that $g(x)$ will be continuous, but the function $y = y(x)$ yielding the maximum need not be continuous. We have already seen an example of this in connection with the functional equation of § 15.

Suppose, however, that we restrict $f(x, y)$ to be a strictly concave function of y for all y in $[0, x]$, for $x \geq 0$.

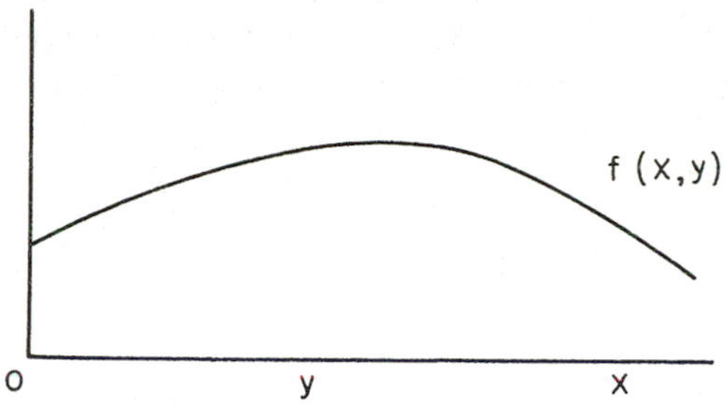

Figure 5

It is clear that as x varies, the maximizing y will now be a continuous function of x.

Let us see how we can utilize this information to simplify the memory problem for computing machines. Consider the equations

$$f_{N+1}(x) = \max_{0 \leq y \leq x} [g(y) + h(x - y) + f_N(ay + b(x - y))], \tag{2}$$
$$N = 1, 2, \ldots .$$

If we have no information concerning the location of a maximizing y, we must have available all values of $f_N(z)$ for $0 \leq z \leq ax$ in order to determine $f_{N+1}(x)$. Suppose, however, we take $g(x)$ and $h(x)$ to be strictly concave as well as continuous. In this case, $f_N(x)$ is strictly concave for each N and the function $g(y) + h(x - y) + f_N(ay + b(x - y))$ is strictly concave for $0 \leq y \leq x$, and what is most important the function $y_N(x)$ which yields the maximum in (2) is unique and continuous as a function of x.

It follows than that if we are using an x-grid of values $0, \Delta, 2\Delta, \ldots$, to compute $f(x)$, the complete set of values of $f_N(z)$ for $0 \leq z \leq ax$ is not required to compute $f_{N+1}(x)$, but only the values of $f_N(z)$ in a relatively small neighborhood of $z = y_N(x - \Delta)$.

The same idea extended to multi-dimensional equations can result in a considerable saving of memory space in computing machines. Reciprocally, we will be able to solve problems using existing machines which might otherwise escape them. In any case, a great saving in running time will result, once again increasing the feasibility of a solution by these means.

§ 24. Stochastic allocation processes

In the preceding pages of the chapter, we have considered, in greater and lesser detail, various multi-stage allocation processes characterized by the property that the outcome of any decision was uniquely determined by the choice of this decision. Processes of this type we call *deterministic*.

Not all multi-stage processes, however, possess this property, and, as a matter of fact, many of the most interesting are quite definitely not of this type. Let us consider here one important class of non-deterministic processes in which the effect of a decision is to determine a distribution of outcomes in the sense of probability theory. Processes of this type we shall call *stochastic*.

We shall limit ourselves in this book to processes of these two types. The discussion of the origin of processes of more complicated nature, and their treatment, we shall defer to another place.

From the mathematical point of view, stochastic processes furnish varied classes of fascinating analytic problems, and throw unexpected light upon many processes of supposedly deterministic nature. Applications of the theory are furnished by scores of processes drawn from biologic, economic, engineering, and physical fields.

Returning to our domain of decision processes, a fundamental problem confronting us is that of defining what we mean by an optimal policy in the face of uncertain outcomes. What is crystal clear, but so often overlooked in a posteriori comment, is the fact that a lack of complete control over a process effectively prevents a guarantee of a maximum return.

On the other hand, despite this Damoclean sword of uncertainty, there must exist some means of comparing policies, taking into account the possible fluctuation of outcomes.

What causes a major difficulty in applications is not that it is hard to find such a measure, but rather that is is hard to find a unique measure. In short, it must be emphasized that there is no one method which can have any pretensions to the title of "best." Whatever method is used depends to a large extent upon various analytic and arithmetic aspects of the process, and, it must be confessed, upon the philosophical and psychological attitudes of the decision-makers.

Having thus dwelt upon the dismal side of the matter, to assuage our consciences, let us now proceed more constructively.

The general idea, and this is fairly unanimously accepted, is to use some *average* of the possible outcomes as a measure of the value of a policy. It is in the choice of this average that the difficulties arise.

Let us point out in passing that there is a definite lack of unanimity concerning the use of averages in determining policies for stochastic processes which may be carried through once, or at best, only a few times. In some cases, "distribution-free" policies can be obtained. In general, however, there seems to be no other approach to these questions than the usual one we present here.

The first average, or criterion, we shall employ is the common arithmetic weighted average, or *expected value*. Due to the linearity of this average, it possesses a most important invariant property which greatly simplifies the functional equations which describe the process. This property enables the future decisions to be based solely upon the present state of the system, independently of the past history of the process.

The second criterion, which is far less frequently used, is the probability of achieving at least a certain level of return. This also possesses the proper invariant structure as far as multi-stage processes are concerned. We will discuss this criterion in greater detail in a subsequent chapter.

§ 25. Functional equations

Let us now consider a simple stochastic version of the deterministic process considered in § 2, and show that the same functional equation technique is applicable.

In place of assuming that the outcome of a division of x into y and $x - y$ is a return of $g(y) + h(x - y)$, leaving a new quantity $x_1 = ay + b(x - y)$, let us assume that with probability p_1 there is a return of $g_1(y) + h_1(x - y)$ and a remaining quantity $a_1 y + b_1(x - y)$, and with probability $p_2 = 1 - p_1$ a return of $g_2(y) + h_2(x - y)$ and a new quantity $a_2 y + b_2(x - y)$

Let us define

(1) $f_N(x) =$ the *expected* total return of an N-stage process, obtained using an optimal policy, starting with an initial quantity x.

Then, as before, we obtain the equations

$$(2) \qquad f_1(x) = \operatorname*{Max}_{0 \leq y \leq x} [p_1(g_1(y) + h_1(x - y)) + p_2(g_2(y) + h_2(x - y))],$$

$$f_{N+1}(x) = \underset{0 \leq y \leq x}{\text{Max}} \; [p_1 [g_1(y) + h_1(x-y) + f_N(a_1 y + b_1(x-y))] + p_2 [g_2(y) + h_2(x-y) + f_N(a_2 y + b_2(x-y))]],$$

for $N \geq 1$.

The equations have the same analytic structure as those obtained from the deterministic process. By agreeing to use the "expected value" as the measure of the value of a policy, we have eliminated the stochastic aspects of the process, at least as far as the analysis is concerned.

§ 26. Stieltjes integrals

For those who are familiar with the Riemann-Stieltjes integral, there is a much more compact way of writing the above equations. Let

(1) $dG(u, v; x, y)$ = distribution function of a return of u and a remaining quantity of v, starting with an initial quantity x and making an allocation of y.

Taking $f_N(x)$ to be defined as above, we obtain the equations

$$\text{(2)} \qquad f_1(x) = \underset{0 \leq y \leq x}{\text{Max}} \int u \, dG(u, v; x, y),$$

$$f_{N+1}(x) = \underset{0 \leq y \leq x}{\text{Max}} \int [u + f_N(v)] \, dG(u, v; x, y)$$

It is much simpler to describe the processes, to establish existence and uniqueness theorems for the resultant functional equations, and to derive analytic properties of the solution, using this short-hand notation. The basic mathematical ideas are, however, the same.

Equations of this type will be discussed again in Chapter III within a more general framework.

Exercises and Research Problems for Chapter I

1. Let us define the function

$$f_N(a) = \underset{R}{\text{Max}} \, [x_1 x_2 \ldots x_N]$$

where R is the region determined by the conditions

a. $x_1 + x_2 + \ldots + x_N = a, \; a > 0$.

b. $x_i \geq 0$.

Prove that $f_N(a)$ satisfies the recurrence relation

$$f_N(a) = \max_{0 \le x \le a} x f_{N-1}(a - x), \ N \ge 2,$$

with $f_1(a) = a$.

2. Show inductively that $f_N(a) = a^N/N^N$, and hence establish the arithmetic-geometric mean inequality,

$$\left(\frac{x_1 + x_2 + \ldots + x_N}{N}\right)^N \ge x_1 x_2 \ldots x_N,$$

for $x_i \ge 0$, with equality only if $x_1 = x_2 = \ldots = x_N$.

3. Let us define the function

$$f_N(a) = \min_R \sum_{i=1}^{N} x_i^p, \ p > 0,$$

where R is the region defined by

a. $\sum_{i=1}^{N} x_i \ge a, \ a > 0.$

b. $x_i \ge 0.$

Show that $f_N(a)$ satisfies the recurrence relation

$$f_N(a) = \min_{0 \le x \le a} [x^p + f_{N-1}(a - x)], \ N \ge 2,$$

with $f_1(a) = a^p$.

4. Show that $f_N(a) = a^p c_N$, where c_N depends only upon N and p, and thus that

$$c_N = \min_{0 \le x \le 1} [x^p + (1 - x)^p c_{N-1}].$$

Determine c_N for the ranges $0 \le p < 1$, $1 < p$, respectively.

5. Consider the problem of minimizing the function

$$F(x_1, x_2, \ldots, x_N) = \sum_{i=1}^{N} p_i s_i/(s_i + x_i),$$

where the p_i and s_i are parameters subject to the conditions $p_i > 0$, $\sum_i p_i = 1$, $s_i > 0$, and the x_i range over the region defined by $x_i \ge 0$, $\sum_{i=1}^{N} x_i = a$.

Obtain the corresponding recurrence relations and show that the solution is of the form

$$x_j = 0, \quad 0 \leq j \leq t,$$
$$x_j > 0, \quad t + 1 \leq j \leq N$$

under a suitable reordering of the x_i's.

6. Consider the problem of maximizing the function

$$F(x_1, x_2, \ldots, x_N) = \sum_{i=1}^{N} \varphi(x_i),$$

subject to the constraints $x_i \geq 0$, $\sum_{i=1}^{N} x_i = c$. Show that the maximum is $\varphi(c)$, under the assumption that $\varphi(x)$ is convex.

7. Consider the case where $\varphi(x)$ is a monotonically increasing function which is strictly concave. Show that the solution of the corresponding functional equation,

$$f_N(c) = \underset{0 \leq y \leq c}{\text{Max}} \, [\varphi(y) + f_{N-1}(c - y)], \; N \geq 2,$$
$$f_1(c) = \varphi(c),$$

has the form

$$y_N = 0, \; 0 \leq c \leq c_N,$$
$$= z_N, \; c > c_N,$$

where z_N is the unique solution of

$$\varphi'(y) = f_{N-1}(c - y),$$

for $N \geq 2$, and show how to determine the sequence $\{c_N\}$.

8. Obtain explicit recurrence relations, and the analytic form of the sequence for the case where

$$\varphi(y) = y - by^2, \; b > 0,$$

and c is restricted to the range $0 \leq c \leq 1/2\,b$.

9. What are the analogues of these result for the case where the function F has the form $\sum_{i=1}^{N} \varphi_i(x_i)$, where each function $\varphi_i(x)$ satisfies the same conditions as above?

10. Carry through the corresponding analysis for the problem of minimizing $F(x_1, x_2, \ldots, x_N) = \sum_{i=1}^{N} \varphi(x_i)$, subject to $x_i \geq 0$, $\sum_{i=1}^{N} x_i = a$ in the

case where $\varphi(x)$ is a non-negative monotonically increasing function which is strictly convex. Consider, in particular, the case where

$$\varphi(x) = x + bx^2, b > 0.$$

11. Consider the problem of maximizing

$$F(x_1, x_2, \ldots, x_N; y_1, y_2, \ldots, y_N) = \sum_{i=1}^{N} \varphi(x_i, y_i),$$

subject to

$$x_i, y_i \geq 0, \sum_{i=1}^{N} x_i = c_1, \sum_{i=1}^{N} y_i = c_2,$$

where $\varphi(x, y)$ is a strictly concave function, monotone increasing in x and y.

Show that the corresponding functional equation

$$f_N(c_1, c_2) = \max_{\substack{0 \leq x \leq c_1 \\ 0 \leq y \leq c_2}} [\varphi(x,y) + f_{N-1}(c_1 - x, c_2 - y)],$$

possesses for each $N \geq 2$ a solution of the form

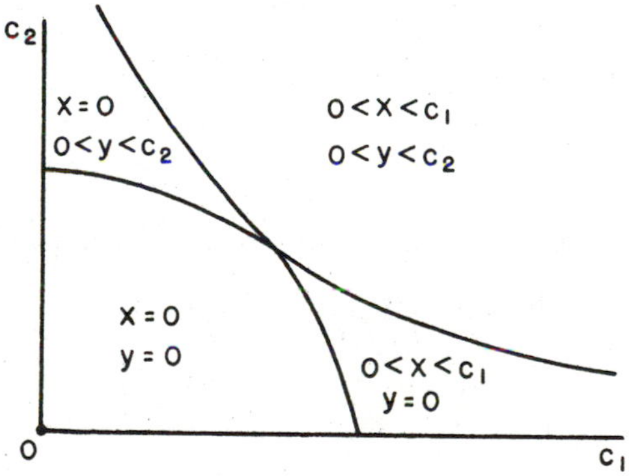

Figure 6

and show how to determine the boundary curves.

Consider, in particular, the case where

$$\varphi(x,y) = u_1 x + v_1 y + u_2 x^2 + 2u_3 xy + u_4 y^2,$$

12. Under the assumption that $\varphi(x)$ is a monotonically increasing strictly concave function, determine the maximum of $F(x_1, x_2, \ldots, x_N) = \sum_{i=1}^{N} \varphi(x_i)$ over the region determined by

a. $\sum_{i=1}^{N} x_i \leq c_1,\ x_i \geq 0$

b. $\sum_{i=1}^{N} x_i^p \leq c_2,$

for $p > 1$ and $p < 1$ respectively.

13. Obtain the recurrence relations arising from the problem of minimizing $\sum_{i=1}^{N} \varphi_i(x_i)$ subject to the restrictions

a. $0 \leq x_i \leq r_i,$

b. $\sum_{i=1}^{N} \psi_i(x_i) \geq a,$

under the assumptions that each $\psi_i(x)$ is a non-negative monotone increasing function of x, with $\sum_{i=1}^{N} \psi_i(r_i) \geq a.$

14. Consider the corresponding multi-dimensional problem of minimizing $\sum_{i=1}^{N} \varphi_i(x_i, y_i)$ subject to the constraints

a. $0 \leq x_i \leq r_i,\ 0 \leq y_i \leq s_i,$

b. $\sum_{i=1}^{N} \psi_i(x_i, y_i) \geq a,$

under appropriate assumptions concerning the sequence $\{\psi_i\}$.

15. Determine the maximum of the function $x_1 x_2 \ldots x_N$ over the region defined by

a. $\sum_{i=1}^{N} x_i = 1,\ x_i \geq 0,$

b. $bx_k \leq x_{k+1},\ b > 1,\ k = 1, 2, \ldots, N-1.$

Consider the same problem for the function $\sum_{i=1}^{N} x_i^p$, for different ranges of p.

16. Consider the recurrence relations

$$f_1(x) = \max_{0 \leq y \leq x} [g(y) + h(x-y)],$$

$$f_{N+1}(x) = \max_{0 \leq y \leq x} [g(y) + h(x-y) + f_N(ay + b(x-y))],$$

where $g(y) = c_1 y^d$, $h(y) = c_2 y^d$, with $c_1, c_2, d > 0$. Show that $f_N(x) = u_N x^d$, where

$$u_1 = \underset{0 \le v \le 1}{\text{Max}} \left[c_1 v^d + c_2 (1 - v)^d \right],$$

$$u_{N+1} = \underset{0 \le v \le 1}{\text{Max}} \left[c_1 v^d + c_2 (1 - v)^d + u_N (av + b(1 - v))^d \right].$$

Show that

$$\lim_{N \to \infty} u_N = \underset{0 \le v \le 1}{\text{Max}} \left[\frac{c_1 v^d + c_2 (1 - v)^d}{1 - (av + b(1 - v))^d} \right].$$

17. Consider the process described in § 2 under the assumption that it is not required to use all the resources available at each stage. Show that the functional equation obtained in this way has the form

$$f(x) = \underset{y_1 + y_2 \le x}{\text{Max}} \left[g(y_1) + h(y_2) + f(ay_1 + by_2 + x - y_1 - y_2) \right].$$

Does this equation have a solution if $g(x)$ and $h(x)$ are both concave functions of x? Does it have a solution if they are both convex? Under what conditions upon $g(x)$ and $h(x)$ does it have a solution with a corresponding optimal policy?

18. Show that if there is a solution with $y_1 + y_2 < x$, $y_1, y_2 > 0$, then $g'(y_1)/(1 - a) = h'(y_2)/(1 - b)$ under suitable assumptions concerning g and h. What is the interpretation of this solution?

19. Consider the process described in § 2 under the assumption that additional resources are added at each stage, either externally or from the conversion of all or part of the return $g(y) + h(x - y)$ into resources, and obtain the corresponding recurrence relations.

20. Consider the process described in § 2. Define $g_N(z)$ as the minimum cost required to obtain a total return of z at the end of N stages. Show that

$$g_1(z) = \underset{\substack{g(y_1) + h(y_2) = z \\ y_1, y_2 \ge 0}}{\text{Min}} \left[(1 - a) y_1 + (1 - b) y_2 \right],$$

$$g_{N+1}(z) = \underset{y_1, y_2 \ge 0}{\text{Min}} \left[(1 - a) y_1 + (1 - b) y_2 + g_N(z - g(y_1) - h(y_2)) \right]$$

21. There are N different types of items, with the ith item having weight w_i and a value v_i. It is desired to load a ship having a total capacity of w pounds with a cargo of greatest possible value. Show that this problem leads to the problem of determining the maximum over the n_i of the linear form $L = \sum_{i=1}^{N} n_i v_i$, subject to the constraints, $n_i = 0, 1, 2, \ldots, N$,

$\sum_{i=1} n_i w_i \leq w$, and thus that this problem leads to the recurrence relations

$$f_1(w) = v_1 [w/w_1], \quad ([a] \text{ denotes the greatest integer contained in } a)$$

$$f_{N+1}(w) = \max_{0 \leq x \leq \left[\frac{w}{w_{N+1}}\right]} [x v_{N+1} + f_N (w - x w_{N+1})],$$

where x can assume only zero or integral values.

22. Suppose that we have a herd of cattle and the prerogative, at the end of the year, of sending one part of the herd to market, and retaining the other part for breeding purposes. Assume that the dollar value of y cattle sent to market is $\varphi(y)$, and that z retained for breeding purposes yield az, $a > 1$, at the beginning of the next year.

Show that the problem of determining a breeding policy which maximizes the total return over an N-year period leads to the recurrence relation

$$f_1(x) = \max_{0 \leq y \leq x} \varphi(y)$$

$$f_N(x) = \max_{0 \leq y \leq x} [\varphi(y) + f_{N-1}(a(x - y))].$$

23. Determine the structure of the optimal policies in the following cases:

a. $\varphi(y) = ky$, $k > 0$
b. $\varphi(y)$ is quadratic in y
c. $\varphi(y)$ is strictly convex
d. $\varphi(y)$ is strictly concave

24. Formulate the equations under the additional restriction that cattle must be 2 years old before they can be sold. Take into account feeding cost and mortality rates.

25. Consider the case in which there are probability distributions for the price and demand.

26. In problem 22, let $\varphi(x) = cx^d$, $c, d > 0$. Show that $f_N(x) = c_N x^d$, where $c_1 = c$ and $c_{N+1} = \max_{0 \leq r \leq 1} [r^d + c_N a^d (1 - r)^d]$, $N = 1, 2, \ldots$. Determine the asymptotic behavior of c_{N+1}/c_N and r_{N+1}/r_N.

27. Suppose that we have a quantity x of money, and that portions of this money can be used for common goods, invested in bonds, or invested in stocks. The return from y dollars invested in bonds is ay dollars, $a > 1$, over a period of one year; the return from z dollars invested in stocks is

bz dollars, $b > 1$, over a period of one year. The utility of w dollars spent is $\varphi(w)$. How should the capital be utilized so as to derive a maximum utility over an N year period?

28. Consider the same problem under the assumption that the return from stocks is a stochastic quantity.

29. A sophomore has three girl friends, a blonde, a brunette, and a redhead. If he takes one of the three to the Saturday night dance, the other two take umbrage, with the result that the probability that they will refuse an invitation to next week's dance increases. Furthermore, as a result of his invitation, there is a certain probability that the young lady of his choice will be more willing to accept another invitation and a certain probability that the young lady will be less willing.

Assuming that feminine memories do not extend back beyond one week, what dating policy maximizes the expected number of dances the sophomore attends—with a date?

30. Obtain a sequence of recurrence relations equivalent to determining the minimum of the linear form $L = \sum_{i=1}^{N} x_i$, subject to the constraints $x_i \geq 0$, $x_i + x_{i+1} \geq a_i$, $i = 1, 2, \ldots, N-1$. Thus, or otherwise, show that $\text{Min } L = \max_i a_i$, granted that one a_i is positive.

31. Solve the corresponding problem for the case where the constraints are $x_i + x_{i+1} + x_{i+2} \geq a_i$, $i = 1, 2, \ldots, N-2$.

32. Determine the recurrence relations for the problem of minimizing $L = \sum_{i=1}^{N} c_i x_i$, $c_i \geq 0$, subject to the constraints

$$x_i \geq 0, \; b_i x_i + d_i x_{i+1} \geq a_i, \; i = 1, 2, \ldots, N-1.$$

33. Solve the problem formulated above in (32) for the case where the constraints are

a. $x_i + x_{i+1} \geq a_i$, $i = 1, 2, \ldots, N-1$, $x_N \geq a_N$, or

b. $x_i + x_{i+1} \geq a_i$, $i = 1, 2, \ldots, N-1$, $x_1 \geq a$, $x_N \geq a_N$, or

c. $x_i + x_{i+1} + x_{i+2} \geq a_i$, $i = 1, 2, \ldots, N-2$, $x_{N-1} + x_N \geq a_{N-1}$,
$$x_N \geq a_N.$$

plus the usual constraint $x_i \geq 0$.

34. Show how to approximate to $f(x)$ in the interval $[a, b]$ by means of a

linear function $ux + v$ according to the following measures of deviation

a. $\int_a^b (f(x) - ux - v)^2 \, dx$

b. $\underset{a \leq x \leq b}{\text{Max}} \; |f(x) - ux - v|$

35. Suppose that it is necessary to traverse a distance x. If we travel at a speed v there is a probability $p(v)\,ds$ of being stopped in the interval $(s, s + ds)$ and incurring a delay of d time units. At what fixed speed should we travel in order to minimize the expected time required to cover a distance x? (Greenspan)

36. Under the same conditions as those of Problem 35, at what speed should we travel in order to minimize the probability of requiring more than a time T to cover the distance x?

37. Assume that there is a penalty of p dollars when stopped and that actual travelling time costs c dollars per unit time. How do we proceed to minimize expected cost?

38. Obtain a recurrence relation equivalent to the problem of minimizing the quadratic form $Q_N = \sum_{k=1}^{N} (x_k - x_{k-1})^2$ over all sets of values for the x_k for which $\sum_{k=1}^{N} x_k^2 = 1$, $x_0 = c$.

39. We are informed that a particle is in either of two states, which we shall call S and T, and are given the initial probability x that it is in state T. If we use an operation A we reduce this probability to ax, where a is a positive constant less than 1, whereas operation L, which consists of observing the particle, will tell us definitely which state it is in. It is desired to transform the particle into state S in a minimum time, with certainty.

If $f(x)$ is defined to be the expected number of operations required to achieve this goal, show that $f(x)$ satisfies the equation

$$f(x) = \text{Min} \begin{Bmatrix} L: & 1 + xf(1) \\ A: & 1 + f(ax) \end{Bmatrix}, \quad 0 < x \leq 1,$$

$$f(0) = 0.$$

40. Show that there is a number x_0 in the interval $(0,1)$ with the property that

$$\begin{aligned} f(x) &= 1 + xf(1), \; 0 < x \leq x_0 \\ &= 1 + f(ax), \; 1 \geq x \geq x_0. \end{aligned}$$

Show that

$$f(1) = \text{Min}\left(\frac{k+1}{1-a^k}\right), \; k = 1, 2, \ldots,$$

$$x_0 = \frac{1}{(1-a)\,f(1)} = \frac{1-a^k}{(1-a)\,(k+1)},$$

for the minimizing value of k.

41. At each stage of a sequence of actions, we are allowed our choice of one of two actions. The first has associated a probability p_1 of gaining one unit, a probability p_2 of gaining two units, and a probability p_3 of gaining nothing and terminating the process. The second has a similar set of probabilities p_1', p_2', p_3'. We wish to determine a sequence of choices which maximizes the probability of attaining at least n units before the process is terminated.

Let $p(n)$ denote the this probability for $n = 1, 2, 3, \ldots$. Show that $p(n)$ satisfies the equation

$$p(n) = \text{Max}\begin{bmatrix} p_1 p(n-1) + p_2 p(n-2), \\ p_1' p(n-1) + p_2' p(n-2) \end{bmatrix},$$

for $n = 2, 3, 4, \ldots$, with $p(0) = 1$, and

$$p(1) = \text{Max}\,(p_1, p_1').$$

42. With reference to § 7, show that if $g(x)$ and $h(x)$ are quadratic in x, then $f_N(c) = \alpha_N + \beta_N c + \gamma_N c^2$ where α_N, β_N, γ_N are independent of c.

43. Show that there exist recurrence relations of the form

$$\alpha_{N+1} = R_1(\alpha_N, \beta_N, \gamma_N),$$
$$\beta_{N+1} = R_2(\alpha_N, \beta_N, \gamma_N),$$
$$\gamma_{N+1} = R_3(\alpha_N, \beta_N, \gamma_N),$$

where the R_i are rational functions.

44. Treat in a similar way the problem of minimizing the function

$$f(x_1, x_2, \ldots, x_N) = \sum_{k=1}^{N} [g(x_k - f_k) + h(x_k - x_{k-1})$$
$$+ m(x_k - 2x_{k-1} + x_{k-2})],$$

where $g(x)$, $h(x)$ and $m(x)$ are quadratic.

45. Suppose that we have a machine whose output per unit time is $r(t)$ as a function of t, its age measured in the same units. Its upkeep cost per unit time is $u(t)$ and its trade-in value at any time t is $s(t)$. The purchase price of a new machine is $p > s(0)$. At each of the times $t = 0, 1, 2, \ldots$, we have the option of keeping the machine, or purchasing a new one. Consider an unbounded process where the return one stage away is discounted by a factor a, $0 < a < 1$. Let $f(t)$ represent the total overall return obtained using an optimal policy.

Show that $f(t)$ satisfies the equation

$$f(t) = \text{Max} \begin{bmatrix} r(t) - u(t) + af(t+1), \\ s(t) - p + r(0) - u(0) + af(1) \end{bmatrix}$$

46. Using the fact that an optimal policy, starting with a new machine, is to retain the machine for a certain number of time periods, and then purchase another one, determine the solution of the above equation.

47. Is it uniformly true that, if given an over-age machine, the optimal policy is to turn it in immediately for a new one?

48. How does one formulate the problem to take into account technological improvement in machines and operating procedures?

49. A secretary is looking for a single piece of correspondence, ordinarily a carbon on thin paper. She usually has 6 places she can look

	Folder Number k
Three folders of about 30 sheets each	1,2,3
One folder of about 50 sheets	4
One folder of about 100 sheets	5
Elsewhere	6

The initial probabilities of the letter being in the various places are usually

k	p_k Probability of letter in folder	$1-2_k$ Probability of being found on one examination if in folder	t_k Time for one examination
1	.11	.95	1
2	.11	.95	1
3	.11	.95	1
4	.20	.85	2
5	.37	.70	3
6	.10	.10	100

How shall the secretary look through the folders so as to

a. Minimize the expected time required to find a particular letter?
b. Maximize the probability of finding it in a given time? (F. Mosteller)

50. Let the function $a(x)$ satisfy the constraint $a(x) \leq d < 1$ for all x. Show that the solution of the equation

$$u = \operatorname*{Max}_{x} [b(x) + a(x)u],$$

if it exists, is unique, and is given by the expression

$$u = \operatorname*{Max}_{x} b(x)/(1 - a(x)).$$

Under what conditions does the solution exist?

If $a(x)$ does not satisfy the above condition, show that the number of solutions is either 0, 1, 2 or a continuum, and give examples of each occurrence.

51. We are given a quantity $x > 0$ that is to be utilized to perform a certain task. If an amount y, $0 \leq y \leq x$, is used on any single attempt, the probability of success is $a(y)$. If the task is not accomplished on the first try, we continue with the remaining quantity $x - y$. Show that if $f(x)$ represents the over-all probability of success using an optimal policy, then $f(x)$ satisfies the functional equation

$$f(x) = \operatorname*{Sup}_{0 \leq y \leq x} [a(y) + (1 - a(y))f(x - y)].$$

52. Derive the corresponding equation for $1 - f(x)$, the probability of failure.

53. Consider the two cases where $a(y)$ is convex or concave, and obtain the explicit solutions for these cases. Observe that in one case there is *no* optimal policy.

54. Consider the process discussed in § 2 under the assumption that the total return from an N-stage process is

$$R'_N = g(y) + h(x - y) + g(y_1) + h(x_1 - y_1) + \dots + g(y_{N-1}) + h(x_{N-1} - y_{N-1}) + k(x_N),$$

where $k(x)$ is a given function.

55. Consider the functional equation

$$f(x) = \operatorname*{Max}_{0 \leq y \leq x} [g(y) + h(x - y) + f(ay + b(x - y))],$$

under the assumption that

a. $g(y) \sim c_1 y^d,\ h(y) \sim c_2 y^d,\ c_1, c_2, d > 0$, as $y \to \infty$

or

b. $g(y) \sim c_1 y^{d_1},\ h(y) \sim c_2 y^{d_2},\ c_1, c_2, d_1, d_2 > 0$ as $y \to \infty$.

In both cases, determine the asymptotic behavior of $f(x)$ as $x \to \infty$.

56. Determine a recurrence relation for

$$\min_{x_i \geq 0} \left[\frac{x_1}{x_2 + x_3} + \frac{x_2}{x_3 + x_4} + \cdots + \frac{x_{n-1}}{x_n + x_1} + \frac{x_n}{x_1 + x_2} \right],$$

with the introduction of suitable additional parameters.

57. Consider the problem of determining the minimum of the function

$$\sum_{k=1}^{N} g_k(r_k, r_{k+1}) + \sum_{k=1}^{N} h_k(r_k),$$

where $r_{N+1} = r_1$, and the r_k are subject to the constraint

a. $0 \leq r_k \leq b_k$,

b. $\sum_{k=1}^{N} \varphi_k(r_k) \geq c$,

with each $\varphi_k(x)$ a known monotone increasing function of x, $\varphi_k(0) = 0$.
Introduce the auxiliary problem:
Minimize

$$g(u, r_2) + g(r_2, r_3) + \cdots + g(r_{N-1}, r_N) + g(r_N, v) + \sum_{k=2}^{N} h_k(r_k),$$

with $r_2, r_3, \ldots, r_N$ subject to the constraints

a. $0 \leq r_k \leq b_k$

b. $\sum_{k=2}^{N} \varphi_k(r_k) \geq c$.

Show that if we designate the above minimum by $F(u, v, c)$, then the minimum in the original problem is given by

$$\min_{0 \leq r_1 \leq b_1} F(r_1, r_1, c - \varphi_1(r_1)).$$

58. Introduce the sequence of functions, $R = 2, 3, \ldots, N-1$,

$$F_R(u, v, c) = \operatorname*{Min}_{r_k} [g(u, r_R) + g(r_R, r_{R+1}) + \ldots + g(r_{N-1}, r_N) + g(r_N, v) + \sum_{k=R}^{N} h_k(r_k)],$$

with

$$F_N(u, v, c) = \operatorname*{Min}_{r_N} [g(u, r_N) + g(r_N, v) + h_N(r_N)].$$

For each R, admit only c-values satisfying the restriction $\sum_{k=R}^{N} \varphi_k(b_k) \geq c$, where the b_k are fixed positive constants.

Show that we have the recurrence relation

$$F_R(u, v, c) = \operatorname*{Min}_{r_R} [g(u, r_R) + h_R(r_R) + F_{R+1}(r_R, v, c - \varphi_R(r_R))],$$

where r_R varies over the interval defined by

a. $0 \leq r_R \leq b_R$,

b. $\sum_{k=R+1}^{N} \varphi_k(b_k) \geq c - \varphi_R(r_R)$.

59. Consider in a similar fashion the problem of minimizing a function such as

$$R_N = g(r_1, r_2, r_3) + g(r_2, r_3, r_4) + \ldots + g(r_{N-1}, r_N, r_1) + g(r_N, r_1, r_2).$$

60. Suppose that we have a quantity of capital x, and a choice of the production in varying quantities of N different products. Assume initially that there is an unlimited supply of labor and machines for the production of any items we choose, in any quantities we wish.

If we decide to produce a quantity x_i of the ith item, we incur the following costs:

a. a_i = unit cost of raw materials required for the ith item
b. b_i = unit cost of machine production of ith item
c. c_i = unit cost of labor required for i^{th} item.
d. C_i = a fixed cost, independent of the amount produced of the ith item, if $x_i > 0$.

The cost of producing a quantity x_i of the ith item is then

$$\begin{aligned} g_i(x_i) &= (a_i + b_i + c_i)\, x_i + C_i, && x_i > 0 \\ &= 0, && x_i = 0. \end{aligned}$$

Let p_i be the selling price per unit of the ith item. The problem is to choose the x_i so as to maximize the total profit

$$P_N = \sum_{i=1}^{N} p_i x_i,$$

subject to the constraints

(a) $\sum_{i=1}^{N} g_i(x_i) \leq x,$

(b) $x_i \geq 0.$

Let

$$f_N(x) = \operatorname*{Max}_{x_i} P_N.$$

Show that

$$f_1(x) = p_1(x - C_1)/(a_1 + b_1 + c_1),\ x \geq C_1,$$
$$= 0, \qquad 0 \leq x \leq C_1,$$

and

$$f_N(x) = \operatorname*{Max}_{\substack{x_N \geq 0 \\ g_N(x_N) \leq x}} [p_N x_N + f_{N-1}(x - g_N(x_N))].$$

Show that $x_N \geq 0$ can be replaced by

$$x_N \geq \frac{f_{N-1}(x) - f_{N-1}(x - C_N)}{p_N}.$$

61. Assume that the demand for each item is stochastic. Let $G_k(z)$ represent the cumulant function for the demand z for the kth item. Show that the expected return from the manufacture of x_k of the kth item is

$$p_k \int_0^{x_k} z\, dG_k(z) + p_k \int_{x_k}^{\infty} x_k\, dG(z)$$

$$= p_k \int_0^{x_k} z\, dG_k(z) + p_k x_k (1 - G_k(x_k)),$$

and obtain the recurrence relation corresponding to the problem of maximizing the total expected return.

62. Consider the problem of maximizing the probability that the return exceed r.

63. Consider the above problem in the deterministic and stochastic versions when there are restrictions upon the quantity of machines available and the labor supply.

64. Obtain the recurrence relations corresponding to the case where we have "complementarity" constraints such as

a. $x_1 x_2 = 0, \quad x_7 x_8 = 0, \quad x_9 x_{10} x_{11} = 0,$

and so on, or

b. $x_i x_{i+1} = 0, \quad i = 1, 2, \ldots, \quad N - 1.$

65. Suppose that we have a complicated mechanism consisting of N interacting parts. Let the ith part have weight W_i, size S_i, and let us assume that we know the probability distribution for the length of time that any particular part will go without a breakdown, necessitating a new part. Assume also that we know the time and cost required for replacement, and the cost of a breakdown. Assuming that there are weight and size limitations on the total quantity of spare parts we are allowed to stock, how do we stock so as to minimize

a. the expected time lost due to breakdowns,
b. the expected cost of breakdowns,
c. a given function of the two, time and cost,
d. the probability that the time lost due to breakdowns will exceed T,
e. the probability that the cost due to breakdowns will exceed C?

66. Determine the possible modes of asymptotic behavior of the sequence $\{u_n\}$ determined by the recurrence relation

$$u_{n+1} = \text{Max}\,[au_n + b, cu_n + d],$$

and generally by the recurrence relation

$$u_{n+1} = \text{Max}_i\,[a_i u_n + b_i], \quad i = 1, 2, \ldots, k.$$

(cf. Problem 50).

67. Determine the minimum of

$$F(x_1, x_2, \ldots, x_N) = \sum_{i=1}^{N} g_i(x_i) + \text{Max}\,(x_1, x_2, \ldots, x_N),$$

subject to the constraints $x_i \geq 0$.

68. Suppose that we have N different activities in which to invest capital. Let $g_i(x_i)$ be the return from the ith activity due to an investment of x_i. Given an initial quantity of capital x, we are required to invest in at most k activities so as to maximize the total return.

Denote the maximum return by $f_{k,N}(x)$. Show that we have the recurrence relation

$$f_{k,N}(x) = \text{Max} \begin{bmatrix} \underset{0 \le y \le x}{\text{Max}} \ [g_N(y) + f_{k-1,N-1}(x-y)], \\ f_{k,N-1}(x) \end{bmatrix},$$

for $1 \le k \le N-1$.

69. Two corporations, with interlocking directorate, are forbidden by anti-monopoly statutes from investing in the same enterprise. The first corporation has capital x to invest, the second capital y, with known returns $g_i(z)$ from an investment of a quantity of capital z in the ith of N different enterprises.

Show that if the directors wish to maximize the total return from the two corporations, they must maximize

$$F_N(x_i, y_i) = \sum_{i=1}^{N} g_i(x_i) + \sum_{i=1}^{N} g_i(y_i),$$

subject to the constraints

a. $\sum_{i=1}^{N} x_i = x, \ x_i \ge 0,$

b. $\sum_{i=1}^{N} y_i = y, \ y_i \ge 0.$

c. $x_i y_i = 0.$

Let

$$f_N(x, y) = \underset{\{x_i, y_i\}}{\text{Max}} F_N(x_i, y_i)$$

Show that

$$f_N(x, y) = \text{Max} \begin{bmatrix} \underset{0 \le y_N \le y}{\text{Max}} \ [g_N(y_N) + f_{N-1}(x, y - y_N)] \\ \underset{0 \le x_N \le x}{\text{Max}} \ [g_N(x_N) + f_{N-1}(x - x_N, y)) \end{bmatrix}$$

Consider the case where the different corporations derive different returns from the same enterprise.

70. It is decided to employ a policy of replacing all light bulbs in an office building at one time. Assume that the cost of replacing the bulbs is a, and that $g(x)$ represents the cost due to lack of lighting if a time interval x elapses between replacements. Over a time interval T, it is decided to make replacements at times $x_1, x_1 + x_2, \ldots, x_1 + x_2 + \ldots + x_n = T$, where n is to be determined.

The efficiency of the program is to be measured by the average loss sustained

$$F(x_1, x_2, \ldots, x_n) = \frac{\sum_{i=1}^{n} (a + g(x_i))}{T}$$

What is the optimal policy?

(I. R. Savage)

71. Let the functions $g_i(x)$ be such that the maximum of

$$F_N(x_1, x_2, \ldots, x_N) = \sum_{i=1}^{N} g_i(x_i)$$

over the region $x_i \geq 0$, $\sum_{i=1}^{N} x_i = c$ may be obtained by use of a Lagrange multiplier λ, considering the expression

$$G_N = \sum_{i=1}^{N} g_i(x_i) - \lambda \sum_{i=1}^{N} x_i.$$

On the other hand, let $f_N(c) = \underset{\{x\}}{\text{Max}}\, F_N$. Show that

$$\lambda = f_N'(c)$$

Obtain the corresponding result for the maximum of $\sum_{i=1}^{N} g_i(x_i, y_i)$ subject to

$$\sum_{i=1}^{N} x_i = c_1, \quad \sum_{i=1}^{N} y_i = c_2, \quad x_i, y_i \geq 0.$$

72. Let

$M_r(x_1, x_2, \ldots, x_N) =$ the rth largest of the quantities $x_1, x_2, \ldots, x_N$,

$N_r(x_1, x_2, \ldots, x_N) =$ the rth smallest of the quantities $x_1, x_2, \ldots, x_N$,

for $r = 1, 2, \ldots, N$. Obtain recurrence relations connecting the members of the sequences

$$\{M_r(x_1, x_2, \ldots, x_N)\}, \{N_r(x_1, x_2, \ldots, x_N)\}, r = 1, 2, \ldots, .$$

73. Consider the problem of maximizing $\sum_{i=1}^{N} 2^{-x_i}$

subject to the constraints $x_i \geq 0$, $\sum_{i=1}^{N} 1/(1 + x_i) \leq x$.

(J. V. Whittaker)

74. A gambler has a capital of x dollars and wishes to bet on the outcomes of N different events. There is a probability p_k that he can predict the kth outcome correctly. The only constraint on the total amount that he bets is the condition that he be able to pay off his losses.

Show that the problem of maximizing his expected return may be converted into the problem of maximizing

$$L_N(x) = \sum_{k=1}^{N} p_k x_k \text{ subject to the constraints}$$

(a) $x_i \geq 0,$

(b) $\sum_{i=1}^{N} x_i \leq x + x_j,\ j = 1, 2, \ldots, N.$

75. Consider the problem of maximizing

$$L_N(x) = \sum_{k=1}^{N} p_k x_k$$

subject to the constraints

(a) $x_i \geq 0$

(b) $\sum_{i=1}^{N} x_i \leq u + x_j$

(c) $\sum_{i=1}^{N} x_i \leq v.$

Define $f_N(u, v) = \text{Max } L_N(x)$. Show that

$$f_N(u, v) = \text{Max}\,[p_N x_N + f_{N-1}(u - x_N, \text{ Min}(v - x_N, u))]$$

76. The problem of designing an efficient water distillation plant for heavy water production involves the minimization of

$$V_N = g(a_1) + \frac{g(a_2)}{a_1} + \frac{g(a_3)}{a_1 a_2} + \ldots + \frac{g(a_m)}{a_1 a_2 \ldots a_{m-1}},$$

where the a_i are subject to the constraints

(a) $a_i \geq 1$

(b) $a_1 a_2 \ldots a_m = x.$

Show that this may be reduced to the functional equation

$$f_{k+1}(x) = \underset{a_1 \geq 1}{\text{Min}} \left[g(a_1) + \frac{1}{a_1} f_k\left(\frac{x}{a_1}\right) \right],$$

and find the solution in the case where $g(y) = y^b,\ b > 0.$

77. Consider the case where

$$V_N = g_1(a_1) + \frac{g_2(a_2)}{a_1} + \ldots + \frac{g_m(a_m)}{a_1 a_2 \ldots a_{m-1}}.$$

(E. Cerri, M. Silvestri and S. Villan, "The Cascading Problem in a Water Distillation plant and Heavy Water Production," *Z. Naturforschg.*, 11a, **694** (1956).)

78. Consider the problem of allocating resources to N different activities, leading to the problem of maximizing a function
$\Sigma\, g_i(x_i)$ subject to the constraints $\Sigma\, x_i = c$, $x_i \geq 0$.
Show that the function $f_N(c)$ obtained via the usual recurrence relations does *not* depend upon the way in which the activities are numbered.

Bibliography and Comments for Chapter I

§ 1. A fairly complete bibliography of papers up to 1954 plus some remarks which complement the text may be found in R. Bellman, "The Theory of Dynamic Programming," *Bull. Amer. Math. Soc.*, vol. 60 (1954), pp. 503–516.

§ 2. This process was first discussed in *Econometrica*, vol. 22 (1954), pp. 37–48.

§ 7. Further discussion of this problem may be found in R. Bellman, "A Class of Variational Problems," *Quart. of Appl. Math.*, 1956. An interesting discussion of general "smoothing" problems may be found in I. J. Schoenberg, "On Smoothing Functions and their Generating Functions," *Bull. Amer. Math. Soc.*, vol. 59 (1953), pp. 199–230, where a number of further references may be found.

§ 11. The importance of the concept of approximation in policy space was stressed in R. Bellman, "On Computational Problems in the Theory of Dynamic Programming," Symposium on Numerical Methods, *Amer. Math. Soc.* Santa Monica, 1953.

§ 12. The elegant proof of Lemma 1 was found independently by I. Glicksberg and W. Fleming to whom the author posed the problem of finding a better proof than that given in the opening lines of the section.

§ 17. The results in this section were derived by D. Anderson.

§ 18. A more complete discussion of the concept of the stability of solutions of functional equations may be found in R. Bellman, *Stability Theory of Differential Equations*, McGraw-Hill, 1954.

§ 19. The reduction of the sequence $\{f_{k,N}(x)\}$ to a sequence $\{f_k(x)\}$ is an important piece of mathematical legerdemain as far as computational solutions are concerned; cf also § 6 and § 7. The limited storage capacity of computing machines makes one quite stingy with subscripts and parameters.

§ 22. The proof in the text follows a paper of S. Johnson, "Optimal Search is Fibonaccian," 1955 (to appear).

An equivalent result was found earlier by J. Kiefer, unbeknownst to Johnson, using a much more difficult argument: J. Kiefer, "Sequential Minimax Search for a Maximum," *Proc. Amer. Math. Soc.*, vol. 4 (1953), pp. 502–6.

The problem of determining a corresponding result for higher dimensions seems extraordinarily difficult, and nothing is known in this direction at the present time.

§ 24. An excellent introduction to the study of stochastic processes is given in the book by W. Feller, *Probability Theory*, John Wiley and Sons, 1948. A number of important physical processes are discussed in the book by M. S. Bartlett, *An introduction to stochastic processes with special reference to methods and applications*, Cambridge, 1955.

Exercise 76. See R. Bellman, *Nuclear Engineering*, 1957

CHAPTER II

A Stochastic Multi-Stage Decision Process

§ 1. Introduction

In the preceding chapter we considered in some detail a multi-stage decision process in both deterministic and stochastic guises. In this chapter we shall discuss a stochastic multistage decision process of an entirely different type which possesses a number of interesting features. In particular, in obtaining the solution of some simple versions of processes of this type, we shall encounter the important concept of "decision regions".

We shall follow essentially the same lines pursued in the previous chapter, first a statement of the problem, then a brief discussion in classical terms. Following this, the problem will be formulated in terms of a functional equation, the required existence and uniqueness theorems will be proved, and then the remainder of the chapter devoted to a discussion of various properties of the solution, such as stability and analytic structure.

For the simple process used as our model, we are fortunate enough to obtain a solution which has a very interesting interpretation. Equally fortunately as far as the mathematical interest of the problem is concerned, this solution does not extend to more general processes of the same type. This forces us to employ techniques of an entirely different type which we shall discuss in a later chapter, Chapter 8.

The failure of the elementary solution is not due solely to the inadequacy of the analysis. A counter-example has been constructed showing that the solution of a multi-stage decision process of this class cannot always have the simple form of the solution given in § 8 below. Another proof of this fact is furnished by Lemma 8 of Chapter 8.

A number of interesting results which we do not wish to discuss in detail are given as exercises at the end of the chapter.

§ 2. Stochastic gold-mining

We shall cast the problem in the mold of a gold-mining process.

Suppose that we are fortunate enough to own two gold mines, Anaconda and Bonanza, the first of which possesses within its depths an amount of gold x, and the second an amount of gold y. In addition, we have a single, rather delicate, gold-mining machine with the property that if used to mine gold in Anaconda, there is a probability p_1 that it will mine a fraction r_1 of the gold there and remain in working order, and a probability $(1 - p_1)$ that it will mine no gold and be damaged beyond repair. Similarly, Bonanza has associated the corresponding probabilities p_2 and $1 - p_2$, and fraction r_2.

We begin the process by using the machine in either the Anaconda or Bonanza mine. If the machine is undamaged after its initial operation, we again make a choice of using the machine in either of the two mines, and continue in this way making a choice before each operation, until the machine is damaged. Once the machine is damaged, the operation terminates, which means that no further gold is obtained from either mine.

What sequence of choices maximizes the amount of gold mined before the machine is damaged?

§ 3. Enumerative treatment

Since we are dealing with a stochastic process, it is not possible to talk about *the* return from a policy, a point we have already discussed in § 24 of the previous chapter, nor can we choose a policy which guarantees a maximum return. We must console ourselves with measuring the value of a policy by means of some average of the possible returns, and choosing an optimal policy on this basis. As before, the simplest such average is the *expected* value.

Let us then agree that we are interested in the policies (since there may be many) which maximize the *expected* amount of gold mined before the machine is damaged. A policy here will consist of a choice of A's and B's, A for Anaconda and B for Bonanza. However, any such sequence such as

$$S = AABBBABB \ldots \tag{1}$$

must be read: A first, then A again if the machine is undamaged, then B if the machine is still undamaged, and so on.

Let us initially, to avoid the conceptual difficulties inherent in unbounded processes, consider only mining operations which terminate automatically after N steps regardless of whether the machine is undamaged or not. In this case it is quite easy, in theory, to list all feasible policies, and to compute all possible returns.[1] It is possible to use this idea to some extent in certain problems. However, in general, this procedure is rather limited in application, unrevealing as to the structure of an optimal policy, and, as a brute force method, a betrayal of one's mathematical birthright.

[1] To quote numbers again, a 10-stage policy would require the listing of $2^{10} = 1024$ possible policies; if three choices at each stage, then 59,049 different policies.

§ 4. Functional equation approach

In place of the above enumerative approach, we shall once again employ the functional equation approach. Let us define

(1) $f_N(x, y) =$ expected amount of gold mined before the machine is damaged when A has x, B has y and an optimal policy which can last at most N stages is employed.

Considering the one-stage process, we see that an A-choice yields an expected amount $p_1 r_1 x$, while a B-choice yields $p_2 r_2 y$. Hence

(2) $$f_1(x, y) = \text{Max}\,[p_1 r_1 x, p_2 r_2 y].$$

Let us now consider the general $(N + 1)$-stage process. Whatever choice is made first, the continuation over the remaining N stages must be optimal if we wish to obtain an optimal $(N + 1)$-stage policy. Hence the total expected return from an A-choice is

(3) $$f_A(x, y) = p_1(r_1 x + f_N((1 - r_1)x, y)),$$

and the total expected return from a B-choice is

(4) $$f_B(x, y) = p_2(r_2 y + f_N(x, (1 - r_2)y)).$$

Since we wish to maximize our total $(N + 1)$-stage return, we obtain the basic recurrence relation

(5) $$\begin{aligned} f_{N+1}(x, y) &= \text{Max}\,[f_A(x, y), f_B(x, y)], \\ &= \text{Max}\,[p(r_1 x + f_N((1 - r_1)x, y)), p_2(r_2 y + \\ &\qquad f_N(x, (1 - r_2)y))]. \end{aligned}$$

§ 5. Infinite stage approximation

The same argumentation shows that the return from the unbounded process, which we call $f(x, y)$, assuming that it exists, satisfies the functional equation

(1) $$f(x, y) = \text{Max}\,[p_1(r_1 x + f((1-r_1)x, y)), p_2(r_2 y + f(x, (1-r_2)y))].$$

Once again, the infinite process is to be considered as an approximation to a finite process with large N. In return for the advantage of having only a single function to consider, we face the necessity of establishing the existence and uniqueness of a solution of the equation in (1). This we proceed to do in the next section.

§ 6. Existence and uniqueness

Let us now prove the following result:

THEOREM 1. *Assume that*

(1) $$\text{a.}\quad |p_1|,\ |p_2| < 1,^2$$
$$\text{b.}\quad 0 \leq r_1, r_2 < 1.$$

Then there is a unique solution to (5.1) *which is bounded in any rectangle* $0 \leq x \leq \bar{X}, 0 \leq y \leq \bar{Y}$.

This solution $f(x, y)$ *is continuous in any finite part of the region* $x, y \geq 0$.

PROOF: Let us, to simplify the notation, set

(2) $$T_1(f) = p_1[r_1 x + f((1 - r_1)x, y)],$$
$$T_2(f) = p_2[r_2 y + f(x, (1 - r_2)y)].$$

Then the functional equation in (5.1) has the form

(3) $$f(x, y) = \text{Max}\,[T_1(f), T_2(f)].$$

Define the sequence of functions

(4) $$f_1(x, y) = \text{Max}\,[p_1 r_1 x, p_2 r_2 y],$$
$$f_{N+1}(x, y) = \text{Max}\,[T_1(f_N), T_2(f_N)],$$
$$= \underset{i=1,2}{\text{Max}}\,[T_i(f_N)]$$

precisely as in the recurrence relation of (4.5).

Let $i = i(N) = i(N, x, y)$ be an index which yields the maximum in the expression $\underset{i=1,2}{\text{Max}}\,[T_i(f_N)]$, for $N = 1, 2, \ldots$

Then we have,

(5) $$f_{N+1}(x, y) = T_{i(N)}(f_N) \geq T_{i(N+1)}(f_N)$$
$$f_{N+2}(x, y) = T_{i(N+1)}(f_{N+1}) \geq T_{i(N)}(f_{N+1}),$$

using the same device we employed in the course of the existence and uniqueness proof for the solution of the functional equation in (8.1) of Chapter 1.

[2] In the equation arising from the process described above, the p_i are non-negative. The proof we give covers the more general equation as well.

Hence

$$(6) \quad |f_{N+1}(x, y) - f_{N+2}(x, y)| \leq \text{Max}\,[\,|T_{i(N)}(f_N) - T_{i(N)}(f_{N+1})|,\; |T_{i(N+1)}(f_N) - T_{i(N+1)}(f_{N+1})|\,]$$

$$\leq \underset{i=1,2}{\text{Max}}\,[\,|T_i(f_N) - T_i(f_{N+1})|\,]$$

$$\leq \text{Max}\,[\,|p_1|\,|f_N((1-r_1)x, y) - f_{N+1}((1-r_1)x, y)|,\; |p_2|\,|f_N(x, (1-r_2)y) - f_{N+1}(x, (1-r_2)y)|\,].$$

Let us now define

$$(7) \qquad u_N(x, y) = \underset{\substack{0 \leq s \leq x \\ 0 \leq t \leq y}}{\text{Max}} |f_N(s, t) - f_{N+1}(s, t)|$$

From (6) we obtain

$$(8) \qquad u_{N+1}(x, y) \leq q\, u_N(x, y),$$

where $q = \text{Max}\,(|p_1|, |p_2|)$. Since $0 \leq q < 1$, we see that the series $\sum_{N=1}^{\infty} u_N(x, y)$ converges uniformly in any bounded rectangle $0 \leq x \leq \bar{X}$, $0 \leq y \leq \bar{Y}$. Hence $f_N(x, y)$ converges uniformly to a function $f(x, y)$ which satisfies the relation (5.1), and which is continuous in any bounded rectangle in the (x, y)-plane.

The uniqueness proof follows the same lines as the proof of Theorem 1 of Chapter 1 and is left as an exercise for the reader.

As we see from the above proof, the choice of $f_1(x, y)$ is arbitrary provided only that it be bounded in any finite rectangle. It is interesting to note that the limit function will be continuous even if the initial function is not, as a consequence of the uniqueness of the solution.

§ 7. Approximation in policy space and monotone convergence

As before, it is easily seen that we can ensure monotone convergence by approximation in policy space, in the case where $p_1, p_2 \geq 0$. The two simplest approximations are those corresponding to A^∞ and B^∞.[3] From the first policy we obtain the expected return

$$(1) \qquad f_A(x, y) = p_1 r_1 x/(1 - p_1(1 - r_1)),$$

and from the second, the return

$$(2) \qquad f_B(x, y) = p_2 r_2 y/(1 - p_2(1 - r_2)).$$

[3] It is interesting to observe the following difference between the process and the functional equation obtained from it. The sequence A^∞ is *conditional* as far as the process is concerned, but *deterministic* as far as the equation is concerned.

As we shall see below in § 8 and § 9, we actually possess a far more sophisticated technique for obtaining a first approximation in the discussion of more complicated processes, at the expense, of course, of the above simplicity of expression. The guiding principle is, however, quite simple.

§ 8. The solution

Let us now turn to the solution of the equation in (5.1) for the case where p_1 and p_2 are real numbers satisfying the inequality $0 \leq p_1, p_2 < 1$. It is intuitively clear that an A-choice is made when $x/y \gg 1$ and a B-choice is made when $y/x \gg 1$ [4].

It is also easily seen that the choice at each stage depends only on the ratio x/y, since $f(kx, ky) = kf(x, y)$ for $k > 0$. Perhaps the quickest way to prove this is to invoke the uniqueness theorem, although it is intuitively clear from the description of the process.

It follows then that if we examine the positive (x, y)-quadrant, and divide it into an A-set and a B-set, which is to say those values of x and y at which an A-decision is the optimal first choice and those at which the B-decision is optimal, then (x, y) in the A-set implies that (kx, ky) is in the A-set for all $k > 0$, and similarly for the B-set.

If these sets are well-behaved, it follows that their boundaries must be straight lines,

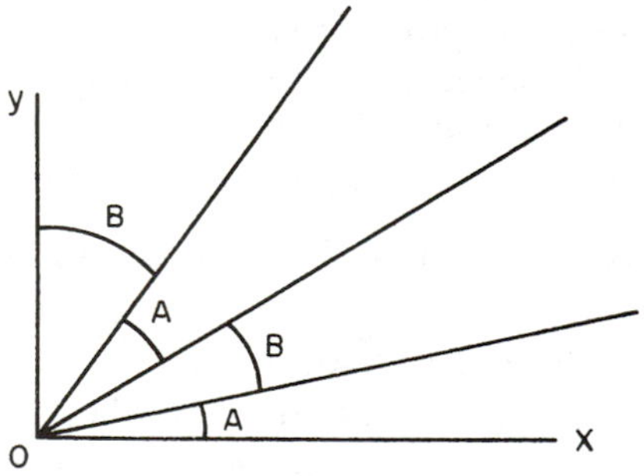

Figure 1

as conceivably in the figure above. The regions where A and B are used are called *decision regions*.

Let us now boldly conjecture that there are only two regions, as in Figure 2,
and see if we can determine the boundary line, L, if this is the case.

[4] The notation $a \gg 1$ signifies that a is very large compared to 1.

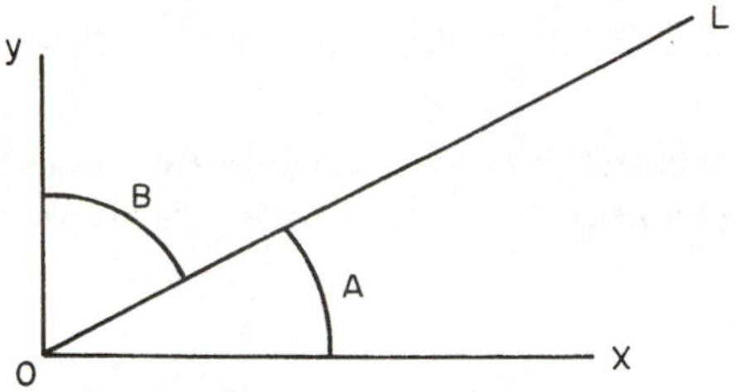

Figure 2

What is the essential feature of the boundary line which will enable us to determine its equation? It is this: it is the line on which A or B choices are equally optimal.

If we use A at a point (x, y), with an optimal continuation from the first stage on, we have

$$f_A(x, y) = p_1 r_1 x + p_1 f((1 - r_1) x, y), \tag{1}$$

while similarly B at (x, y), and an optimal continuation, yield

$$f_B(x, y) = p_2 r_2 y + p_2 f(x, (1 - r_2) y). \tag{2}$$

Equating these two expressions we obtain the equation for L. Unfortunately, this equation as it stands is of little use since it involves the unknown function f.

In order to complete the analysis successfully we must make a further observation. When at a point on L we employ A, we decrease x while keeping y constant and hence enter the B region; similarly, if we use B at a point on L we enter the A region (see Figure 2 above). It follows that for a point on L an initial first choice of A is equivalent to an initial first and second choice of A and then B, while, conversely, an initial first choice of B is equivalent to an initial first and second choice of B and then A.

If we use A and then B and continue optimally, we have

$$f_{AB}(x, y) = p_1 r_1 x + p_1 p_2 r_2 y + p_1 p_2 f((1 - r_1) x, (1 - r_2) y), \tag{3}$$

and similarly

$$f_{BA}(x, y) = p_2 r_2 y + p_1 p_2 r_1 x + p_1 p_2 f((1 - r_1) x, (1 - r_2) y). \tag{4}$$

Equating f_{AB} and f_{BA}, the unknown function f disappears[5] and we obtain the equation

[5] The meaning of this is that having survived both an A choice and a B choice, it is no longer of any importance in the *continuation* of the process as to the original order of these choices.

(5) $$p_1 r_1 x/(1-p_1) = p_2 r_2 y/(1-p_2),$$

for L.

It remains to establish this equation rigorously. Let us begin by proving that there is a region near the x-axis where A is always the optimal first choice.

If $y = 0$, we have

(6) $$f(x, 0) = \text{Max}\begin{bmatrix} p_1 r_1 x + p_1 f((1-r_1)x, 0) \\ p_2 f(x, 0) \end{bmatrix}$$
$$= p_1 r_1 x + p_1 f((1-r_1)x, 0).$$

Since $f(x, y)$ is continuous in y, it follows that

(7) $$f(x, y) > p_2 (r_2 y + f(x, (1-r_2) y)),$$

for $0 \leq y \leq kx$, where k is some small positive constant, since the strict inequality holds for $y = 0$.

Thus we have a region in which A is used first, shown below in Figure 3.

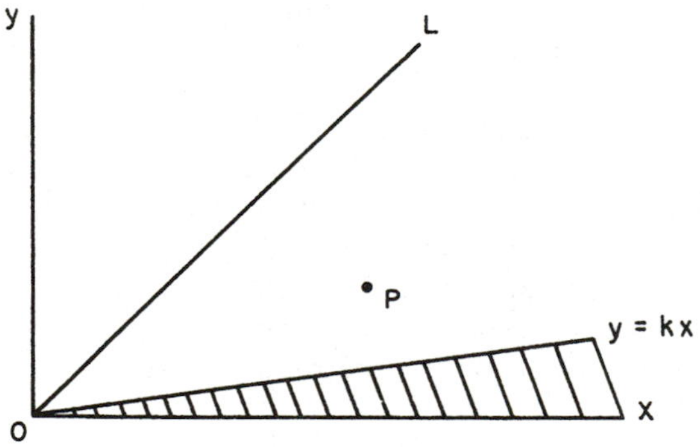

Figure 3

Let us now take a point $P = P(x, y)$, in the region between L and $y = kx$, with the property that $(x, (1-r_2)y)$ is in the shaded region. In other words, use of B at P must result in an A-choice next, provided that machine is undamaged. (This proviso is necessary when discussing the process, but not when discussing the equation, as we have noted above.)

If B is optimal at P, we obtain

(8) $$f(x, y) = f_{BA}(x, y),$$

as given by (4). However, we know that below L, $f_{BA}(x, y) < f_{AB}(x, y)$. Hence B cannot be optimal at P. Proceeding inductively in this fashion we extend the shaded region up to L. Since precisely the same argument shows that the region between L and the y-axis is a B-region, we have completed the proof of

THEOREM 2. *Consider the equation*

$$(9) \qquad f(x, y) = \text{Max} \begin{bmatrix} p_1 [r_1 x + f((1 - r_1) x, y)], \\ p_2 [r_2 y + f(x, (1 - r_2) y)] \end{bmatrix}, \quad x, y \geq 0,$$

where $0 \leq p_1, p_2 < 1, 0 \leq r_1, r_2 \leq 1$.

The solution is given by

$$(10) \qquad \begin{aligned} f(x, y) &= p_1 [r_1 x + f((1 - r_1) x, y)], \textit{for} \\ &\qquad p_1 r_1 x/(1 - p_1) > p_2 r_2 y/(1 - p_2) \\ &= p_2 [r_2 y + f(x, (1 - r_2) y)], \textit{for} \\ &\qquad p_1 r_1 x/(1 - p_1) < p_2 r_2 y/(1 - p_2). \end{aligned}$$

For $p_1 r_1 x/(1 - p_1) = p_2 r_2 y/(1 - p_2)$ *either choice is optimal.*

§ 9. Discussion

The solution has a very interesting interpretation. We may consider $p_1 r_1 x$ to be the immediate expected gain and $(1 - p_1)$ to be the immediate expected loss. The theorem then asserts that the solution consists of making the decision which at each instant maximizes the ratio of immediate expected gain to immediate expected loss. As we shall see, this intriguing criterion occurs from time to time throughout the theory of dynamic programming.

§ 10. Some generalizations

The same methods suffice to prove the two results below.

THEOREM 3. *Consider the equation*

$$(1) \qquad f(x, y) = \text{Max} \begin{bmatrix} A: \sum_{k=1}^{N} p_k [c_k x + f(c'_k x, y)], \\ \\ B: \sum_{k=1}^{N} q_k [d_k y + f(x, d'_k y)] \end{bmatrix}$$

where $x, y \geq 0$ *and*

$$(2) \qquad \begin{aligned} &\text{(a)}\ p_k \geq 0, q_k \geq 0, \sum_{k=1}^{N} p_k, \sum_{k=1}^{N} q_k < 1, \\ &\text{(b)}\ 1 \geq c_k, d_k \geq 0, c'_k + c_k = d'_k + d_k = 1. \end{aligned}$$

The optimal choice of operations is the following: If

$$(3)\qquad \frac{\sum_{k=1}^{N} p_k c_k}{1-\sum_{k=1}^{N} p_k}\, x > \frac{\sum_{k=1}^{N} q_k d_k}{1-\sum_{k=1}^{N} q_k}\, y$$

choose A; if the reverse inequality holds, choose B. In case of equality, either choice is optimal.

THEOREM 4. *Consider the functional equation*

$$(4)\quad f(x_1, x_2, \ldots, x_N) = \operatorname*{Max}_{i} \Big[\sum_{k=1}^{K} p_{ik}[c_{ik} x_i + f(x_1, x_2, \ldots, c'_{ik} x_i, \ldots, x_n)]\Big]$$

where $x_i \geq 0$ *and*

$$(5)\qquad \text{(a)}\quad p_{ik} \geq 0,\ \sum_{k=1}^{K} p_{ik} < 1,\ i = 1, 2, \ldots, n.$$

$$\text{(b)}\quad 1 \geq c_{ik} \geq 0,\ c_{ik} + c_{ik}' = 1.$$

The decision functions are

$$D_i(x) = \frac{\sum_{k=1}^{K} p_{ik} c_{ik}}{1-\sum_{k=1}^{K} p_{ik}}\, x_i$$

in the sense that the index which yields the maximum of $D_i(x)$ *for* $i = 1, 2, \ldots, n$ *is the index to be chosen in* (4). *In case of equality, it is a matter of indifference as to which is used.*

It is clear that we can combine Theorems 3 and 4 into one more comprehensive result, which in turn can be generalized by the use of the Stieltjes integral. Thus a version of (1) arising from a continuous distribution of outcomes is

$$f(x, y) = \operatorname{Max} \begin{bmatrix} \int_0^1 [zx + f((1-z)\,x, y)]\, dG(z), \\ \int_0^1 [wy + f(x, (1-w)\,y)]\, dH(w). \end{bmatrix}$$

We leave the derivation of the extensions of Theorems 3 and 4, and the statements and proof of the corresponding existence and uniqueness theorem, as exercises for the reader.

§ 11. The form of $f(x, y)$

Having obtained a very simple characterization of the optimal policy, let us now turn our attention to the function $f(x, y)$. In general, no simple analytic representation will exist. If, however, we consider the equation

$$(1) \qquad f(x, y) = \text{Max} \begin{bmatrix} a_1 x + a_2 y + p_2 f(c_2 x, y) \\ b_1 x + b_2 y + q_2 f(x, d_2 y) \end{bmatrix}$$

we can show that if c_2 and d_2 are connected by a relation of the type $c_2{}^m = d_2{}^n$, with m and n positive integers, a piece-wise linear representation for $f(x, y)$ may be obtained.

It is sufficient, in order to illustrate the technique, to consider the simplest case where the relation is $c_2 = d_2$.

Let (x, y) be a point in the A-region. If A is applied to (x, y), this point is transformed into $(c_2 x, y)$, which may be in either an A- or a B-region. Let L_1 be the line that is transformed into L[6] when (x, y) goes into $(c_2 x, y)$, let L_2 be the line transformed into L_1, and so on. Similarly, let M_1 be the line transformed into L when (x, y) goes into $(x, d_2 y)$, and so on. In the sector LOL_1, A is used first, followed by B, as shown below.

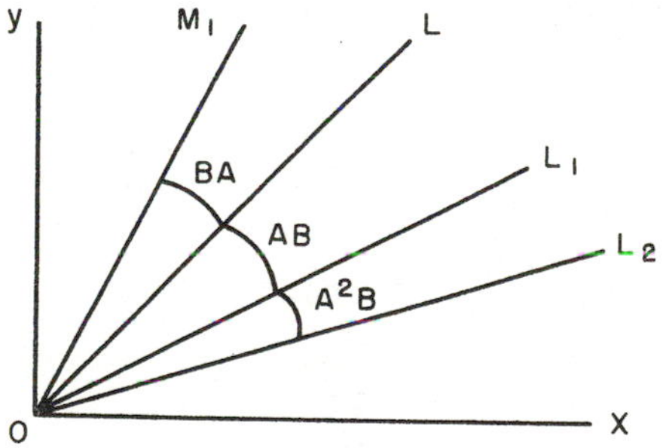

Figure 4

Hence, for (x, y) in this sector we obtain

$$(2) \qquad \begin{aligned} f(x, y) &= a_1 x + a_2 y + p_2 f(c_2 x, y) \\ &= a_1 x + a_2 y + p_2 (b_1 c_2 x + b_2 y) + q_2 p_2 f(c_2 x, c_2 y) \\ &= (a_1 + p_2 b_1 c_2) x + (a_2 + p_2 b_2) y + p_2 q_2 c_2 f(x,y) \end{aligned}$$

[6] The boundary line, whose equation obtained as above, is $[a_1(1 - q_2) + b_1 (p_2 c_2 - 1)]x = [b_2 (1 - p_2) + a_2 (q_2 d_2 - 1)]y$

This yields

$$(3) \qquad f(x, y) = \frac{(a_1 + p_2 b_1 c_2)\, x + (a_2 + p_2 b_2)\, y}{1 - p_2 q_2 c_2}$$

for (x, y) in LOL_1. Similarly, we obtain a linear expression for f in LOM_1. Having obtained the representations in these sectors, it is clear that we obtain linear expressions in $L_1 OL_2$, and so on.

§ 12. The problem for a finite number of stages

Let us first establish

THEOREM 5. *Consider the recurrence relations*

$$(1) \qquad f_1(x, y) = \operatorname{Max} \{p_1 r_1 x,\, p_2 r_2 y\}$$

$$f_{N+1}(x, y) = \operatorname{Max} \begin{bmatrix} A: & p_1 [r_1 x + f_N((1 - r_1)\, x, y)], \\ B: & p_2 [r_2 y + f_N(x, (1 - r_2)\, y)] \end{bmatrix},$$

$N = 1, 2, \ldots$.

For each N, there are two decision regions.

PROOF. For each $N \geq 2$, the points determined by the condition that AB plus an optimal continuation for the remaining $(N\text{-}2)$ moves is equivalent to BA plus an optimal continuation for the remaining $(N\text{-}2)$ moves lie on the same line L we have determined above, namely

$$(2) \qquad L: \quad \frac{p_1 r_1 x}{1 - p_1} = \frac{p_2 r_2 y}{1 - p_2}.$$

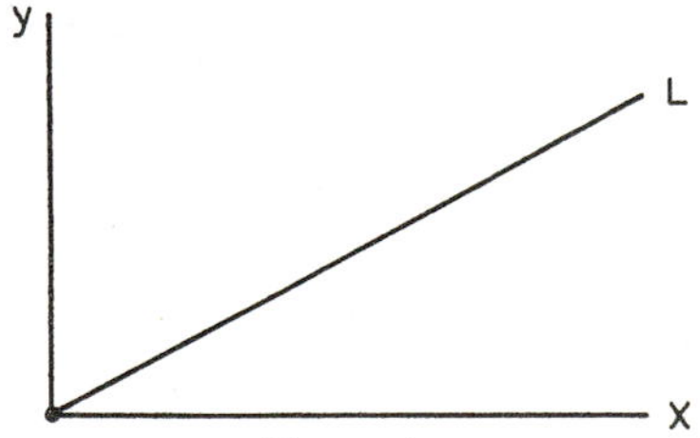

Figure 4a

For the N-stage process, any policy, and consequently, any optimal policy has the form

$$(3) \qquad S_N: \; A^{a_1} B^{b_1} \ldots A^{a_N} B^{b_N},$$

where the a_i and b_i are positive integers or zero, restricted by the condition, $\sum_i (a_i + b_i) = N$.

Let us now consider a point $P = P(x, y)$ lying above L. If A is used at P, there are two possibilities: either A is used k times in succession, and then followed by B,

$$S_N = A^k B \ldots, 1 \leq k \leq N - 1, \tag{4}$$

or $S_N = A^N$. Let us consider the first case. If A is used $(k - 1)$ times in succession, we reach a point P' further above L. At P', AB cannot be the first two moves in an optimal $(N - k + 1)$-stage policy, since BA plus an optimal continuation is superior.

Consequently above L, either B is used first, or the optimal policy is A^N. Let us now show that if A^N is optimal at P, then it is optimal in the region between OP and the x-axis.

To demonstrate this we begin with the observation that it is permissible to assume that $x + y = 1$, $0 \leq x, y \leq 1$, because of the homogeneity of $f_N(x, y)$ as a function of x and y. Considering the N-stage process, we see that there are 2^N possible policies, say $P_1, P_2, \ldots, P_{2N}$. Each of these policies used at a point (x, y) yields a N-stage return which is a linear function of x and y, say $L_k(x, y)$. For $x + y = 1$, we may plot these functions obtaining a set of 2^N straight lines,

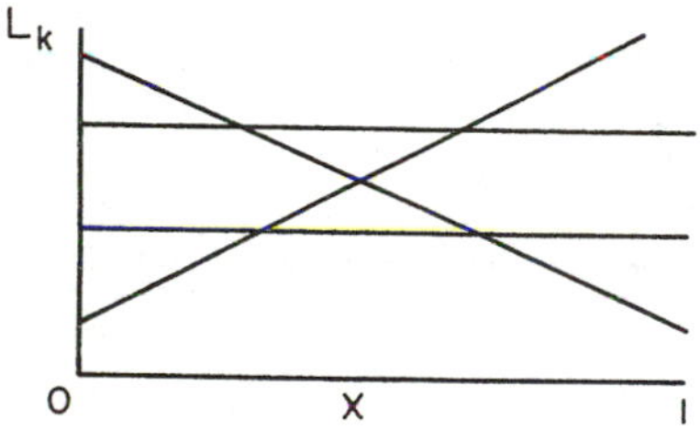

Figure 5

If N were 2, so that the four policies AA, AB, BA, BB yielded four lines as above, the maximum return as a function of x would have the form

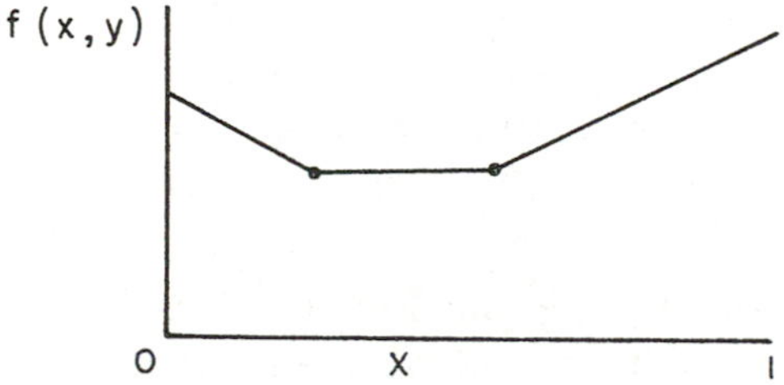

Figure 6

It is clear that A^N is an optimal policy for $y = 0$, $x = 1$. It follows that if A^N is optimal at $(\bar{x}, \bar{y})$, $0 < y < 1$, the line corresponding to A^N will dominate all other lines for $x \leq \bar{x} \leq 1$.

Combining the above results we see that for any N, the boundary between the A-region and the B-region will either be $AB = BA$ or $A^N = M_1$, where M_1 is a policy of complicated form, or $B^N = M_2$ is also a complicated policy.

We can now establish a sharper result:

THEOREM 6. *The decision regions for f_N converge towards those of f as $N \to \infty$ in a monotone fashion. There is always an integer N_0 with the property that for $N \geq N_0$ the regions for f_N are identical with those of f.*

PROOF: Consider the situation for $N = 3$. Let L_2 be the boundary line for the two-stage process, and assume that the relative positions of L_2 and L are as shown below.

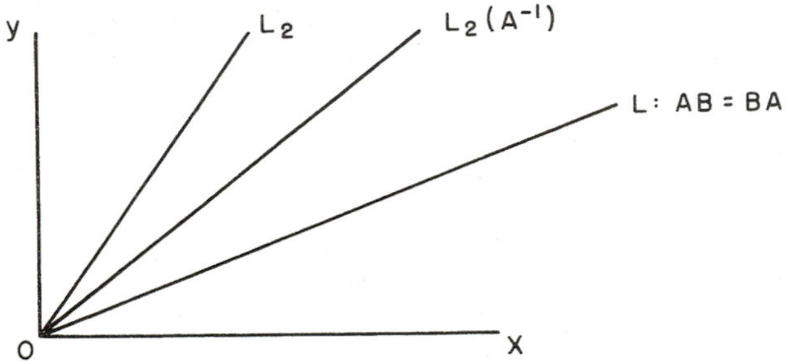

Figure 7

Let $L_2\,(A^{-1})$ denote the line transformed into L_2 when A is used at a point on $L_2\,(A^{-1})$, which is to say when (x, y) is transformed into (cx, y). Let Q be a point in the sector between L_2 and $L_2\,(A^{-1})$. If A is used at Q as the first move in a three-stage policy, B is used next, since the transformed point is in the B-region for a two stage process. However, if Q is above L, we know that AB cannot be the first two moves of an optimal policy. Hence B is used at Q. This shows that the B-region for the three-stage process is at least that containing the region above $L_2\,(A^{-1})$. This process may be continued for larger and larger N until $L_k\,(A^{-1})$, for some finite k, lies below L. At this point, the boundary line becomes $AB = BA$, and remains so for all larger N.

§ 13. A three-choice problem

Let us now assume that in addition to the two A and B choices already discussed, we have a third choice which is a compromise between the A

and B choices. The equation we obtain in this case takes the form

$$(1)\quad f(x, y) = \text{Max} \begin{bmatrix} \text{A:}\ p_1 [r_1 x + f((1 - r_1)\, x, y)] \\ \text{B:}\ p_2 [r_2 y + f(x, (1 - r_2)\, y)] \\ \text{C:}\ p_3 [r_3 x + r_4 y + f((1 - r_3)\, x, (1 - r_4)\, y)] \end{bmatrix}$$

where $0 \leq r_3, r_4 \leq 1$ and $0 \leq p_3 < 1$, and the quantities p_1, p_2, r_1, r_2 satisfy the previous inequalities.

On the basis of what we know concerning the solution of the equation where the C-term is missing, it might be suspected that the solution of this equation would be determined in the following way: There are three decision regions, as in the figure below, with A, B and C each optimal first choices in these regions

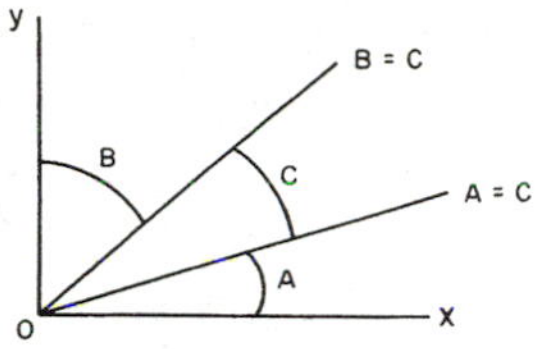

Figure 8

Unfortunately, a counter-example has been constructed showing that this is not true generally. It shows, by means of a fairly complicated but straightforward calculation, that the solution can, for suitable values of the parameter, take the form shown in Figure 9 below.

The solution of (1) above seems to be quite a difficult problem, and very little is known concerning the character of the solution.

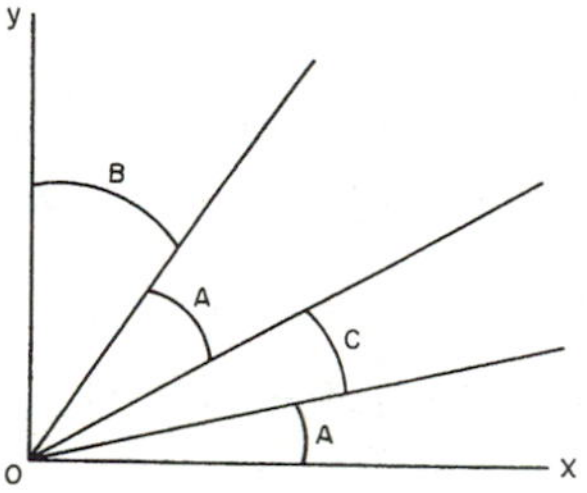

Figure 9

It is not even known whether or not the number of decision regions is always finite and whether the number is uniformly bounded if finite. To obtain some information about this problem in a part of the parameter space, we shall consider a continuous version in Chapter 8, where with the aid of variational techniques the decision regions may be determined.

For the continuous version they do assume the simple form shown in the first figure above, Figure 8.

§ 14. A stability theorem

Let us now derive a stability theorem for the solution [7] of the equation

$$f(x, y) = \text{Max} \begin{bmatrix} \text{A}: & p_1 [r_1 x + f((1 - r_1) x, y)] \\ \text{B}: & p_2 [r_2 y + f(x, (1 - r_2) y)] \end{bmatrix} \tag{1}$$

THEOREM 7. *Let $g(x, y)$ be the solution of*

$$g(x, y) = \text{Max} \begin{bmatrix} \text{A}: & p_1 [r_1 x + g((1 - r_1) x, y)] \\ \text{B}: & p_2 [r_2 y + g(x, (1 - r_2) y)] \end{bmatrix} + h(x, y). \tag{2}$$

Then, in any rectangle R: $0 \leq x \leq \bar{X}, 0 \leq y \leq \bar{Y}$

$$|f(x, y) - g(x, y)| \leq \text{Max}_R |h(x, y)|/q, \tag{3}$$

where $q = \text{Min}((1 - p_1), (1 - p_2))$.

PROOF. The proof proceeds by successive approximations, as in the corresponding section in Chapter 1. Consequently, we shall merely sketch the details. Set

$$\begin{aligned} f_1(x, y) &= \text{Max}[p_1 r_1 x, p_2 r_2 y] \\ g_1(x, y) &= \text{Max}[p_1 r_1 x, p_2 r_2 y] + h(x, y). \end{aligned} \tag{4}$$

and, generally,

$$\begin{aligned} f_{n+1}(x, y) &= \text{Max} \begin{bmatrix} \text{A}: & p_1 [r_1 x + f_n((1 - r_1) x, y)] \\ \text{B}: & p_2 [r_2 y + f_n(x, (1 - r_2) y)] \end{bmatrix} \\ g_{n+1}(x, y) &= \text{Max} \begin{bmatrix} \text{A}: & p_1 [r_1 x + g_n((1 - r_1) x, y)] \\ \text{B}: & p_2 [r_2 y + g_n(x, (1 - r_2) y)] \end{bmatrix} + h(x, y). \end{aligned} \tag{5}$$

It is clear that

$$|f_1(x, y) - g_1(x, y)| \leq \text{Max}_R |h(x, y)|. \tag{6}$$

[7] By the term "solution", here and in the following pages, we shall mean the unique solution in the appropriate function class.

Applying the techniques used repeatedly above, we see that

$$(7)\quad \operatorname*{Max}_R |f_{n+1}(x, y) - g_{n+1}(x, y)| \leq \operatorname*{Max}_R p_3 |f_n(x, y) - g_n(x, y)| + \operatorname*{Max}_R |h|$$

where $p_3 = \text{Max}(p_1, p_2)$. Iteration of this inequality yields

$$(8)\quad \operatorname*{Max}_R |f_n(x, y) - g_n(x, y)| \leq \operatorname*{Max}_R |h| (1 + p_3 + \ldots p_3^{n-1}),$$

for $n = 2, \ldots,$. Letting $n \to \infty$ we obtain the stated result.

Exercises and Research Problems for Chapter II

1. With reference to the process described in § 2, consider the case where the purpose of the process is to maximize the expected value of $\varphi(R)$, where R is the total return, and $\varphi(z)$ is a given function of z. Define the function

$f(x, y, a) =$ expected value of $\varphi(R)$ obtained employing an optimal policy with initial quantities x and y in the respective mines and a quantity a already mined.

Show that $f(x, y, a)$ satisfies the following functional equation

$$f(x, y, a) = \text{Max} \begin{bmatrix} A: & p_1 f(r_1' x, y, a + r_1 x) + p_1' \varphi(a) \\ B: & p_2 f(x, r_2' y, a + r_2 y) + p_2' \varphi(a) \end{bmatrix}, \quad x, y \geq 0$$

$$f(0, 0, a) = \varphi(a).$$

Here $p_1' = 1 - p_1$, $p_2' = 1 - p_2$, $r_1' = 1 - r_1$, $r_2' = 1 - r_2$

2. Establish an existence and uniqueness theorem for this equation.

3. Consider the case where $\varphi(z)$ is defined as follows: $\varphi(z) = 0$, $0 \leq z < u$, $\varphi(z) = 1$, $z \geq u$, where $u < x + y$.

4. Let $g(x, y) = \operatorname*{Max}_P \text{Exp}(e^{br})$, $b > 0$, where Exp stands for expected value and we maximize over all policies P. Show that $g(x, y)$ satisfies the equation

$$g(x, y) = \text{Max} \begin{bmatrix} A: & p_1 e^{br_1 x} g(r_1' x, y) + p_1' \\ B: & p_2 e^{br_2 y} g(x, r_2' y) + p_2' \end{bmatrix}$$

5. Show that the solution of the above equation is determined by the relation between the functions $p_1 (e^{br_1 x} - 1)/p_1'$ and $p_2 (e^{br_2 y} - 1)/p_2'$

6. Show that Theorem 2 is the limiting case of this result as $b \to 0$.

7. The function $g(x, 0)$ satisfies the equation

$$g(x, 0) = p_1 e^{br_1 x} g(r_1' x, 0) + p_1'.$$

Obtain its asymptotic behavior as $x \to \infty$.

8. Referring to Problem 1 obtain some sufficient conditions upon $\varphi(x)$ which will ensure precisely two decision regions.

9. Solve the equation

$$f(x, y) = \text{Max} \begin{bmatrix} A: & p_1 [r_1 x + f(r_1' x, y)] \\ B: & p_2 [r_2 y + f(x, r_2' y)] \\ C: & p_3 [r_3 x + r_4 y + f(tx, ty)] \end{bmatrix}$$

10. Solve the equation

$$f(x, y) = \text{Max} \begin{bmatrix} A: x + f(ax, by) \\ B: y + f(cy, dx) \end{bmatrix}$$

assuming that $0 \leq a, b, c, d < 1$. (Gross-Shapiro)

11. Consider the process described in § 2 under the assumption that there is a probability p_1 of obtaining $r_1 x$ and continuing, a probability p_2 of obtaining nothing and continuing, and a probability p_3 of obtaining nothing and terminating, if A is chosen, with $p_1 + p_2 + p_3 = 1$, with similar probabilities q_1, q_2, q_3 if B is chosen. Show that the corresponding functional equation is

$$f(x, y) = \text{Max} \begin{bmatrix} A: p_1 [r_1 x + f((1 - r_1) x, y)] + p_2 f(x, y) \\ B: q_1 [s_1 y + f(x, (1 - s_1) y)] + q_2 f(x, y) \end{bmatrix}$$

and that this may be written in the simpler form

$$f(x, y) = \text{Max} \begin{bmatrix} A: & \dfrac{p_1}{1 - p_2} [r_1 x + f((1 - r_1) x, y)] \\ \\ B: & \dfrac{q_1}{1 - q_2} [s_1 y + f(x, (1 - s_1) y)] \end{bmatrix}$$

12. Consider the process described in § 2 in which it is not possible to observe the effect of any of the decisions once the process has started.

Discuss the problem of determining the policies maximizing the expected return in the following situations:

a. when the machine is undamaged, it mines a fixed fraction of the gold in any particular mine.
b. when the machine is undamaged, there is a distribution of returns.

Suppose that we wish to maximize the probability that the return exceeds a fixed quantity R_0.

13. Consider the process described in § 2 under the assumption that the machine mines a fixed quantity in each mine, dependent upon the mine, in place of a fixed fraction, as long as the amount remaining in the mine exceeds the fixed amount.

14. Show that the equation in (5.1) is equivalent to

$$f(z) = \text{Max} \begin{Bmatrix} \text{A:} & p_1 [r_1 + (1 - r_1) f(z/(1 - r_1))] \\ \text{B:} & p_2 [r_2 z + f((1 - r_2) z)] \end{Bmatrix}$$

for $0 \leq z < \infty$.

15. Consider the equation

$$f(x, y) = \text{Max} \begin{bmatrix} \text{A:} & rx + f((1 - r) x, y) \\ \text{B:} & q [sy + f(x, (1 - s) y)] \end{bmatrix}$$

for $x, y \geq 0$, $0 \leq r, s, q < 1$.

Show that a solution is

$$f(x, y) = x + \frac{sqy}{1 - q(1 - s)}.$$

16. Show that the gold-mining process generating this equation possesses *no* optimal policy, i.e. no policy yielding this return, but that there are arbitrarily many policies yielding a return of more than

$$x + \frac{sqy}{1 - q(1 - s)} - \delta \text{ for any } \delta > 0.$$

17. Prove that the solution above is not unique in the class of bounded functions over any bounded rectangle, but that it is unique over the class of functions $f(x, y)$ for which $f(0, 0) = 0$, $f(x, y)$ is continuous at $x = y = 0$.

Bibliography and Comments for Chapter II

§ 1. The concept of "decision regions" is a very important one in the study of decision processes. We shall meet it again in Chapter VIII, where it guides us to the solution of the variational problems treated there, and again in Chapter IX, in connection with variational problems with constraints. An interesting paper in this connection is K. D. Arrow, D. Blackwell, M. Girshick, "Bayes and Minimax Solutions of Sequential Decision Problems," *Econometrica*, vol. 17 (1949), pp. 213–214.

§ 8. The result of § 8 was obtained in conjunction with M. Shiffman in the summer of 1950.

§ 12. The type of geometric argument used here was extensively developed by S. Karlin and H. N. Shapiro to give an alternative proof of Theorem 2 and other results.

§ 13. The first counter-example was obtained by S. Karlin and H. N. Shapiro after a great deal of fruitless effort had been expended attempting to establish a result based upon Figure 8. See S. Karlin and H. N. Shapiro, "Decision Processes and Functional Equations," RM–933, Sept. 1952, The RAND Corporation.

CHAPTER III

The Structure of Dynamic Programming Processes

§ 1. Introduction

In this chapter we wish to examine and compare the essential features of the two processes we have considered in some detail in the first and second chapters. Disparate as these processes may seem at first glance, one being of deterministic type with a stochastic version and the other of a stochastic type with no deterministic version, we shall see that from an abstract point of view they are examples of the same general type of process. It is therefore no accident that they are governed by functional equations of a similar form.

After a discussion and analysis of these similarities, we shall consider the formulation of the more general decision processes and from these derive a number of functional equations possessing a common structure. We could, if we so desired, condense these into one all-embracing functional equation. However, since extreme generality is only gained at the expense of fine detail, it seems decidedly better, from both a conceptual and analytic point of view, to consider separately a number of important sub-categories of processes, each of which possesses certain distinctive mathematical and physical features.

We shall close the chapter with a further discussion of the concept of approximation in function space, which we have already encountered in the previous chapters, and a demonstration of its most important property, that of monotone convergence.

§ 2. Discussion of the two preceding processes

Let us begin by observing that the processes discussed in Chapters I and II have the following features in common:

a. In each case we have a physical system characterized at any stage by a small set of parameters, the *state variables*.
b. At each stage of either process we have a choice of a number of decisions.
c. The effect of a decision is a transformation of the state variables.
d. The past history of the system is of no importance in determining future actions.

e. The purpose of the process is to maximize some function of the state variables.

We have purposely left the description a little vague, since it is the spirit of the approach to these processes that is significant rather than the letter of some rigid formulation. It is extremely important to realize that one can neither axiomatize mathematical formulation nor legislate away ingenuity. In some problems, the state variables and the transformations are forced upon us; in others there is a choice in these matters and the analytic solution stands or falls upon this choice; in still others, the state variables and sometimes the transformations must be artificially constructed. Experience alone, combined with often laborious trial and error, will yield suitable formulations of involved processes.

Let us now identify the two processes discussed in the foregoing chapters with the description given above.

In the unbounded multi-stage allocation process, the state variables are x, the quantity of resources, and z the return obtained up to the current stage. The decision at any stage consists of an allocation of a quantity y to the first activity where $0 \leq y \leq x$. This decision has the effect of transforming x into $ay + b\,(x - y)$ and z into $z + g\,(y) + h\,(x - y)$. The purpose of the process is to maximize the final value of z.

In the stochastic gold-mining process, the state variables are x and y, the present levels of the two mines, and z the amount of gold mined to date. The decision at any stage consists of a choice of Anaconda or Bonanza. If Anaconda is chosen, (x, y) goes into $((1 - r_1)\,x, y)$ and z into $z + r_1\,x$, and if Bonanza, (x, y) goes into $(x, (1 - r_2)\,y)$ and z into $z + r_2 y$. The purpose of the process is to maximize the expected value of z obtained before the machine is defunct.

In the finite versions of both processes, we have the additional parameter of time, manifesting itself in the form of the number of stages remaining in the process. It is, however, very useful to keep this state variable distinct from the others, since, as usual, time plays a unique role.

Let us now agree to the following terminology: A *policy* is any rule for making decisions which yields an allowable sequence of decisions; and an *optimal policy* is a policy which maximizes a preassigned function of the final state variables. A more precise definition of a policy is not as readily obtained as might be thought. Although not too difficult for deterministic processes, stochastic processes require more care. For any *particular* process, it is not difficult to render the concept exact. The key word is, of course, "allowable".

A convenient term for this preassigned function of the final state variables is *criterion function*. In many applications, the determination of a proper criterion function is a matter of some difficulty. From the analytic point of view, a solution may be quite easy to obtain for one criterion function, and quite difficult for a closely related one. It is well, consequently, to retain a certain degree of flexibility in the choice of such functions.

§ 3. The principle of optimality

In each process, the functional equation governing the process was obtained by an application of the following intuitive:

PRINCIPLE OF OPTIMALITY. *An optimal policy has the property that whatever the initial state and initial decision are, the remaining decisions must constitute an optimal policy with regard to the state resulting from the first decision.*

The mathematical transliteration of this simple principle will yield all the functional equations we shall encounter throughout the remainder of the book. A proof by contradiction is immediate.

§ 4. Mathematical formulation—I. A discrete deterministic process

Let us now consider a *deterministic* process, by which we mean that the outcome of a decision is uniquely determined by the decision, and assume that the state of the system, apart from the time dependence, is described at any stage by an M-dimensional vector $p = (p_1, p_2, \ldots, p_M)$, constrained to lie within some region D. Let $T = \{T_q\}$ where q runs over a set S which may be finite, enumerable, composed of continua, or a combination of sets of this type, be a set of transformations with the property that $p \varepsilon D$ implies that $T_q(p) \varepsilon D$ for all $q \varepsilon S$, which is to say that any transformation T_q carries D into itself.

The term "discrete" signifies here that we have a process consisting of a finite or denumerably infinite number of stages.

A policy, for the finite process which we shall consider first, consists of a selection of N transformations in order, $P = (T_1, T_2, \ldots, T_N)$,[1] yielding successively the sequence of states

$$
\begin{aligned}
p_1 &= T_1(p), \\
p_2 &= T_2(p_1), \\
&\;\cdot \\
&\;\cdot \\
&\;\cdot \\
p_N &= T_N(p_{N-1}).
\end{aligned} \tag{1}
$$

These transformations are to be chosen to maximize a given function, R, of the final state p_N.

[1] where we write T_1 for T_{q_1}, T_2 for T_{q_2}, and so on.

There are a number of cases in which it is easy to see that a maximum will exist, in which case an optimal policy exists. The simplest is that where there are only a finite number of allowable choices for q at each stage. Perhaps next in order of simplicity is where we assume that D is a finite closed region, with $R(p)$ continuous in p for $p \varepsilon D$, $T_q(p)$ jointly continuous in p and q for all $p \varepsilon D$ and all q belonging to a finite closed region S.

These two cases cover the most important of the finite processes, while their limiting forms account for the unbounded processes.

Observe that the maximum value of $R(p_N)$, as determined by an optimal policy, will be a function only of the initial vector p and the number of stages N. Let us then define our basic auxiliary functions

$$(2) \qquad f_N(p) = \operatorname*{Max}_{P} R(p_N)$$

= the N-stage return obtained starting from an initial state p and using an optimal policy.

This sequence is defined for $N = 1, 2, \ldots$, and for $p \varepsilon D$.

Simple as this step is, it represents a fundamental principle in analysis, the *principle of continuity*. In order to solve our original problem involving one initial vector, p, and a multi-stage process of a definite number of stages, N, we consider the entire set of maximization problems arising from arbitrary values of p and from an arbitrary number of stages.

The original process has thus been imbedded within a family of similar processes. In place of attempting to determine the characteristics of an optimal policy for an isolated process, we shall attempt to deduce the common properties of the set of optimal policies possessed by the members of the family.

This procedure will enable us to resolve the original problem in a number of cases where direct methods fail.

To derive a recurrence relation connecting the members of the sequence $\{f_N(p)\}$, let us employ the principle of optimality stated above in 3. Assume that we choose some transformation T_q as a result of our first decision, obtaining in this way a new state vector, $T_q(p)$. The maximum "return"[2] from the following $(N - 1)$ stages is, by definition, $f_{N-1}(T_q(p))$. It follows that if we wish to maximize the total N-stage return q must now be chosen so as to maximize this $N - 1$ stage return. The result is the basic recurrence relation

$$(3) \qquad f_N(p) = \operatorname*{Max}_{q \varepsilon S} f_{N-1}(T_q(p)),$$

for $N \geq 2$, with

[2] i.e. the value of the criterion function.

$$f_1(p) = \operatorname*{Max}_{q \in S} R(T_q(p)) \tag{4}$$

Observe that $f_N(p)$ is unique, but that the q which maximizes is not necessarily so. Thus the maximum return is uniquely determined, but there may be many optimal policies which yield this return.

For the case of an unbounded process, the sequence $\{f_N(p)\}$ is replaced by a single function $f(p)$, the total return obtained using an optimal policy starting from state p, and the recurrence relation is replaced by the functional equation

$$f(p) = \operatorname*{Max}_{q} f(T_q(p)). \tag{5}$$

§ 5. Mathematical formulation—II. A discrete stochastic process

Let us once again consider a discrete process, but one in which the transformations which occur are *stochastic* rather than deterministic.

A decision now results in a distribution of transformations, rather than a single transformation. The initial vector p is transformed into a stochastic vector z with an associated distribution function $dG_q(p, z)$, dependent upon p and the choice q.

Two distinct types of processes arise, depending upon whether we assume that z is known after the decision has been made and before the next decision has to be made, or whether we assume that only the distribution function is known. We shall only consider processes of the first type in this volume, since processes of the second type require in general the concept of functions of functions, which is to say *functionals*.

It is clear, as we have stated several times before, that it is now on the whole meaningless to speak of maximizing *the* return. Rather we must agree to measure the value of a policy in terms of some average value of the function of the final state. Let us call this expected value the *return*.

Beginning with the case of a finite process, we define $f_N(p)$ as in (4.2). If z is a state resulting from any initial transformation T_q, the return from the last $N - 1$ stages will be $f_{N-1}(z)$, upon the employment of an optimal policy. The expected return as a result of the initial choice of T_q is therefore

$$\int_{z \in D} f_{N-1}(z)\, dG_q(p, z) \tag{1}$$

Consequently, the recurrence relation for the sequence $\{f_N(p)\}$ is

$$f_N(p) = \operatorname*{Max}_{q \in S} \int_{z \in D} f_{N-1}(z)\, dG_q(p, z),\ N \geq 2, \tag{2}$$

with

(3) $$f_1(p) = \underset{q \varepsilon S}{\text{Max}} \int_{z \varepsilon D} R(z)\, dG_q(p, z).$$

Considering the unbounded process, we obtain the functional relation

(4) $$f(p) = \underset{q \varepsilon S}{\text{Max}} \int_{z \varepsilon D} f(z)\, dG_q(p, z)$$

§ 6. Mathematical formulation—III. A continuous deterministic process

There are a number of interesting processes that require that decisions be made at each point of a continuum, such as a time interval. The simplest examples of processes of this character are furnished by the calculus of variations. As we shall see in Chapter IX below, this conception of the calculus of variations leads to a new view of various parts of this classical theory.

Let us define

(1) $f(p; T) =$ the return obtained over a time interval $[0, T]$ starting from the initial state p and employing an optimal policy.

Although we consider the process as one consisting of choices made at each point t on $[0, T]$, it is better to begin with the concept of choosing policies, which is to say functions, over intervals, and then pass to the limit as these intervals shrink to points. The analogue of (4.3) is

(2) $$f(p; S + T) = \underset{D\,[0, S]}{\text{Max}}\ f(p_D; T)$$

where the maximum is taken over all allowable decisions made over the interval $[0, S]$.

As soon as we consider infinite processes, occurring as the result of either unbounded sequences of operations, or because of choices made over continua, we are confronted with the difficulty of establishing the existence of an actual maximum rather than a supremum. In general, therefore, in the discussion of processes of continuous type, it is better to use initially the equation

(3) $$f(p; S + T) = \underset{D}{\text{Sup}}\, f(p_D; T),$$

which is usually easy to establish, and then show, under suitable assumptions that the maximum is actually attained.

As we shall see in Chapter IX, the limiting form of (2) as $S \to 0$ is a non-linear partial differential equation. This is the important form for actual analytic utilization. For numerical purposes, S is kept non-zero but small.[3]

[3] We shall show, in Chapter IX, that it is possible to avoid many of the quite difficult rigorous details involved in this limiting procedure if we are interested only in the computational solution of variational processes.

§ 7. Continuous stochastic processes

An interesting and challenging question which awaits further exploration is the formulation and solution of general classes of continuous stochastic decision processes of both one-person and two-person variety. Although we shall discuss a particular process in Chapter VIII, we shall not discuss the general formulation of continuous stochastic decision processes here, since a rigorous treatment requires delicate and involved argumentation based upon sophisticated concepts.

§ 8. Generalizations

It will be apparent to the reader that the functional equations we have derived above for the case where the state variables and the decision variables were constrained to finite dimensional Euclidean spaces can be extended to cover the case where the state variables and decision variables are elements of more general mathematical spaces, such as Banach spaces.

Rather than present this extension abstractly we prefer to wait until a second volume where we will discuss examples of these more general processes. The theory of integral equations and variational problems involving functions of several variables, as well as more general stochastic processes, all afford examples of processes which escape the finite dimensional formulation to which we have restricted ourselves in this volume, and require for their formulation in the foregoing terms the theory of functionals and operations.

§ 9. Causality and optimality

Consider a multi-stage process involving no decisions, say one generated by the system of differential equations,

$$dx_i/dt = g_i(x_1, x_2, \ldots, x_N),\ x_i(0) = c_i,\ i = 1, 2, \ldots, N, \tag{1}$$

which may, more compactly, be written in vector form

$$dx/dt = g(x),\ x(0) = c. \tag{2}$$

The state of the system at time t, taking for granted existence and uniqueness of the solution, is a function only of c and t, thus we may write

$$x(t) = f(c, t). \tag{3}$$

The uniqueness of the solution leads to the functional equation

$$f(c, s + t) = f(f(c, s), t), \tag{4}$$

for $s, t \geq 0$, an analytical transliteration of the law of causality. This equation expresses the fundamental semi-group property of processes of this type.

Comparing (4) above with (6.2), we see that we may regard multi-stage decision processes as furnishing a natural extension of the theory of semi-groups. Any further discussion here along these lines would carry us beyond our self-imposed limits, and we shall consequently content ourselves with the above observation.

§ 10. Approximation in policy space

In solving a functional equation such as (4.4) or (5.3), we shall in Chapter IV make use of that general factotum of analysis, the method of successive approximations. The method very briefly, consists of choosing an initial function $f_0(p)$, and then determining a sequence of functions, $\{f_N(p)\}$, by means of the algorithm

$$f_N(p) = \operatorname*{Max}_q f_{N-1}(T_q(p)), \; N = 1, 2, \ldots \tag{1}$$

as, for instance, in (4.4) We have already employed this method in dealing with the equations of Chapters I and II.

In many important cases, this method after a suitable preliminary preparation of the equation actually leads to a convergent sequence whose limit yields the solution of the functional equation.[4] We shall make extensive use of it in the following chapter.

In the theory of dynamic programming, however, we have an alternate method of approximation which is equally important in its own right, a method which we call "approximation in policy space".

Before discussing this method of approximation, let us observe that there is a natural duality existing in dynamic programming processes between the function $f(p)$ measuring the overall return, and the optimal policy (or policies) which yields this return. Each can be used to determine the other, with the additional feature that a knowledge of $f(p)$ yields *all* optimal policies, since it determines all maximizing indices q in an equation such as (4.4), while a knowledge of *any* particular optimal policy yields $f(p)$.

The maximizing index q can be considered to be a function of p. If the index is not unique, we have a multi-valued function. Whereas we call $f(p)$ an element in function space, let us call $q = q(p)$ an element of *policy space*. Both spaces are, of course, function spaces, but it is worth distinguishing between them, since their elements are quite different in meaning.

It follows now that we have two ways of making an initial approxima-

[4] It is interesting to observe that in many theories, as, for example, partial differential equations, the preliminary transformation of the equation is of such a nature that the principal difficulty of the existence proof resides in the demonstration that the limit function actually satisfies the *original* equation.

tion. We may approximate to $f(p)$, as we do ordinarily in the method of successive approximation, or we may, and this is a feature of the functional equations belonging to dynamic programming processes, approximate initially in policy space.[5]

Choosing an initial approximation $q_o = q_o(p)$, we compute the return from this policy by means of the functional equation

$$f_o(p) = f_o(T_{q_o}(p)). \tag{2}$$

We have already given an example of this in § 11 of Chapter I.

There are now two ways we can proceed. Taking the function of q, $f_o(T_q(p))$, we can determine a function $q(p)$ which maximizes. Call this function $q_1(p)$. Using this new policy, we determine $f_1(p)$, the new return, by means of the functional equation

$$f_1(p) = f_1(T_{q_1}(p)). \tag{3}$$

This equation is solved iteratively, as in (11.3) and (11.4) of Chapter I.

Continuing in this way, we obtain two sequences $\{f_N(p)\}$ and $\{q_N(p)\}$.

In place of this procedure, we can define

$$f_1(p) = \operatorname*{Max}_q f_o(T_q(p)), \tag{4}$$

and then continue inductively, employing the usual method of successive approximations,

$$f_{N+1}(p) = \operatorname*{Max}_q f_N(T_q(p)). \tag{5}$$

It is immediate that $f_1 \geq f_o$ and thus that the sequence $\{f_N\}$ is monotone increasing. We shall discuss the convergence of this process in the next chapter.

The first procedure, although a more natural one, seems more difficult to treat rigorously and we shall not consider it here. In dealing with various types of continuous processes, such as those furnished by the calculus of variations, it would seem, however, that this technique is required for successive approximations. We shall discuss this topic again in Chapter IX.

[5] Actually this type of approximation is tacitly encountered in other branches of analysis as, for instance, in the theory of differential equations, where a differential equation is frequently replaced by a difference equation for approximation purposes. This replaces the space of general functions by the subspace of step-functions.

Exercises and Research Problems for Chapter III

1. Suppose that we are given the information that a ball is in one of N boxes, and the a priori probability, p_k, that it is in the kth box. Show that the procedure which minimizes the expected time required to find the ball consists of looking in the most likely box first.

2. Consider the more general process where the time consumed in examining the kth box is t_k, and where there is a probability q_k that any particular examination of the kth box will yield no information concerning its contents. When this happens, we continue the search operation with the information already available.

Let $f(p_1, p_2, \ldots, p_N)$ be the expected time required to obtain the ball using an optimal policy. Show that this function satisfies the equation

$$f(p_1, p_2, \ldots, p_N) = \operatorname*{Min}_k \left[\frac{t_k}{(1-q_k)} + (1-p_N) f(p_1^*, p_2^*, \ldots, 0, \ldots, p_N^*) \right]$$

where $p_i^* = p_i / (1 - p_k)$ and the 0 occurs in the kth place.

3. Prove that if we wish to *obtain* the ball, the optimal policy consists of examining the box for which $p_k (1 - q_k)/t_k$ is a maximum first. On the other hand, if we merely wish to locate the box containing the ball in the minimum expected time, the box for which this quantity is a maximum is examined first, or not at all.

4. Consider the situation in which we can simultaneously perform operations which locate the ball within given sets of boxes.

5. We have a number of coins, all of the same weight except for one which is of different weight, and a balance. Determine the weighing procedures which minimize the maximum time required to locate the distinctive coin in the following cases

a. The coin is known to be heavier

b. It is not known whether the coin is heavier or lighter.

6. Determine the weighing procedures which minimize the expected time required to locate the coin.

7. Consider the more general problem where there are two or more distinctive coins, under various assumptions concerning the properties of the distinctive coins. (Cairns)

8. We are given n items, not all identical, which must be processed through a number of machines, m, of different type. The order in which the machines are to be used is not immaterial, since some processes must be

carried out before others. Given the times required by the ith item on the jth machine, a_{ij}, $i = 1, 2, \ldots, n$, $j = 1, 2, \ldots, m$, we wish to determine the order in which the items should be fed into the machines so as to minimize the total time required to complete the lot.

Consider the case where there are only two stages with $a_{i1} = a_i$ and $a_{i2} = b_i$, and where the machines must be used in this order. Let

$f(a_1, b_1, a_2, b_2, \ldots, a_N, b_N; t) =$ time consumed processing the N items with required times a_i, b_i on the first and second machines when the second machine is committed for t hours ahead, and an optimal scheduling procedure is employed.

Prove that f satisfies the functional equation

$$f(a_1, b_1, a_2, b_2, \ldots, a_N, b_N; t) = \operatorname*{Min}_i [a_i + f(a_1, b_1, a_2, b_2, \ldots, 0, 0, \ldots, a_N, b_N; b_i + \max(t - a_i, 0)],$$

where the (0, 0) combination is in place of (a_i, b_i).

9. Show that an optimal ordering is determined by the following rule: Item i precedes item j if $\min(a_i, b_j) < \min(a_j, b_i)$. If there is equality, either ordering is optimal, provided that it is consistent with all the definite preferences. (Johnson)

What is the solution if either machine can be used first?

10. Let x_i be the inactive time in the second machine immediately before the ith item is processed on the second machine. Let a_i, b_i be the times required to process the ith item on the first and second machines respectively and assume that the items are arranged in numerical order. Then

$$\sum_{i=1}^{n} x_i = \operatorname*{Max}_{1 \leq u \leq n} \left[\sum_{i=1}^{u} a_i - \sum_{i=1}^{u-1} b_i\right] \text{ (Johnson)}$$

11. For the three-stage process the corresponding expression for the total idle time on the third machine is

$$\operatorname*{Max}_{1 \leq u \leq v \leq n} \left[\sum_{i=1}^{u} a_i - \sum_{i=1}^{u-1} b_i + \sum_{i=1}^{v} b_i - \sum_{i=1}^{v-1} c_i\right] \text{ (Johnson)}$$

12. Consider the following problem arising in the production of many-part items, or alternately in the maintenance of a complex system. There are N different stages of production involved in turning out the final item. The probability that the item is processed correctly at the ith stage is p_i.

Assume that k machines are available which can be used to increase the accuracy of any particular stage of the process in the following way. If one machine is added to the ith stage, p_i becomes p_{i1}, if two machines, then p_{i2}, and so on.

How should we distribute the machines to maximize the overall accuracy of the process? Consider the same problem under the following alternative assumptions.

(a) At most m_i machines are allowed at the ith stage

(b) A machine at the ith stage costs d_i dollars and we have at most d dollars to spend.

(c) A machine at the ith stage requires h_i operators at the ith point, and at most h men are available.

13. A mistake found at the ith point requires a time t_i and a cost c_i to rectify. Taking into account laboring costs, machine costs, and the cost of turning out a defective item, say z, how much money should be spent on checking equipment and how should it be used?

14. Consider the problem of maximizing the function

$\sum_{i=1}^{n} \varphi_i (x_i)$ under the constraints

a. $x_i \geq 0$

b. $\sum_{i=1}^{n} x_i = c$

c. $x_{i_k} x_{i_{k+1}} = 0$ for a set of integers $i_1 < i_2 < i_3 < \ldots < i_m$,

$$m \leq n - 1.$$

Consider, in particular the cases

a. $x_i x_{i+1} = 0, \qquad i = 1, 2, \ldots, n - 1$

b. $x_i x_{i+1} x_{i+2} = 0, \qquad i = 1, 2, \ldots, n - 2$

Consider the reverse situation, where we have constraints of the form

a. $x_{i_k} x_{i_{k+1}} \geq d_k.$

Discuss the special cases

a. $x_i x_{i+1} \geq 1,$

b. $x_i x_{i+1} x_{i+2} \geq 1.$

15. A manager of a restaurant has two types of laundry service available for napkins, a quick service which requires q days, and costs c cents per napkin, and a slow service which requires $p > q$ days and costs d cents, $d < c$, per napkin. Assuming that he knows in advance the number of

customers he will have on any given day of an N-day period and that he prides himself on providing every customer with a napkin, how many napkins should he purchase and how should he launder them so as to minimize the total cost over the N-day period? Consider first the cases where $p = q + 1$, and $p = q + 2$.

16. Consider the analogous problem under the assumption that k launderings wear out a napkin.

17. Consider the above problem under the assumption that number of customers on each day is a stochastic quantity.

18. We have a resource x which may be utilized in a number of ways. If y is a parameter specifying a particular use, let $R(x, y)$ be the immediate return, and $D(x, y)$ the cost in resources. If $f(x)$ is the total return from repeated use of an initial resource x, obtained using an optimal allocation policy, we derive the functional equation

$$f(x) = \operatorname*{Max}_{y} [R(x, y) + f(x - D(x, y))].$$

Assuming that $D(x, y)$ is small compared to x, for all y, show that we obtain the formal approximate equation

$$f'(x) = \operatorname*{Max}_{y} \frac{R(x, y)}{D(x, y)},$$

and give the interpretation of this result.

19. Consider the stochastic case. Show that the corresponding functional equation has the form

$$f(x) = \operatorname*{Max}_{y} \left[\int_0^\infty z\,dR(y, z, x) + f\left(x - \int_0^\infty w\,dD(y, w, x)\right)\right],$$

and the approximate equation has the form

$$f'(x) = \operatorname*{Max}_{y} \frac{\int_0^\infty z\,dR(y, z, x)}{\int_0^\infty w\,dD(y, w, x)},$$

and give the interpretation of the result.

20. Consider the application of approximation in policy space to the functional equation

$$f(x) = \operatorname*{Max}_{0 \le y \le x} [g(y) + h(x - y) + f(ay + b(x - y))].$$

We choose an initial $y_0(x)$ and compute $f_0(x)$. Then determine $y_1(x)$ by the condition that y_1 maximize the function $g(y) + h(x - y) + f_0(ay + b(x - y))$, compute $f_1(x)$ using $y_1(x)$, and so on. When are the elements of the sequences $\{y_n(x)\}$ and $\{f_n(x)\}$ continuous in x, and when do they converge? Consider, in particular, the cases where g and h are both convex, or both concave.

21. Assume that we have two machines, unimaginatively called I and II, with the following properties. If machine I is used there is a probability r of receiving a gain of one unit; if machine II is used, there is a probability s of receiving a gain of one unit. We shall assume that s is known, but that r is determined only by an a priori probability distribution. The problem is to determine a selection policy which maximizes the expected return obtained over N trials, or alternatively the discounted return from an unbounded process, discounting the return one stage hence by a factor $a < 1$.

Assume that the distribution function for r after m successes and n failures on the first machine is given by

$$dF_{m,n}(r) = \frac{r^m (1 - r)^n \, dF(r)}{\int_0^1 r^m (1 - r)^n \, dF(r)} .$$

Let $f_{m,n}$ equal the expected return obtained using an optimal policy for an unbounded process after the first machine has had m successes and n failures. Show that $f_{m,n}$ satisfies the recurrence relation

$$f_{m,n} = \text{Max} \begin{bmatrix} \text{I:} \quad \int_0^1 r dF_{m,n}(r) \, [1 + af_{m+1,n}] \\ \quad + \int_0^1 (1 - r) \, dF_{m,n}(r) \, [af_{m,n+1}], \\ \text{II:} \quad s/(1 - a) \end{bmatrix}$$

22. Prove that there is a unique bounded solution to this equation, which may be obtained by successive approximations.

23. Prove that for each $m, n \geq 0$ there is a unique quantity $s(m, n)$ with the property that the sequence $\{f_{mn}\}$ is determined by the equations

(a) $f_{mn} = s/(1 - a), \; 1 \geq s \geq s(m, n)$,

$$f_{mn} = \int_0^1 r dF_{mn}(r) \, [1 + af_{m+1,n}]$$
$$+ a\left(1 - \int_0^1 r dF_{m,n}(r)\right) f_{m,n+1}, \; 0 \leq s < s(m, n).$$

The sequence $s(m, n)$ has the following properties

$$\text{(b)}\quad s(m+1, n) > s(m, n) > s(m, n+1),$$

and

$$\text{(c)}\quad f_{m+1,n} > f_{m,n} > f_{m,n+1}.$$

How can the sequence $s(m, n)$ be calculated?

24. Prove the corresponding results for the process allowing only a finite number of trials.

25. Consider the following situation. We have a warehouse with fixed capacity and an initial stock of a certain product which is subject to known seasonal price and cost variations. The problem is to determine the optimal pattern of purchasing (or production), storage and sales.

Let B denote the fixed warehouse capacity, and A the initial stock in the warehouse. Consider a seasonal product bought (or produced) and sold for each of $i = 1, 2, \ldots, n$ periods. For the ith period, let

(1)
$$\begin{aligned} c_i &= \text{cost per unit} \\ p_i &= \text{selling price per unit} \\ x_i &= \text{amount bought (or produced)} \\ y_i &= \text{amount sold} \end{aligned}$$

The constraints are as follows:

(2)
(a) *Buying Constraints:* The stock on hand at the end of the ith period cannot exceed the warehouse capacity.
(b) *Selling Constraints:* The amount sold in the ith period cannot exceed the amount available at the end of the $(i-1)^{\text{st}}$ period.
(c) *Non-negativity:* Amounts purchased or sold in any period are non-negative.

The problem is to determine the policy which maximizes the over-all profit.

Show that it may be converted into the problem of determining the x_i and y_i which maximize

$$P = \sum_{j=1}^{n} (p_j y_j - c_j x_j), \tag{3}$$

subject to the constraints

(4)
$$\text{(a)}\quad A + \sum_{j=1}^{i} (x_i - y_i) \leq B, \qquad i = 1, 2, \ldots, n,$$

$$\text{(b)}\quad y_i \leq A + \sum_{j=1}^{i-1} (x_j - y_j), \qquad i = 1, 2, \ldots, n,$$

$$\text{(c)}\quad x_i, y_i \geq 0.$$

26. For fixed B, define

$$f_n(A) = \text{Max}\ P, \quad n = 1, 2, \ldots.$$

Show that $f_1(A) = p_1 A$, and that

$$(1) \qquad f_n(A) = \underset{x_1,\, y_1}{\text{Max}}\ [p_1 y_1 - c_1 x_1 + f_{n-1}(A + x_1 - y_1)],$$

for $n \geq 2$, where the maximum is over the region

$$(2) \qquad \text{(a)}\quad 0 \leq y_1 \leq A$$

$$\text{(b)}\quad x_1 - y_1 \leq B - A, \qquad x_1 \geq 0.$$

27. Prove that the function $f_N(v)$ is linear in v, namely

$$f_N(v) = K_N(p_1, p_2, \ldots, p_N, c_1, c_2, \ldots, c_N) + L_N(p_1, p_2, \ldots, p_N, c_1, c_2, \ldots, c_N)\, v,$$

and thus that the optimal policy is independent of v.

(Dreyfus)

28. Consider the following idealized transportation system

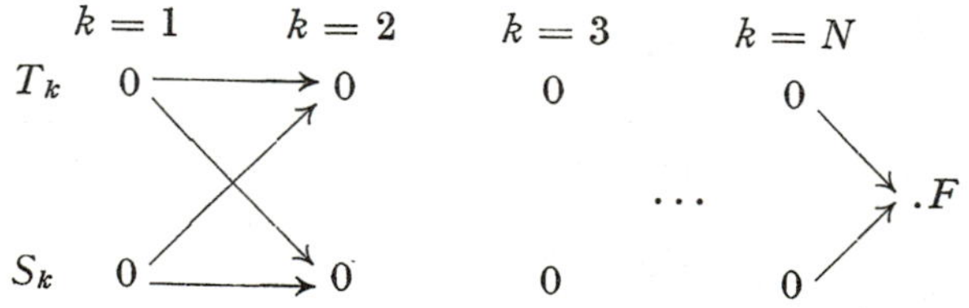

At each stage we have two terminals T_k and S_k. From either T_k or S_k we can ship materials to T_{k+1} or S_{k+1}.

The maximum amounts we can ship along these routes are the following

a. $T_k \to T_{k+1} = R_{k,\, k+1}, \qquad T_N \to F = R_N$

b. $T_k \to S_{k+1} = R_{k,\, k+1} \qquad S_N \to F = S_N$

c. $S_k \to S_{k+1} = S_{k,\, k+1}$

d. $S_k \to T_{k+1} = S_{k,\, k+1}$

Starting with initial quantitites x at T_k and y at S_k, denote by $F_k(x, y)$ the quantity arriving at F using an optimal shipping policy. Show that

$$F_N(x, y) = \text{Min}(x, R_N) + \text{Min}(y, S_N),$$

$$F_k(x, y) = \text{Max}\ F_{k+1}(z_1 + w_2, z_2 + w_1),$$

where the maximum is taken over the region

$$z_1 + z_2 \leq x, \qquad w_1 + w_2 \leq y,$$
$$0 \leq z_1 \leq R_{k,\, k+1}, \qquad 0 \leq z_2 \leq R_{k,\, k+1},$$
$$0 \leq w_1 \leq S_{k,\, k+1}, \qquad 0 \leq w_2 \leq S_{k,\, k+1}$$

29. Formulate the corresponding problem for the case where the terminals have maximum capacities.

30. Consider the stochastic case where the capacities are random variables with known distribution functions. Obtain a recurrence relation for the maximum expected quantity arriving at F, under various assumptions concerning the information pattern.

31. Consider the following transportation problem. We are given a number of "sources", $S_1, S_2, \ldots, S_M$, and a number of "sinks" or "terminals", $T_1, T_2, \ldots, T_N$. Each source S_i has a quantity x_i of resources which must be transported to various terminals in such a way that the total quantity arriving at T_j fulfills an a priori demand y_j. It is assumed that $\sum_i x_i = \sum_j y_j$. Given the distances, d_{ij}, between the sources and the terminals, and assuming that the cost of shipping a unit quantity of resources between S_i and T_j is equal to d_{ij}, we wish to determine the routing which minimizes the total cost of supplying the demands.

Show that the problem above is equivalent to minimizing the sum

$$C = \sum_{i,j} d_{ij}\, x_{ij}$$

subject to the constraints

$$\sum_j x_{ij} = x_i, \quad \sum_i x_{ij} = y_j, \quad x_{ij} \geq 0. \quad \text{(Hitchcock-Koopmans)}$$

32. Write, for fixed $y_1, y_2, \ldots, y_N$,

$$\min_{x_{ij}} C = f_N(x_1, x_2, \ldots, x_M).$$

Show that

$$f_1(x_1, x_2, \ldots, x_M) = d_{1N}\, x_1 + d_{2N}\, x_2 + \ldots + d_{MN}\, x_M,$$

$$f_N(x_1, x_2, \ldots, x_M) = \min_{\{x_{i1}\}} \Big[\sum_{i=1}^{M} d_{i1}\, x_{i1} + f_{N-1}(x_1 - x_{11}, x_2 - x_{21}, \ldots, x_M - x_{M1}) \Big]$$

where the minimum is over the region

$$\sum_{i=1}^{M} x_{i1} = y_1, \quad 0 \leq x_{i1} \leq x_i.$$

33. Show that as a consequence of the relation $\sum_{i=1}^{M} x_i = \sum_{j=1}^{N} y_j$, we may always reduce the dimension of the problem by one, by writing

$$f_N(x_1, x_2, \ldots, x_M) = f_N(x_1, x_2, \ldots, x_{M-1}).$$

34. Consider the stochastic case where the d_{ij} represent random variables with given distributions.

35. Assuming that the cost of transportation from an i-port to a j-port is quadratically nonlinear, $d_{ij} x_{ij} + e_{ij} x_{ij}^2$, $e_{ij} > 0$, show that there is now a *unique* minimizing schedule. (Prager)

36. Consider a similar multi-stage process where resources at (A_i, B_i, C_i) must be transported to $(A_{i+1}, B_{i+1}, C_{i+1})$ and so on, until reaching assigned destinations, T_1, T_2, T_3, as indicated below

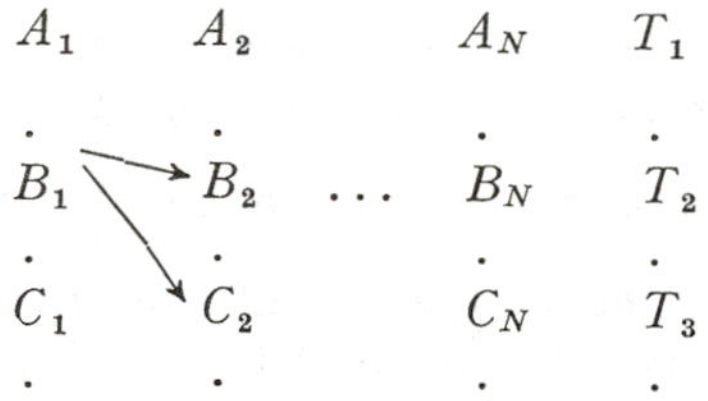

37. Consider the problem of determining the minimum of

$$L(x) = \sum_{i=1}^{N} c_i x_i,$$

subject to the constraints

$$\sum_{j=1}^{N} a_{ij} x_j \leq b_i, \quad i = 1, 2, \ldots, M,$$
$$x_i \geq 0,$$

where we assume that $a_{ij} \geq 0$.

Denote $\operatorname{Min}_x L(x)$ by $f_N(b_1, b_2, \ldots, b_M)$. Show that

$$f_N(b_1, b_2, \ldots, b_M) = \operatorname{Min}_{x_N} [c_N x_N + f_{N-1}(b_1 - a_{1N} x_N, b_2 - a_{2N} x_N, \ldots, b_M - a_{MN} x_N)],$$

where x_N is constrained by the relations

$$x_N \geq 0, \; x_N \leq \operatorname{Min}_i (b_i / a_{iN}).$$

38. Suppose that we have an empty five-gallon jug, J_1, an empty two-gallon jug, J_2, and unlimited supplies of usquebaugh and water. The allowable operations are

- A_1 Fill J_1
- A_2 Empty J_1 of any contents
- A_3 Fill J_2
- A_4 Empty J_2 of any contents
- A_5 Pour contents of J_1 into J_2, as much as allowable
- A_6 Pour contents of J_2 into J_1, as much as allowable.

After any finite number of operations, the state of the system may be described as follows:

1. There are $i = 0, 1, 2$ gallons of liquid in J_2, with a ratio $r: (1 - r)$ of usquebaugh to water.
2. There are $j = 0, 1, 2, 3, 4, 5$ gallons in J_1 with a ratio $s: (1 - s)$ of usquebaugh to water.

Starting in some initial state $(i, j; r, s)$, let $f(i, j; r, s)$ denote the minimum number of operations required to attain a given state, say a fifty-fifty mixture of water and usquebaugh in J_2.

Show that

$$f(i, j; r, s) = 1 + \min_{1 \le k \le 6} A_k f.$$

Is $f(i, j; r, s)$ finite for all rational r, with $j = 0$, and all rational s, with $i = 0$? If not, what final combinations of water and usquebaugh can be attained in J_2 in a finite number of operations?

39. Consider the following problem: At each stage of sequence of actions we are allowed our choice of one of two actions. The first has associated a probability p_1 of gaining one unit, a probability p_2 of gaining two units, and a probability p_3 of terminating the process. The second has a similar set of probabilities p_1', p_2', p_3'. What sequence of choices maximizes the probability of attaining at least n units before the process is terminated?

Let $u(n)$ be the maximum probability. Then

$$u(n) = \text{Max} \begin{bmatrix} p_1 u(n-1) + p_2 u(n-2) \\ p_1' u(n-1) + p_2' u(n-2) \end{bmatrix}, \quad n \geq 1,$$

$$u(n) = 1, \qquad n \leq 0.$$

40. Prove that if

$$u(n) = \max_{1 \le i \le K} \left[\sum_{j=1}^{R} a_{ij} u(n-j) \right], \quad n \geq R,$$

$$u(l) > 0, \; l = 0, 1, 2, \ldots, R-1,$$

and if

(a) $a_{ij} \geq 0$

(b) there is one equation, $r^R = \sum_{j=1}^{R} a_{kj} r^{R-j}$ whose largest positive root is greater than the corresponding roots of the other equations of this type;

(c) for this index k, $a_{k1} \neq 0$,

under these circumstances, the solution of (1) is given by

$$u(n) = \sum_{j=1}^{R} a_{kj} u(n-j)$$

for n sufficiently large.

What happens if at least two characteristic equations have the same maximum root?

41. Consider the equation

$$u(n) = \underset{1 \leq i \leq M}{\text{Max}} \left[\sum_{j=1}^{R} a_{ij} u(n-j) + g_i \right]$$

where $a_{ij} \geq 0$, $\sum_{j=1}^{R} a_{ij} = 1$, $g_i \geq 0$, $u(l) \geq 0$, $l = 0, 1, 2, \ldots, R-1$.

Let $c = \underset{i}{\text{Max}}\, g_i / \sum_{j=1}^{R} j a_{ij}$ be attained for the single value $i = s$. If $a_{s1} > 0$, the solution is given by

$$u(n) = \sum_{j=1}^{R} a_{sj} u(n-j) + g_s$$

for $n \geq n_o$ where n_o depends upon the initial conditions and coefficients.

42. Is the result true if $a_{s1} = 0$? Construct a counter-example.

43. Given a finite set $\{A_i\}$ of non-negative square matrices, let C_N be the matrix $B_1 B_2 \ldots B_N$, where each B_i is an A_j, which possesses the characteristic root of largest absolute value. Let r_N be this root. Prove that $\mu = \lim_{N \to \infty} r_N^{1/N}$ exist. Let M_N denote the smallest majorant of the products $P_N = B_1 B_2 \ldots B_N$; i.e., the ij^{th} element in M_N is greater than or equal the ij^{th} element in any P_N. Let m_N be the characteristic root of M_N of largest absolute value. Prove that $\lambda = \lim_{N \to \infty} M_N^{1/N}$ exists as $N \to \infty$.

44. Prove or disprove that $\mu = \lambda$.

45. Consider the following problem: We are given initially x dollars and a quantity y of a serum, together with the prerogative of purchasing additional amounts of the serum at specified times $t_1 < t_2 < \ldots$. At the kth purchasing opportunity, t_k, a quantity $c_k z$ of serum may be purchased for z dollars, where c_k is a monotone-increasing function of k. Given the probability that an epidemic occurs between t_k and t_{k+1}, and the condition that if an epidemic occurs we may only use the amount of serum on hand, the problem is to determine the purchasing policy that maximizes the over-all probability of successfully combating an epidemic, given the probability of success with a quantity w of serum available.

The condition $c_k > c_{k-1}$ is imposed to indicate the cheaper cost of serum at a later date because of technological improvement. Let

p_k = probability that the epidemic occurs between t_k and t_{k+1}, assuming that it has not occurred previously,

$\varphi(w)$ = probability of combating the epidemic successfully with a quantity w of serum,

$f_k(x, y)$ = over-all probability of success using an optimal purchasing policy from t_k on, given x dollars and a quantity y of serum on hand.

Show that $f_k(x, y)$ satisfies the functional equation

$$f_k(x, y) = \max_{0 \le z \le x} [p_k \varphi(y + c_k z) + (1 - p_k) f_{k+1}(x - z, y + c_k z)]$$

46. Show that if $\varphi(w)$ is convex for all values of w which occur, the optimal policy consists of purchasing no serum at $t_1, t_2, \ldots, t_{k-1}$ and then using all available money at t_k where k is chosen so as to maximize

$$[1 - (1 - p)^{k-1}]\varphi(y) + (1 - p)^{k-1}\varphi(y + c_k x),$$

if $p_k \equiv p$. Find the corresponding expression for general p_k.

47. Let

$$F_N(f) = \min_{\{a_k\}} \int_0^1 |f - \sum_{k=1}^{N} a_k \varphi_k|^R dx.$$

Show that

$$F_N(f) = \min_{a_N} F_{N-1}(f - a_N \varphi_N)$$

48. Show that if we let

$m(x_1, x_2, \ldots, x_N)$ = the minimum of N quantities, $x_1, x_2, \ldots, x_N$, we have the functional equation

$$m\,(m\,(x_1, x_2, \ldots, x_{N-1}), x_N) = m\,(x_1, x_2, \ldots, x_N),$$

and similarly for $M\,(x_1, x_2, \ldots, x_N)$, the maximum of N quantities.

49. Show that

$$\operatorname*{Max}_{\{x_i\}} [(1 - x_1)\, e^{x_1} + (1 - x_2)\, e^{x_1 + x_2} + \ldots + (1 - x_N)\, e^{x_1 + x_2 + \ldots + x_N}] = e_N,$$

where $e_1 = e$, $e_N = e^{e_{N-1}}$.

50. Set

$$f_N\,(b, k) = \operatorname*{Min}_{a_i} \int_I \left[e^{-k\,(x-b)^2 - \sum_{i=1}^{N} (x - a_i)^2} \right] dV\,(x).$$

Show that

$$f_N\,(b, k) = \operatorname*{Min}_{a_N} \left[e^{\frac{-k}{k+1}(a_N - b)^2} \quad f_{N-1}\left(\frac{kb + a_N}{k+1}, k+1\right)\right],$$

$$f_1\,(b, k) = \operatorname*{Min}_{a} \int_I \,[e^{-k\,(x-b)^2 - (x-a)^2}]\, dV\,(x).$$

51. Obtain recurrence relations for the problem of determining the minimum and maximum of

(a) $Q_N = (ax_1)^2 + (x_1 + ax_2)^2 + \ldots (x_1 + x_2 + \ldots + x_{N-1} + ax_N)^2$,

subject to $x_1^2 + x_2^2 + \ldots + x_N^2 = 1$,

(b) $Q_N = x_1^2 + (x_1 + ax_2)^2 + \ldots (x_1 + ax_2 + a^2 x_3 + \ldots + a^{N-1} x_N)^2$,

subject to $x_1^2 + x_2^2 + \ldots + x_N^2 = 1$,

(c) $Q_N = x_1^2 + (x_1 + ax_2)^2 + (x_1 + ax_2 + (a + b)\, x_3)^2 + \ldots$

$(x_1 + ax_2 + (a + b)\, x_3 + \ldots (a + (N - 2)\, b)\, x_N)^2$,

subject to $x_1^2 + x_2^2 + \ldots + x_N^2 = 1$.

52. Suppose that a piece of candy is to be shared by two children. Show that an optimal procedure is to let one child divide the candy, and the other choose the piece he wants. Show that this leads to the equation

$$r = \operatorname*{Max}_{0 \le y \le x} \operatorname{Min}\,(y, x - y) = x/2,$$

for the share of the first child.

53. What is the corresponding procedure for N children? (Steinhaus)

54. Suppose that we have a vehicle which can carry enough gasoline to go a distance of d miles. In order to traverse a distance of $2d$ miles over barren territory, it is necessary to establish intermediate caches of gasoline. How should these be located so as to minimize the total expenditure of gasoline required to traverse this distance, and what is the total distance travelled by the vehicle before it reaches its destination?

(N. J. Fine, "The Jeep Problem," *Amer. Math. Monthly*, Vol. LIV, Jan. 1947)

55. Consider the following more realistic versions:

a. Use of more than one vehicle
b. Transportation of an additional cargo
c. Use of some fixed caches, established in advance
d. Delivery to more than one destination
e. Establishment of a rate of delivery
f. Minimization of total cost, including cost of gasoline, cost of purchasing vehicles, cost of establishing caches.
g. Arbitrary distance $x > 2d$. (Helmer)

56. Prove that, in general, the problem of determining

$$\operatorname*{Max}_{\{x_k\}} \operatorname*{Min}_{\{y_k\}} \left[\sum_{k=1}^{N} F_k(x_k, y_k) \right], \text{ where } \sum_{k=1}^{N} x_k \leq x, \ \sum_{k=1}^{N} y_k \leq y, \ x_k, y_k \geq 0$$

cannot be reduced to a recurrence relation of the form

$$f_N(x, y) = \operatorname*{Max}_{x_N,} \operatorname*{Min}_{y_N} [F_N(x_N, y_N) + f_{N-1}(x - x_N, y - y_N)].$$

57. Suppose that the requirements of a system at time n are r_n. Let x_n be the actual level, and let it be required to have $x_n \geq r_n$ for all n.

Furthermore, the restriction on the level at any time is

$$x_{n+1} - x_n \leq \lambda (x_n - x_{n-1}), \quad n \geq 1,$$

an "expansion-limitation".

We wish to chose the x_i so as to minimize

$$J(\{x\}) = \sum_{n=1}^{N} (x_n - r_n).$$

Show that the x_i are given by

$$\begin{aligned}
x_1 &= \varphi_1 \\
x_2 - x_1 &= \operatorname{Min}[\lambda \varphi_1, \lambda \varphi_2] = \operatorname{Min}[\lambda x_1, \varphi_2] \\
x_3 - x_2 &= \operatorname{Min}[\lambda^2 \varphi_1, \lambda \varphi_2, \varphi_3] = \operatorname{Min}[\lambda (x_2 - x_1), \varphi_3] \\
&\vdots \\
x_n - x_{n-1} &= \operatorname{Min}[\lambda^{n-1} \varphi_1, \lambda^{n-2} \varphi_2, \ldots, \varphi_n] = \\
&\quad \operatorname{Min}[\lambda (x_{n-1} - x_{n-2}), \varphi_n],
\end{aligned}$$

where

$$\varphi_k = \underset{j}{\text{Max}} \left[\frac{r_j - r_{k-1}}{\sum_{i=k}^{j} \lambda^{i-k}} \right] \qquad \text{(Shepherd)}.$$

58. Determine the maximum of $\sum_{i=1}^{N} a_i x_i$ subject to the constraints $\sum_{i=1}^{N} x_i^2 = 1$, $0 \leq x_1 \leq x_2 \leq \ldots \leq x_N$, where the a_i are non-negative.

59. Consider the problem of determining the maximum of $\prod_{i=1}^{N} (x_i - a)$ subject to the restrictions $0 \leq x_i \leq b$, where $b > a$, and $\sum_{i=1}^{N} x_i = c$. Show that to obtain a functional equation we must consider also the problem of determining the minimum of $\prod_{i=1}^{N} (x_i - a)$, and obtain the functional equations governing the problem.

Show that this problem does not arise if we consider

$$\prod_{i=1}^{N} | x_i - a |.$$

60. Assume that we are a contestant on a quiz program where we have an opportunity to win a substantial amount of money provided that we answer a series of questions correctly.

Let r_k be the amount of money obtained if the kth question is answered correctly, and let p_k be the a priori probability that we can answer the kth question where $k = 1, 2, \ldots, N$. Let $\varphi(x)$ be the utility function measuring the value to us of winning an amount x.

Assume that we have a choice at the end of each question of attempting to answer the next question, or of stopping with the amount already won. Determine the optimal policies to pursue under the following conditions:

a. Any wrong answer terminates the process with a total return of zero.

b. A total of two wrong answers is allowed.

c. Having answered k_o questions correctly, we must win at least $\sum_{k=1}^{k_o} r_k$, no matter what happens subsequently.

d. We are competing with other contestants. The contestant obtaining the largest total has an opportunity to answer a "jackpot question" worth much more than $\sum_{k=1}^{N} r_k$.

e. At each stage of the process, we have a choice of a hard question or an easy question with the proviso that a miss on an easy question terminates the process with a return of zero, and a miss on a hard question terminates the process with a total return of one-half the amount won to date.

61. Let the quantities b_i, a_{ij} be stochastic variables, subject to known distributions. Obtain a recurrence relation for the sequence

$$\{f_N(t, c_1, c_2, \ldots, c_m)\}$$

defined by the equation

$$f_N(t, c_1, c_2, \ldots, c_m) = \operatorname*{Min}_{x_i} \operatorname{Exp} \left[e^t \sum_{i=1}^{N} b_i x_i \right],$$

where the x_i satisfy constraints of the form

a. $x_i \geq 0$,

b. $\sum_{j=1}^{N} a_{ij} x_j \leq c_i$, $i = 1, 2, \ldots, m$,

and for the sequence

$$g_N(c_1, c_2, \ldots, c_m) = \operatorname*{Min}_{x_i} \operatorname{Exp} \left[\sum_{i=1}^{N} b_i x_i \right].$$

In both cases, Exp represents the expected value with respect to the random elements.

62. Consider the Selberg form

$$Q_N(x) = \sum_{n \leq N} \Big(\sum_{k \mid n} x_k \Big)^2,$$

where $x_1 = 1$ and the other x_k are as yet undetermined. The notation $\sum_{k \mid n} x_k$ means that the sum is to be taken over all integers k which divide n, e.g. $\sum_{k \mid 6} x_k = x_1 + x_2 + x_3 + x_6$. With the introduction of suitable state variables, determine recurrence relations for $\operatorname*{Min}_{x_i} Q_N(x)$.

63. The problem of determining the minimum and maximum characteristic roots of the Jacobi matrix

$$J = \begin{pmatrix} a_1 & b_1 & \dots & & & & \\ b_1 & a_2 & b_2 & & & & \cdot \\ & & & & & & \cdot \\ \cdot & b_2 & a_3 & b_3 & & & \cdot \\ \cdot & & & & & & \\ \cdot & & & b_{N-2} & a_{N-1} & b_{N-1} \\ & & & \dots & & b_{N-1} & a_N \end{pmatrix},$$

where the dots signify that all the other elements are zero, is equivalent to determining the minimum and maximum values of the quadratic form,

$$Q_N(x) = \sum_{i=1}^{N} a_i x_i^2 + 2 \sum_{i=1}^{N-1} b_i x_i x_{i+1},$$

on the sphere $\sum_{i=1}^{N} x_i^2 = 1$.

Consider the two sequences

$$f_N(c) = \operatorname*{Max}_S [Q_N(x) + 2cx_N],$$
$$g_N(c) = \operatorname*{Min}_S [Q_N(x) + 2cx_N],$$

where S represents the N-dimensional sphere. Show that recurrence relations may be obtained, connecting $f_N(c)$ with $f_{N-1}(c)$, and $g_N(c)$ with $g_{N-1}(c)$.

64. Obtain analogous results for the quadratic form

$$Q_N(x) = \sum_{i=1}^{N} a_i x_i^2 + 2 \sum_{i=1}^{N-1} b_i x_i x_{i+1} + 2 \sum_{i=1}^{N-2} c_i x_i x_{i+2}.$$

65. Let $A = (a_{ij})$ be a positive definite symmetric matrix. Show that the problem of solving the system of linear equations

$$\sum_{j=1}^{N} a_{ij} x_j = c_i, \; i = 1, 2, \dots, N,$$

is equivalent to determining the absolute minimum of the form

$$Q_N(x) = \sum_{i,j=1}^{N} a_{ij} x_i x_j - 2 \sum_{i=1}^{N} c_i x_i.$$

66. Define this minimum to be $f_N(c_1, c_2, \ldots, c_N)$, and obtain a recurrence relation connecting f_N and f_{N-1}.

Show that f_N is a quadratic form in the variables c_i,

$$f_N(c_1, c_2, \ldots, c_N) = \sum_{i,j=1}^{N} b^{(N)}{}_{ij}\, c_i\, c_j\,,$$

and show how the recurrence relation connecting f_N and f_{N-1} may be utilized to obtain recurrence relations for the sequences $\left\{b_{ij}^{(N)}\right\}$.

67. A television broadcasting company wishes to lease video links so that certain of its stations may be formed into a connected network. Video links exist between all pairs of stations, and the costs, in general different, of links between the various pairs of stations are known. Show that to construct a network at minimal cost, we choose among the links not yet included in the network the lowest price link which does not form any loop with the links already chosen. (Kalaba)

68. Consider the problem of minimizing $\sum_{j=1}^{n} \varphi_j(x_j)$ over all n—tuples of non-negative integers $x = (x_1, x_2, \ldots, x_n)$ which satisfy $\sum_{j=1}^{n} x_j = m$, where $\varphi_1, \varphi_2, \ldots, \varphi_n$ are convex functions for $x_i \geq 0$. Let $I = \{1, 2, \ldots, n\}$ and for any admissible set, $\{x_1, x_2, \ldots, x_n\}$, let $S^+(x)$ denote the set of indices $j \in I$ for which $x_j > 0$. Show that a necessary and sufficient condition that an admissible set of x_i provide the minimum is

$$\min_{j \in I} [\varphi_j(x_{j+1}) - \varphi_j(x_j)] \geq \max_{j \in S^+(x)} [\varphi_j(x_j) - \varphi_j(x_{j-1})]\,,$$

and obtain the corresponding condition when the x_i are restricted merely to be non-negative and satisfy $\sum_{j=1}^{n} x_j = m$. (Gross)

69. Consider a rectangular matrix $A = (a_{ij})$. It is desired to start at the (1, 1) position and proceed to the (m, n) position moving one step to the right or one step down each move, in such a way as to minimize the sum of the a_{ij} encountered. Show how to determine optimal paths. (Dreyfus)

70. Suppose that we have a toaster capable of toasting two slices of bread simultaneously, each on one side. What toasting procedure minimizes the time required to toast three slices of bread, each on two sides?

(J. E. Littlewood)

Solve the generalized problem requiring the processing of N k-sided items by means of M machines which can each process R items on s sides simultaneously.

71. Consider a 3-terminal communication system,

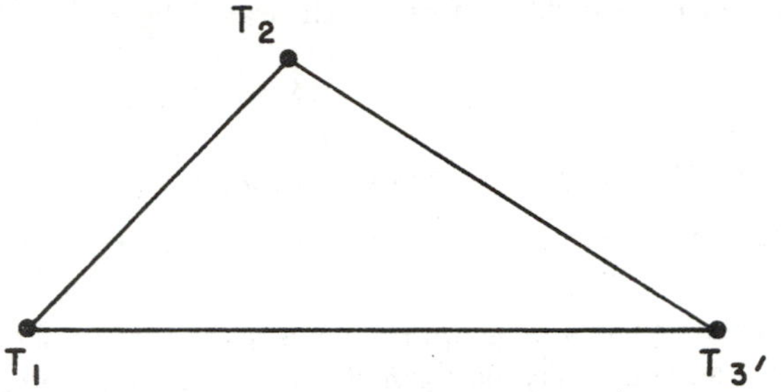

with message loads at each of the terminals for the other terminals. Let r_{ij} denote the maximum number of messages that can be sent from T_i to T_j in unit time, and consider the two cases, first, where there is no interference between signals going from T_i to T_j and those going in the reverse direction from T_j to T_i, and, second, where the total number messages in both directions cannot exceed r_{ij}.

Let x_{ij}, $i, j = 1, 2, 3, i \pm j$, denote the quantity of messages at T_i with ultimate destination T_j, and assume that a unit time is consumed transmitting a message from any T_i to any T_j. Denoting by $f_n(x_{ij})$ the maximum quantity of messages that can be delivered in n time units, derive a recurrence relation for the sequence $\{f_n(x_{ij})\}$.

(Juncosa-Kalaba)

72. A newspaper delivers papers to a number of newsstands. Assuming that the distribution of sales at each of these stands is known, and assuming that a certain quantity of unsold papers may be returned, suitably discounted, how many papers should be published, and how should they be distributed?

73. Consider the problem of minimizing a sum

$$F_N(x_1, x_2, \ldots, x_N) = g_1(x_1) + g_2(x_2) + \ldots + g_N(x_N),$$

where each g_i is a convex function, and the variables are subject to the constraints $a \leq x_1 \leq x_2 \leq \ldots \leq x_N \leq b$. Define

$$f_N(a, b) = \operatorname*{Min}_{\{x_i\}} F_N(x_1, x_2, \ldots, x_N), \text{ for } N = 1, 2, \ldots,$$

and $-\infty < a < b < \infty$. Show that

$$f_{j+1}(a, x) = \operatorname*{Min}_{a \leq y \leq x} [g_{j+1}(y) + f_j(a, x)].$$

74. Let $g(x)$ be continuous and convex for $x \geq d$. Define

$$g(r, s) = \min_{r \leq x \leq s} g(x), \quad d \leq r \leq s.$$

Show that for $d \leq a \leq b \leq c$, we may write

$$g(a, c) = g(a, b) + g(b, c) - g(b, b).$$

In addition, show that $g(a, x)$, as a function of x, is continuous and convex for $x \geq a$. (Karush)

75. Prove that under the above hypotheses that

$$f_N(a, c) = f_N(a, b) + f_N(b, c) - f_N(b, b), \quad -\infty < a \leq b \leq c < \infty.$$

(Karush)

76. Let the $g_i(y)$ be convex functions for $-\infty < y < \infty$ which are bounded from below. Then $f_N(a, b)$ may be written in the form

$$f_N(a, b) = u_N(a) + v_N(b), \quad a \leq b,$$

where $u_N(x)$ and $v_N(x)$ are, respectively, increasing and decreasing convex functions for $-\infty < x < \infty$. (Karush)

77. Let

$$f_N(a_0, a_1, a_2, \ldots, a_N) = \min_{c_i} \max_{0 \leq L \leq N} \left| a_L - \sum_{k=0}^{M} x_k c_{L-k} \right|.$$

Show that

$$f_N(a_0, a_1, a_2, \ldots, a_N) = \min_{c_0} \max \left[|a_0 - x_0 c_0|, f_{N-1}(a_1 - x_1 c_0, \right.$$
$$\left. a_2 - x_2 c_0, \ldots) \right].$$

78. Derive a similar expression for

$$f_N(a_0, a_1, a_2, \ldots, a_N) = \min_{c_i} \sum_{L=0} \left(a_L - \sum_{k=0}^{M} x_k c_{L-k}\right)^2,$$

and obtain thereby recurrence relations for the coefficients in

$$f_N(a_0, a_1, \ldots, a_N) = \sum_{s=0}^{N} q_{rs}^{(N)} a_r a_s.$$

79. A sleuth investigating a murder has N witnesses, one of whom is the murderer, of different degrees of reliability. Let p_i be the probability that the ith witness tells the truth at any particular time to any particular question. The detective interviews the witnesses in some order, asks the first witness a question, and then each succeeding witness a question,

which may be a direct question or a question concerning the truth of the testimony of preceding witnesses. Supposing that he is allowed one question at a time, and that the time required for the ith witness to answer a question is t_i, in what order should the witnesses be interrogated, and what questions should be asked to maximize the probability of determining the murderer in a fixed time T?

80. Consider the problem of minimizing the function

$$F_N(x_1, x_2, \ldots, x_N) = \varphi_1(x_1) + \varphi_2(x_2) + \ldots + \varphi_N(x_N)$$

over all values of the x_i subject to

$$\begin{aligned} &(a) \quad x_i \geq 0 \\ &(b) \quad x_1 \geq r_1, \\ &\qquad x_1 + x_2 \geq r_2, \\ &\qquad \vdots \\ &\qquad x_1 + x_2 + \ldots + x_N \geq r_N . \end{aligned}$$

Define the sequence

$$f_k(z) = \operatorname*{Min}_{x} \sum_{i=k}^{N} \varphi_i(x_i) ,$$

over the region determined by

$$\begin{aligned} &(a) \quad x_i \geq 0 \\ &(b) \quad x_k \geq r_k - z, \\ &\qquad x_k + x_{k+1} \geq r_{k+1} - z, \\ &\qquad \vdots \\ &\qquad x_k + x_{k+1} + \ldots + x_N \geq r_N - z, \end{aligned}$$

for $z \geq 0$, $k = 1, 2, \ldots, N$. Show that

$$f_k(z) = \operatorname*{Min}_{\substack{x_k \geq 0 \\ x_k \geq r_k - z}} [\varphi_k(x_k) + f_{k+1}(r_k)] ,$$

for $k = 1, 2, \ldots, N-1$, and hence that $\operatorname*{Min}_{x} F_N(x_1, x_2, \ldots, x_N) = f_1(0)$.

81. Show that the above problem with the additional restriction that $x_{i+1} - x_i \leq d_{i+1}$ may be reduced to the problem of determining the sequence $\{f_k(z, c)\}$ as defined by

$$f_k(z, c) = \operatorname*{Min}_{R} [\varphi_k(x_k) + f_{k+1}(z + x_k, x_k)] .$$

(*Management Science*, Vol. 3 (1956), p. 111–113).

82. Consider similarly the restriction $x_{i+1} \leq \lambda x_i$.

83. Determine the structure of the optimal policy in the case where the $\varphi_k(x)$ are linear functions of x, $\varphi_k(x) = r_k x$, and we assume

a. $r_{k+1} > r_k$
b. $r_{k+1} < r_k$
c. the r_i steadily increase, then decrease.
d. the r_i steadily decrease, then increase.

(Antosiewicz-Hoffman)

84. Given a continuous convex function, $f(x)$, and two values, one positive, $f(x_1) > 0$, and one negative, $f(x_2) < 0$, $x_1 < x_2$, we wish to determine the position of the zero of the function in $[x_1, x_2]$. The problem is to minimize the maximum length of interval in which we can guarantee that the zero lies after n evaluations of $f(x)$, where the evaluations are performed sequentially.

Define $R_n(s, y)$ to be the minimum length of interval on which we can guarantee locating the zero in $[0, 1]$ of any convex function f, given that $f(0) = 1$, $f(1) = -y$, that we know that the root is between S and 1, and that we have n evaluations to perform. Show that

$$R_0(x, y) = \frac{1}{1+y} - s,$$

$$R_n(s, y) = \operatorname*{Min}_{s \leq x \leq \frac{1}{1+y}} \operatorname{Max} \left[\begin{array}{l} \operatorname*{Max}_{0 \leq v' \leq \frac{y(x-s)}{1-s}} x R_{n-1}\left(\frac{x(y-v')}{xy-v'}, v'\right) \\ \operatorname*{Max}_{0 \leq v \leq 1-x(1+y)} (1-x) R_{n-1}\left(\frac{x}{1-x} \cdot \frac{v}{1-v}, \frac{y}{v}\right) \end{array} \right]$$

(Gross-Johnson)

85. A man is standing on a queue waiting for service, with N people ahead of him. He knows the utility of waiting out the queue, r, and the probability p that a person will be served in unit time. On the other hand, he incurs a cost of c for every unit of time spent waiting. The problem is to determine his waiting policy if he wishes to maximize his expected return.

Let f_N denote the expected return obtained employing an optimal waiting policy when there are N people ahead. Show that

$$f_N = \operatorname{Max}\left[-c + p f_{N-1} + (1-p) f_N, 0\right],$$

$N = 1, 2, \ldots,$ with $f_0 = r$. Hence show that

$$f_N = \text{Max}\left[f_{N-1} - \frac{c}{p}, 0\right]$$

and thus determine the optimal policy. (Haight)

86. Consider the same problem under the assumption that he can wait at most a time T. (Haight)

87. What policy does he pursue if he knows that a probability p exists, but does not know its precise value? (Haight)

88. Consider a forestry firm in which we start with a fixed capital and a certain presence of timber. We assume that

1. There is a fixed initial amount of cash available, and no revenue other than proceeds from selling timber, and from interest on cash on hand. No borrowing is allowed, and all current expenses must be covered by cash and sales.
2. Trees can be grown only from seed; it is impossible to buy young trees from outside the "economy."
3. The annual increment of "timber" depends on the age of the tree (growth rates need not be monotonic).
4. The cost of "carrying" a growing tree for one year depends on the age of the tree.
5. The selling value of a tree depends only on its timber content, i.e. its age.
6. The aim of the process is to maximize the money available after a fixed number of years.

Four activities may be engaged in, lending, planting, carrying, and felling.

1. Money can be lent for a year at interest rate r.
2. Money can be used to plant trees.
3. Money and trees can be used to provide older trees.
4. Trees of a given age can be cut down to provide money.

Over a given time period how does one proceed so as to maximize the total assets, capital plus timber?

(Morton, *Dynamic Programming*, Proceeding of an International Conference on Input-Output Analysis, J. Wiley and Sons, 1956).

89. Consider a multi-component electronic system whose reliability may be taken to be the product of the reliabilities of the individual components.

To improve the reliability of a particular stage, we can put a number of units in parallel. Let $p_k(x_k)$ be the reliability of the kth stage when x_k units are put in parallel at the kth stage, and let $g_k(x_k)$ be the cost of inserting x_k units in parallel.

The problem is to maximize the total reliability

$$P_N(x) = \prod_{k=1}^{N} p_k(x_k),$$

subject to the restrictions

a. $x_k = 1, 2, 3, \ldots,$

b. $\sum_{k=1}^{N} g_k(x_k) \leq c.$

If $f_N(c) = \text{Max}\, P_N(x)$, show that

$$f_N(c) = \text{Max}\,[P_N(x) f_{N-1}(c - g_N(x))],$$

where the maximum is over

a. $x = 1, 2, \ldots,$

b. $g_N(x) \leq c.$ (Nadel)

90. Assume that there are two "costs," one in terms of actual money, and the other in terms of weight.

91. Discuss the connections between the following problems:

a. Maximize $\prod_{k=1}^{N} p_k(x_k)$, subject to $\sum_{k=1}^{N} g_k(x_k) \leq c_1$,

$\sum_{k=1}^{N} h_k(x_k) \leq c_2$, and $x_k = 1, 2, \ldots,$.

b. Maximize $\prod_{k=1}^{N} p_k(x_k) - \lambda_1 \sum_{k=1}^{N} g_k(x_k) - \lambda_2 \sum_{k=1}^{N} h_k(x_k)$,

subject to $x_k = 1, 2, \ldots,$.

c. Maximize $\prod_{k=1}^{N} p_k(x_k) - \lambda_1 \sum_{k=1}^{N} g_k(x_k)$, subject to

$$\sum_{k=1}^{N} h_k(x_k) \leq c_2, \text{ and } x_k = 1, 2, \ldots, .$$

d. Minimize $\sum_{k=1}^{N} g_k(x_k) + \lambda_3 \sum_{k=1}^{N} h_k(x_k)$, subject to

$$\prod_{k=1}^{N} p_k(x_k) \geq r, x_k = 1, 2, \ldots, .$$

92. Obtain the corresponding functional equations, and discuss the question of most convenient computation.

93. The requirement for a machine of a certain type as a function of time is known. It is desired to institute a procurement policy to meet this demand at minimum cost, given the following information.

1. Procurement of new machines cost p dollars per machine.
2. Maintenance of a machine costs m dollars per time period.
3. Cost of upkeep and repair per period is a known function of the number of machines on hand and the number required.

Show that the corresponding functional equation is

$$f_N(x) = \min_{z_1 + x \geq r_1} [Pz_1 + M(z_1 + x) + L_1(z_1 + x) + f_{N-1}(x + z_1)],$$

where z_1 can assume only the values 0, 1, 2, ...

Obtain the solution under the assumption that each function $L_k(x)$ has the form

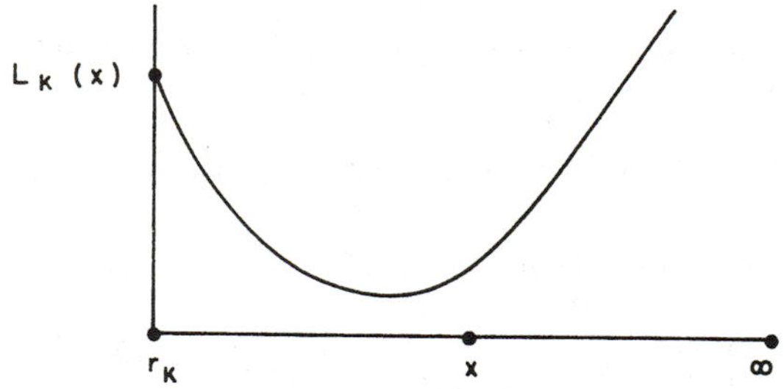

and, as a special case, is parabolic, *i.e.* a quadratic in x.

94. Consider the problem for the case where two distinct types of machines are being procured, with joint maintenance facilities, but independent demand.

Bibliography and Comments for Chapter III

§ 1. The basic ideas of this chapter, together with the "principle of optimality", were first stated in the monograph "An Introduction to the Theory of Dynamic Programming," RAND Corporation, 1953, an outgrowth of a shorter paper written in 1952, but not published then. This paper, in turn, was the result of research done in 1949, 1950 and 1951, and contained in a number of unpublished papers.

§ 3. As we have recently shown in connection with some joint work (R. Bellman and R. Kalaba, "On the Principle of Invariant Imbedding and Propagation Through Inhomogeneous Media," *Proc. Nat. Acad. Sci.*, (1956), the "principle of optimality" is actually a particular application of what we have called the "principle of invariant imbedding." A special form of the invariance principle was used by Ambarzumian "On the Scattering of Light by a Diffuse Medium," *C. R. Doklady, Sci. U.R.S.S.* 38 (1943), p. 257 and extensively developed by S. Chandrasekhar *Radiative Transfer*, Oxford, 1950. An early use of the method is due to G. Stokes (*Mathematical and Physical Papers*, Vol. IV, "On the intensity of the light reflected from or transmitted through a pile of plates," pp. 145–156).

The functional equation technique used throughout is intimately related to the "Point of Regeneration" method used in the study of branching processes, cf. R. Bellman and T. E. Harris, "On Age-Dependent Binary Branching Processes," *Ann. Math.*, Vol. 55 (1952), pp. 280–295.

Actually, we have made no systematic effort to trace the origin and use of invariance principles, and the above references represent only a few of the many that could be cited. One, however, which cannot be ignored is J. Hadamard, "Le principe de Huygens," *Bull. Soc. Math. France*, 52 (1924), pp. 610–640, where there is an interesting discussion of causality, functional equations and Huygens' principle.

The classic reference to semi-group theory is E. Hille, "Functional Analysis and Semi-groups," *Amer. Math. Soc.*, 1948.

§ 6. A detailed discussion of the formulation of variational problems as continuous decison processes will be found in Chapter 9.

§ 9. A discussion of causality and optimality, together with the interrelation with semi-groups may be found in R. Bellman, "Dynamic Programming and A New Formalism in the Theory of Integral Equations," *Proc. Nat. Acad. Sci.*, Vol. 41 (1955), pp. 31–34.

Problem 92. See R. Bellman, "Dynamic Programming and Lagrange Multipliers", *Proc. Nat. Acad. Sci.*, Vol. 42 (1956), pp. 767-769.

CHAPTER IV

Existence and Uniqueness Theorems

§ 1. Introduction

In the previous chapter we outlined the skeletal structure of dynamic programming processes and derived various general classes of functional equations. In this chapter, we shall abstract the particular methods utilized in Chapter I and II to treat the equations occurring therein and derive some existence and uniqueness theorems for the more general equations of Chapter III. Our principal tool will be the method of successive approximations due to Picard.

Although all the proofs follow essentially a common track, each requires its own detour at an appropriate point. Consequently, in place of attempting to frame the hypotheses in such general terms that we can state all our results in a single theorem, at the possible expense of clarity and loss of understanding of the simple mechanism involved, we have divided our results into a number of theorems referring to particular classes of equations. The basic method of proof is, however, the same throughout.

Our first step consists of formalizing the device we have used before to compare the solutions of two equations, cf. § 7 of Chapter I and § 6 of Chapter II. The resulting inequality is essential to our proofs in this chapter, and will be utilized again in our treatment of multi-stage games in a later chapter, and in comparison theorems in the calculus of variations in Chapter IX.

The first class of equations we treat are those where each operation results in a shrinking of resources, which is to say, the point transformations involved are shrinking transformations in the sense of Cacciopoli. Equations of this type we rather unimaginatively call equation of type one.

The next class of equations which we discuss are those where the probability of survival decreases uniformly with each operation. This is equivalent to the functional transformation being a shrinking transformation. These equations we name equations of type two.

Both types have, in particular cases, the form

$$f(p) = \operatorname*{Sup}_{q} \, [g(p, q) + h(p, q) f(T(p, q))] \tag{1}$$

where the quantities occurring are as defined in the previous chapter.

As we shall see these equations are rather readily treated by standard iterative techniques, with the aid of our basic inequality. Equations which do not belong to either of these classes usually require some fancier techniques, as we shall see in our treatment of a particular equation in § 8. All equations not of types one or two, we blithely lump together as those of type three.

Following these results on existence and uniqueness, we shall discuss monotone convergence in a general setting, and state some general stability theorems established in the same fashion as before.

After indicating some directions of generalization, which can be carried quite far, we shall consider a particular equation of type three, as mentioned above. Here we have a combination of two types of shrinking transformations, and the treatment is a bit more involved.

We shall close the chapter with a discussion of an interesting integral equation arising in the theory of "optimal inventory" or "stock control," a subject which we shall treat in greater detail in the following chapter, where particular solutions are obtained.

Apart from their interest in connection with multi-stage decision processes, the equations we consider possess the analytic merit of constituting in many ways a natural extension of linear equations. As such, their study is valuable since they serve as a bridge between the well-regulated preserve of linear equations and the as yet untamed jungle of nonlinear equations.

§ 2. A Fundamental inequality

Let us consider the two functional transformations

$$(1)\qquad S_1(f, p, q) = g(p, q) + \int_{r \varepsilon D} f(r)\, dG(p, q, r),$$

$$S_2(f, p, q) = h(p, q) + \int_{r \varepsilon D} f(r)\, dG(p, q, r),$$

where $dG(p, q, r) \geq 0$, and define two additional transformations as follows:

$$(2)\qquad \text{a. } f_2(p) = \operatorname*{Sup}_q S_1(f_1, p, q)$$

$$\text{b. } F_2(p) = \operatorname*{Sup}_q S_2(F_1, p, q).$$

There is no need to go into a discussion of what we mean by the Stieltjes integral here since we are using it in a purely formal manner. All our results will actually be utilized for the case where $\int_{r \varepsilon D} f(r)\, dG(p, q, r) \equiv$

$h(p, q) f(T(p, q))$, and the reader unfamiliar with the Stieltjes integral need merely make this transformation to reduce all the equations to familiar terms, or he may consider $dG(p, q, r)$ to have the form $H(p, q, r)\, dr$, with $H \geq 0$.

The inequality we wish to prove is

LEMMA 1.

$$(3) \qquad |f_2(p) - F_2(p)| \leq \operatorname*{Sup}_q \left[|g(p, q) - h(p, q)| + \int_{r \varepsilon D} |f_1(r) - F_1(r)|\, dG(p, q, r) \right].$$

PROOF. Let us simplify the notation initially by assuming that both transformations in (2) have the property that the supremum is actually a maximum. Let then $q = q(p)$ be a value of q for which the maximum is assumed in (2a), and $\bar{q} = \bar{q}(p)$ be a value of q for which the maximum is assumed in (2b). Then we have the following set of equalities and inequalities:

$$(4) \qquad \begin{aligned} &\text{a. } f_2(p) = S_1(f_1, p, q) \geq S_1(f_1, p, \bar{q}) \\ &\text{b. } F_2(p) = S_2(F_1, p, \bar{q}) \geq S_2(F_1, p, q), \end{aligned}$$

as in § 7 of Chapter I and § 6 of Chapter II.

From these follow immediately

$$(5) \qquad f_2(p) - F_2(p) \geq \{g(p, \bar{q}) - h(p, \bar{q})\} + \int_{r \varepsilon D} (f_1(r) - F_1(r))\, dG(p, \bar{q}, r),$$

and

$$f_2(p) - F_2(p) \leq \{g(p, q) - h(p, q)\} + \int_{r \varepsilon D} (f_1(r) - F_1(r))\, dG(p, q, r).$$

These, in turn, yield the single inequality

$$(6) \qquad |f_2(p) - F_2(p)| \leq \operatorname{Max} \begin{bmatrix} |g(p, \bar{q}) - h(p, \bar{q})| + \int_{r \varepsilon D} |f_1(r) - F_1(r)|\, dG(p, \bar{q}, r), \\ |g(p, q) - h(p, q)| + \int_{r \varepsilon D} |f_1(r) - F_1(r)|\, dG(p, q, r), \end{bmatrix}$$

from which the result in (3) is immediate.[1]

To obtain the result as stated in terms of the supremum it is only necessary to note that the supremum may be obtained arbitrarily closely by the value of the function for some $q = q(p)$. The argument then proceeds via a limiting procedure.

[1] We are using the simple result that $a \leq x \leq b$ implies $|x| \leq \operatorname{Max}(|a|, |b|)$.

§ 3. Equations of Type One

Let us now impose the following conditions upon the functions entering into the equation of (1.1):

(1) a. *$g(p, q)$ is uniformly bounded for all $q \varepsilon S$ and all $p \varepsilon D$ satisfying the restriction $\|p\| \leq c_1$, where $\|p\| = \left(\sum_{i=1}^{N} p_i^2\right)^{1/2}$. D is the domain of f, it contains the nullvector $p = \theta$, and $T(p, q) \varepsilon D$ for all $p \varepsilon D$.*

b. *$g(\theta, q) = 0$ for all $q \varepsilon S$.*

c. *$|h(p, q)| \leq 1$ for all $p \varepsilon D$ and $q \varepsilon S$.*

d. *$\|T(p, q)\| \leq a\|p\|$, for some $a < 1$, for all $q \varepsilon S$ and all $p \varepsilon D$.*

e. *If $v(c) = \operatorname*{Sup}_{\|p\| \leq c} \operatorname*{Sup}_{q} |g(p, q)|$, then $\sum_{n=0}^{\infty} v(a^n c) < \infty$.*

Equations which satisfy these assumptions are called *equations of Type One*. In many cases, it may be more convenient, and natural, to use the norm $\|p\| = \sum_{i=1}^{N} |p_i|$. It will be clear from the argumentation below that the precise form of the norm is of little importance.

Our principal result concerning these equations is the following:

THEOREM 1. *Consider the equation*

(2)
$$f(p) = \operatorname*{Sup}_{q} [g(p, q) + h(p, q) f(T(p, q))], \quad p \neq \theta$$
$$f(\theta) = 0,$$

assumed to be of Type One.

There is exactly one solution of (2) which is continuous at $p = \theta$ and equal to zero there, and defined over all of D.

This solution may be obtained as the limit of the sequence $\{f_n(p)\}$ defined as follows:

(3) a. $f_0(p) = \operatorname*{Sup}_{q} g(p, q)$

b. $f_{n+1}(p) = \operatorname*{Sup}_{q} [g(p, q) + h(p, q) f_n(T(p, q))], \; n = 0, 1, 2, \ldots$

Alternatively, any initial function $f_0(p)$ which is continuous at $p = \theta$ and equal to zero there, and bounded for $\|p\| \leq c_1$ for any $c_1 > 0$, $p \varepsilon D$, may be used in (3b) to yield a convergent sequence.

If $g(p, q)$, $h(p, q)$, and $T(p, q)$ are continuous in p in any bounded *portion of D, uniformly for all $q \in S$, then $f(p)$ is continuous in any bounded portion of D.*

PROOF. Let us consider the sequence defined by (3). Using Lemma 1, proved in § 2, we have for $n \geq 1$,

$$(4) \quad |f_{n+1}(p) - f_n(p)| \leq \operatorname*{Sup}_q |h(p, q)| \, |f_n(T(p, q)) - f_{n-1}(T(p, q))| \leq \operatorname*{Sup}_q |f_n(T(p, q)) - f_{n-1}(T(p, q))|,$$

and

$$(5) \qquad |f_1(p) - f_0(p)| \leq \operatorname*{Sup}_q |f_0(T(p, q))| = \operatorname*{Sup}_q |g(p, q)|.$$

Let us now define the new sequence

$$(6) \qquad v_n(c) = \operatorname*{Sup}_p |f_{n+1}(p) - f_n(p)|, \quad \|p\| \leq c, \; p \in D.$$

Using the function defined in (1 e), we see that $v_0(c) = v(c)$. Turning to (4) we have, for $p \in D$, $\|p\| \leq c$.

$$(7) \quad \operatorname*{Sup}_q |f_{n+1}(p) - f_n(p)| \leq \operatorname*{Sup}_p \operatorname*{Sup}_q |f_n(T(p, q)) - f_{n-1}(T(p, q))| \leq \operatorname*{Sup}_{\|p\| \leq ac} |f_n(p) - f_{n-1}(p)|,$$

by virtue of our assumption concerning $T(p, q)$. Hence $v_{n+1}(c) \leq v_n(ac)$, $n = 0, 1, 2, \ldots$, or $v_n(c) \leq v_0(a^n c)$. It follows that the series $\sum_{n=0}^{\infty} [f_{n+1}(p) - f_n(p)]$ converges uniformly for $\|p\| \leq c$, and hence that $\{f_n(p)\}$ converges uniformly to a function $f(p)$ for $\|p\| \leq c$.

This completes the proof of existence and the proof of the statements concerning convergence and continuity.

To establish uniqueness, let $f(p)$ and $F(p)$ be two solutions of (1) continuous at $p = \theta$, and hence defined for all $p \in D$. Let

$$(8) \qquad v(c) = \operatorname*{Sup}_p |f(p) - F(p)|, \quad \|p\| \leq c, \; p \in D.$$

Applying Lemma 1, we have

$$(9) \qquad |f(p) - F(p)| \leq \operatorname*{Sup}_q |f(T(p, q)) - F(T(p, q))|,$$

whence

$$(10) \qquad v(c) \leq v(ac) \leq \ldots \leq v(a^n c).$$

Since $f(p)$ and $F(p)$ are continuous for $p = 0$, $v(a^n c) \to 0$ as $n \to \infty$. Hence $v(c) \equiv 0$, and $f(p) \equiv F(p)$.

The utility of Lemma 1 lies in the fact that it enables us to bypass any discussion of the behavior of the maximizing q as a function of p, a subject of great difficulty about which little is known, in general.

§ 4. Equations of Type Two

Let us now consider the equation of (1.1) where we impose the conditions

(1) a. $|g(p, q)|$ *is uniformly bounded for all* $q \varepsilon S$, *and* $\|p\| \leq c_1, p \varepsilon D$.

b. $|h(p, q)| \leq a < 1$ *for all* $q \varepsilon S$ *and uniformly in any region* $\|p\| \leq c_1, p \varepsilon D$.

c. $\|T(p, q)\| \leq \|p\|$ *for all* p *or alternatively* D *is a bounded region, and no condition is imposed upon* T *apart from the condition that* $T(p, q) \varepsilon D$ *for all* $p \varepsilon D$.

Equations satisfying these conditions we shall call *equations of Type Two.* We shall demonstrate.

THEOREM 2. *If*

$$f(p) = \operatorname*{Sup}_q [g(p, q) + h(p, q) f(T(p, q))] \tag{2}$$

is an equation of Type Two, there is a unique solution which is bounded in any finite part of D.

The solution may be found by means of successive approximations as before, and the previous statements concerning continuity of the solution remain valid.

PROOF. Let

$$\begin{aligned} f_0(p) &= \operatorname*{Sup}_q g(p, q) \\ f_{n+1}(p) &= \operatorname*{Sup}_q [g(p, q) + h(p, q) f_n(T(p, q))], \; n = 0, 1, 2, \ldots \end{aligned} \tag{3}$$

Using Lemma 1, we have

$$\begin{aligned} |f_{n+1}(p) - f_n(p)| &\leq \operatorname*{Sup}_q |h(p, q)[f_n(T(p, q)) - f_{n-1}(T(p, q))]| \\ &\leq a \operatorname*{Sup}_q |f_n(T(p, q)) - f_{n-1}(T(p, q))|, \end{aligned} \tag{4}$$

where $a < 1$. From this point on the proof clearly parallels the proof of Theorem 1. The vanishing at $p = \theta$ is now a consequence of the equation itself.

§ 5. Monotone convergence

We have in the preceding sections demonstrated convergence of the successive approximation under assumptions which yielded essentially geometric convergence. Let us now show that, under the assumption that $h(p, q) \geq 0$, which is true in all the applications to date, we have at our disposal a method of choosing an initial approximation which will yield monotone convergence in addition.

In some equations of Type Three, where convergence of geometric type is either difficult to establish, or else non-existent, this is a valuable technique.

Let us consider our equation in the form

$$f(p) = \operatorname*{Max}_q [g(p, q) + h(p, q) f(T(p, q))]. \tag{1}$$

Let $q_0 = q_0(p)$ be an initial approximation to $q(p)$ and let $f_0(p)$ be determined by use of this policy, i.e.,

$$f_0(p) = g(p, q_0) + h(p, q_0) f_0(T(p, q_o)), \tag{2}$$

and the sequence $\{f_n(p)\}$, $n = 1, 2, \ldots$, then be determined recursively,

$$f_{n+1}(p) = \operatorname*{Sup}_q [g(p, q) + h(p, q) f_n(T(p, q))], \quad n = 0, 1, 2, \ldots \tag{3}$$

[Having introduced the concept of approximation in policy space, it is now convenient to use the supremum again to bypass questions of no little difficulty, concerning continuity over q.] Let us assume, as in the case of equations of Types One and Two, that sufficient conditions have been imposed to have the sequence $\{f_n(p)\}$ uniformly bounded in any finite portion of D.

It is immediately seen that $f_1(p) \geq f_0(p)$, and therefore, by virtue of the non-negativity of $h(p, q)$, that $f_{n+1}(p) \geq f_n(p)$ for all n. It follows that $f_n(p)$ converges to a function $f(p)$ as $n \to \infty$, in any finite part of D.

If q is a member of a finite set S, there is no question of the convergence of $\{f_n(p)\}$ to an actual solution of (3), where the supremum is now a maximum. If S contains a continuum, it is perhaps not immediate that $f(p)$ is the bounded solution of

$$f(p) = \operatorname*{Sup}_q [g(p, q) + h(p, q) f(T(p, q))]. \tag{4}$$

To establish this, we observe that by virtue of the monotone convergence, we have

$$f_{n+1}(p) \leq \operatorname*{Sup}_q [g(p, q) + h(p, q) f(T(p, q))], \tag{5}$$

whence

$$(6) \qquad f(p) \leq \operatorname*{Sup}_q [g(p, q) + h(p, q) f(T(p, q))].$$

On the other hand, we have

$$(7) \qquad \begin{aligned} f(p) &\geq \operatorname*{Sup}_q [g(p, q) + h(p, q) f_n(T(p, q))] \\ &\geq g(p, q) + h(p, q) f_n(T(p, q)) \end{aligned}$$

for all $q \varepsilon S$ and all n. Letting $n \to \infty$, we obtain the reverse inequality to (6), and hence equality.

This property of monotone convergence or, at worst, monotone approximation, is particularly useful in other parts of the theory of dynamic programming, and in particular, in applications to the calculus of variations, as we shall see in a later chapter.

§ 6. Stability theorems

In the theory of functional equations a problem of great theoretical interest, with important physical ramifications, is that of the dependence of the solution upon the form of the equation. In particular, a great deal of effort has been devoted to the determination of those equations which have the property that small changes in the form of the equation effect correspondingly small changes in the form of the solution. Equations which do not have this property are in the main of little physical interest.

Let us now consider the two equations,

$$(1) \qquad \begin{aligned} &\text{a. } f(p) = \operatorname*{Sup}_q [g(p, q) + h(p, q) f(T(p, q))], \\ &b.\ F(p) = \operatorname*{Sup}_q [G(p, q) + h(p, q) F(T(p, q))], \end{aligned}$$

and assume, to begin with, that they are both of Type One. We wish to obtain an inequality for $\operatorname*{Sup}_p |f(p) - F(p)|$, $p \varepsilon D$, $\|p\| \leq c$, where f and F are the unique solutions vanishing at $p = \theta$, and continuous there, of their respective equations.

To obtain this inequality, we employ the method of successive approximations in both equations, setting

$$(2) \qquad \begin{aligned} f_1(p) &= \operatorname*{Sup}_q g(p, q) \\ f_{n+1}(p) &= \operatorname*{Sup}_q [g(p, q) + h(p, q) f_n(T(p, q))] \\ F_1(p) &= \operatorname*{Sup}_q G(p, q) \\ F_{n+1}(p) &= \operatorname*{Sup}_q [G(p, q) + h(p, q) F_n(T(p, q))] \end{aligned}$$

We have

$$(3)\qquad | F_1(p) - f_1(p) | \leq \operatorname*{Sup}_q | G(p, q) - g(p, q) | ,$$

and

$$(4)\qquad | F_{n+1}(p) - f_{n+1}(p) | \leq \operatorname*{Sup}_q [\, | G(p, q) - g(p, q) + h(p, q) || F_n(T(p, q)) - f_n(T(p, q))| \,] .$$

Let us define

$$(5)\qquad u(c) = \operatorname*{Sup}_{\|p\| \leq c} \operatorname*{Sup}_q | G(p, q) - g(p, q) | .$$

Then we have

THEOREM 3. *With the above notation, for equations of Type One,*

$$(6)\qquad \operatorname*{Sup}_{\|p\| \leq c} | F(p) - f(p) | \leq \sum_{n=0}^{\infty} u(a^n c) .$$

PROOF. Set

$$(7)\qquad w_n(c) = \operatorname*{Sup}_{\|p\| \leq c} \operatorname*{Sup}_q | F_n(p) - f_n(p) | .$$

It can be shown inductively that we have $w_n(c) \leq \sum_{k=0}^{n-1} u(a^k c)$, $n \geq 1$, using (4), and the hypotheses governing an equation of Type One. Letting $n \to \infty$, we obtain (6), since $F_n(p) \to F(p)$, and $f_n(p) \to f(p)$.

Similarly,

THEOREM 4. *With the above notation, for equations of Type Two,*

$$(8)\qquad \operatorname*{Sup}_{\|p\| \leq c} | F(p) - (p) | \leq u(c)/(1 - a) .$$

The proof follows the same lines as above, and is therefore omitted.

Similar estimates can be obtained in the cases where $h(p, q)$ and $T(p, q)$ are perturbed.

§ 7. Some directions of generalization

A first generalization of (1.1) is the equation

$$(1)\qquad f(p) = \operatorname*{Sup}_q [g(p, q) + \sum_{i=1}^{N} h_i(p, q) f(T_i(p, q))] ,$$

which, in turn, is a particular case of

$$(2)\qquad f(p) = \operatorname*{Sup}_q [g(p, q) + \int_{r \varepsilon D} f(r)\, dG(p, q, r)] .$$

The methods utilized above yield analogues of the preceding theorems concerning existence, uniqueness and stability for the above equations, and systems of the form

$$(3)\quad f_i(p) = \operatorname*{Sup}_q \left[g_i(p, q) + \sum_{j=1}^{N} \int_{r \varepsilon D} f_j(r)\, dG_{ij}(p, q, r)\right], i = 1, 2, \ldots, N,$$

which is equivalent in form to (2) if we employ vector-matrix notation.

An example of (2) is the equation of "optimal inventory,"

$$(4)\quad f(x) = \operatorname*{Inf}_{y \geq x} \left[v(x, y) + a\left[(1 - G(y)) f(0) + \int_0^y f(y - s)\, dG(s)\right]\right],$$

which we shall treat in detail in the next chapter.

§ 8. An equation of the third type

The technique of approximation in policy space which yields monotone convergence, discussed above in § 5, is very useful in establishing the existence of solutions of equations of Type Three, a class, let us recall, defined quite simply as the complementary class of equations of Type One or Type Two.

Establishing the uniqueness of the solution of equations of Type Three is, in general, a problem of a greater level of difficulty, as we shall see below, and in a later chapter on multi-stage games where we discuss "games of survival."

Let us illustrate these remarks by considering the functional equation,

$$(1)\qquad f(p) = \operatorname{Min}\left[1 + \sum_{k=0}^{n} p_k f(x_k), \operatorname*{Min}_l \left[1 + f(T_l p)\right]\right], p \neq x_0,$$
$$f(x_0) = 0,$$

where l runs over the set of integers $1, 2, \ldots, M$. Here we set

$$(2)\qquad p = (p_0, p_1, \ldots, p_n), p_i \geq 0, \sum_{i=0}^{n} p_i = 1;$$
$$T_l p = (p_{0l}, p_{1l}, \ldots, p_{nl}), p_{il} \geq 0, p_{0l} \neq 1, \sum_{i=0}^{n} p_{il} = 1,$$
$$\text{where } p_{il} = p_{il}(p);\ l = 1, 2, \ldots, M;$$
$$x_k = (0, \ldots, 1, \ldots, 0), \text{ the 1 occurring in the } k^{\text{th}} \text{ place},$$
$$k = 0, 1, \ldots, n.$$

The function $f(p)$ is a scalar function of p.

This equation is a greatly extended version of the equation appearing in Exercise 39 of Chapter I.

This equation can be considered to arise in the following way. A system is known to be in one of $(N + 1)$ different states, which we denote by $0, 1, 2, \ldots, N$, with an initial probability $\{p_k\}$ that it is in the kth state. By means of a combination of the following operations, each of which consumes a unit time, we wish to transform the system into the 0-state, with certainty that it is in that state, in a minimum expected time:

L: We observe the actual state of the system and proceed with that knowledge;

A: We perform an operation A_l that converts the original probability distribution $\{p_k\}$ into a new distribution $\{p_{kl}\}$.

Let $p = (p_0, p_1, \ldots, p_N)$, and $f(p)$ denote the expected time required using an optimal policy, when the system is initially in state p. Then $f(p)$ satisfies (1) above.

We shall prove

THEOREM 5. *If for each transformation* T_l, *and for all* p, *it is true that*

$$\sum_{k=1}^{n} p_{kl} \leq c_1, \quad 0 < c_1 < 1, \tag{3}$$

then there exists a unique bounded solution to (1) *above. This function is positive for* $p \neq x_0$.

PROOF. We shall employ the method of successive approximations, using as our first approximation an approximation in policy space. Let us represent by L the choice of $1 + \sum_{k=0}^{n} p_k f(x_k)$, and by T_1 the choice of $l = 1$ in (1). We consider the function $F_1(p)$ determined by the policy symbolized by $LT_1 LT_1 \ldots$, and the function $F_2(p)$ determined by the policy $T_1 LT_1 L \ldots$ It is clear that

$$\begin{aligned} F_1(p) &= 1 + F_2(T_1 p), \quad p \neq x_0, \\ F_2(p) &= 1 + \sum_{k=0}^{n} p_{kl} F_1(x_k), \quad p \neq x_0, \\ F_1(x_0) &= F_2(x_0) = 0. \end{aligned} \tag{4}$$

Hence, for $l = 1, 2, \ldots, n$,

$$F_1(x_l) = 2 + \sum_{k=1}^{n} p_{kl} F_1(x_k), \quad l = 1, 2, \ldots, n. \tag{5}$$

Since, by assumptions $\sum_{k=1}^{n} p_{kl} \leq c_1 < 1$, the determinant of the system does not vanish and the system has a unique solution, necessarily positive, as we see by solving iteratively. Having determined $F_1(x_l)$, the determination of $F_2(p)$ and, hence, $F_1(p)$ for general p, is immediate.

To begin our successive approximations, define

$$(6)\quad f_0(p) = \text{Min}\,[F_1(p), F_2(p)],$$

$$f_{n+1}(p) = \text{Min}\,[[1 + \sum_{k=1}^{n} p_k f_n(x_k)], \underset{l}{\text{Min}}\,[1 + f_n(T_l p)]], p \neq x_0$$

$$f_{n+1}(x_0) = 0.$$

It is readily seen that $f_0(p) \geq f_1(p) \geq \dots f_n(p) \geq 1, p \neq x_0$.

Hence $f_n(p)$ converges monotonically to a function $f(p)$ which clearly satisfies the functional equation. This establishes the existence of a bounded solution.

The uniqueness proof is considerably more complicated and proceeds in a series of steps. Let $f(p)$ and $g(p)$ be two bounded solutions of (1). The first step is

LEMMA 2. $\underset{p}{\text{Sup}}\,|f(p) - g(p)| = \underset{k}{\text{Max}}\,|f(x_k) - g(x_k)|$.

PROOF. The inequality

$$(7)\qquad \underset{k}{\text{Max}}\,|(x_k) - g(x_k)| \leq \underset{p}{\text{Sup}}\,|f(p) - g(p)|$$

is clear. To demonstrate the reverse inequality, we consider four cases:

$$(8)\qquad \begin{aligned}
\text{a.}\quad & f(p) = 1 + \sum_{k=1}^{n} p_k f(x_k) \\
& g(p) = 1 + \sum_{k=1}^{n} p_k g(x_k) \\
\text{b.}\quad & f(p) = 1 + \sum_{k=1}^{n} p_k f(x_k) \\
& g(p) = 1 + g(T_l p) \\
\text{c.}\quad & f(p) = 1 + f(T_l p) \\
& g(p) = 1 + \sum_{k=1}^{n} p_k g(x_k) \\
d.\quad & f(p) = 1 + f(T_l p) \\
& g(p) = 1 + g(T_{l'} p)
\end{aligned}$$

Consider first the case corresponding to (a). We have

(9) $$f(p) - g(p) = \sum_{k=0}^{n} p_k [f(x_k) - g(x_k)],$$

whence

(10) $$|f(p) - g(p)| \leq \underset{k}{\text{Max}} |f(x_k) - g(x_k)|.$$

Therefore for all p for which (8a) holds, the lemma is correct. Equation (8a) will hold whenever p is close to x_0, since $1 + \sum_{k=1}^{n} p_k f(x_k)$ is less than 2 in this case, and $1 + f(T_l p) \geq 2$. Thus $1 + f(T_l p)$ and $1 + g(T_l p)$ will exceed the result of the L-move[2] for p close to x_0, for $l = 1, 2, \ldots, M$.

This is an important point since the crux of our proof is the fact that (8a) will always occur after a finite number of moves, by virtue of the condition in (3).

Now consider case (8b). We have

(11) $$f(p) = 1 + \sum_{k=0}^{n} p_k f(x_k) \leq 1 + f(T_l p)$$

$$g(p) = 1 + g(T_l p) \leq 1 + \sum_{k=0}^{n} p_k g(x_k).$$

Hence

(12) $$|f(p) - g(p)| \leq \text{Max}\,\{\underset{k}{\text{Max}} |f(x_k) - g(x_k)|, \underset{p}{\text{Sup}} |f(T_l p) - g(T_l p)|\},$$

and similarly for (8c).

From (8d) we derive

(13) $$|f(p) - g(p)| \leq \text{Max}\,\{|f(T_l p) - g(T_l p)|, |f(T_{l'} p) - g(T_{l'} p)|\}.$$

We now iterate these inequalities. For any fixed p, $T_{l_1} T_{l_2} \ldots T_{l_n} p$ will be in the region governed by (8a) for n large enough. Consequently, we obtain

(14) $$\underset{p}{\text{Sup}} |f(p) - g(p)| \leq \underset{k}{\text{Max}} |f(x_k) - g(x_k)|.$$

This completes the proof of the lemma.

It remains to show that $\underset{k}{\text{Max}} |f(x_k) - g(x_k)| = 0$. Let k be an index at which the maximum is assumed. It follows from the functional equation for f and g that

(15) $$f(x_k) = 1 + f(T_l x_k), \; l = l(k)$$

$$g(x_k) = 1 + g(T_{l'} x_k), \; l' = l'(k).$$

[2] i.e., L-choice.

As above we have

$$(16)\qquad \begin{aligned} f(x_k) &= 1 + f(T_l\, x_k) \leq 1 + f(T_{l'}\, x_k) \\ g(x_k) &= 1 + g(T_l'\, x_k) \leq 1 + g(T_l\, x_k). \end{aligned}$$

If both inequalities are proper, we have

$$(17)\quad |f(x_k) - g(x_k)| < \text{Max}\,|\,|f(T_l\, x_k) - g(T_l\, x_k)|,\ |f T_l'\, x_k) - - g(T_{l'}\, x_k)|\,| \leq \operatorname*{Sup}_p |f(p) - g(p)|,$$

a contradiction.

Thus, for either l or l', we have

$$(18)\qquad \begin{aligned} f(x_k) &= 1 + f(T_l\, x_k), \text{ or} \\ g(x_k) &= 1 + g(T_l\, x_k). \end{aligned}$$

This means that the first choices from the position x_k can be the same.

Consider now the situation for second moves. Using the same argument, we see that the second moves, i.e., the equations for $f(T_l x_k)$ and $g(T_l\, x_k)$ can also be the same, and so on, inductively.

Let $p_n = p_n(x_k)$ be the distribution achieved after n moves, where the $(n + 1)$st move puts x_k into the region governed by (8a), The argument above shows that f and g land in this region on the same move. Thus,

$$(19)\qquad \begin{aligned} f(x_k) &= (n + 1) + \sum_{k=0}^{n} p_{kn} f(x_k) \\ g(x_k) &= (n + 1) + \sum_{k=0}^{n} p_{kn}\, g(x_k), \end{aligned}$$

and consequently

$$(20)\qquad \begin{aligned} |f(x_k) - g(x_k)| &\leq \sum_{k=1}^{n} p_{kn}\, |f(x_k) - g(x_k)| \\ &\leq |1 - p_{on}|\, \operatorname*{Sup}_k |f(x_k) - g(x_k)|. \end{aligned}$$

Since $1 > p_{on} > 0$, this implies that $|f(x_k) - g(x_k)| = 0$. Hence $\operatorname*{Sup}_p |f(p) - g(p)| = 0$, which completes our uniqueness proof.

§ 9. An "optimal inventory" equation

In this section we shall discuss the equation

$$(1)\quad f(x) = \operatorname*{Inf}_{y \geq x} [k(y - x) + a\,[\int_y^\infty p(s - y)\,\varphi(s)\,ds + f(0)\int_y^\infty \varphi(s)\,ds + \int_0^y f(y - s)\,\varphi(s)\,ds]],$$

for $x \geq 0$, which we have already mentioned above in its more general form involving a Stieltjes integral. As we shall see in the following chapter this is an equation which occurs in the study of "optimal inventory" or "stock level" control. The proof of existence and uniqueness of solution logically appears here since we shall employ the same techniques as in the previous sections.

Consistent with the policy we have followed throughout, we shall not consider the general equation, involving Stieltjes integrals.

To simplify the subsequent notation, set

$$(2) \quad T(y, x, f) = k(y - x) + a\left[\int_y^\infty p(s - y)\varphi(s)\,ds + f(0)\int_y^\infty \varphi(s)\,ds + \int_0^y f(y - s)\varphi(s)\,ds\right].$$

The equation in (1) then has the form

$$(3) \qquad f(x) = \operatorname*{Inf}_{y \geq x} T(y, x, f).$$

Let us impose the following conditions:

(4) a. $\varphi(s) \geq 0, \quad \int_0^\infty \varphi(s)\,ds = 1$

b. $p(s)$ is monotone increasing, continuous, and $\int_0^\infty p(s)\varphi(s)\,ds < \infty$

c. $k(y)$ is continuous for $y \geq 0$, $k(\infty) = \infty$.

d. $0 < a < 1$.

Under these conditions, we have the result

THEOREM 6. *There is a unique solution to* (1) *which is bounded for x in any finite interval. This solution $f(x)$ is continuous.*

Let $f_0(x)$ be any non-negative continuous function defined over $0 \leq x$. Define the sequence $\{f_n(x)\}$ as follows,

$$(5) \qquad f_{n+1}(x) = \operatorname*{Min}_{y \geq x} T(y, x, f_n), \quad n = 0, 1, 2, \ldots.$$

Then $f(x) = \lim_{n \to \infty} f_n(x)$ *exist for $x \geq 0$ and is the solution of*

$$(6) \qquad f(x) = \operatorname*{Min}_{y \geq x} T(y, x, f).$$

PROOF. The proof follows very familiar lines. For each $n \geq 1$, let

$y_n = y_n(x)$ be a value of y for which $T(y, x, f_n)$ attains its minimum. Since $f_1(x)$ is continuous by assumption, we see inductively that each element of the sequence is continuous. Since $T(\infty, x, f_n) = \infty$, the minimum is attained.

We have then,

$$(7) \qquad \begin{aligned} f_{n+1} &= T(y_n, x, f_n) \leq T(y_{n-1}, x, f_n) \\ f_n &= T(y_{n-1}, x, f_{n-1}) \leq T(y_n, x, f_{n-1}) \end{aligned}$$

Combining these inequalities in the usual way, we obtain

$$(8) \qquad \begin{aligned} |f_{n+1} - f_n| \leq \text{Max}\,\{ &|T(y_n, x, f_n) - T(y_n, x, f_{n-1})|, \\ &|T(y_{n-1}, x, f_n) - T(y_{n-1}, x, f_{n-1})| \} \end{aligned}$$

or

$$(9) \qquad \begin{aligned} |f_{n+1} - f_n| \leq \text{Max}\,\{ &a \int_0^{y_n} |f_n(y_n - s) - f_{n-1}(y_n - s)|\, \varphi(s)\, ds \\ &+ a|f_n(0) - f_{n-1}(0)| - \varphi(s)\, ds, \\ &a \int_0^{y_{n-1}} |f_n(y_{n-1} - s) - f_{n-1}(y_n - s)|\, \varphi(s)\, ds + \\ &a\, |f_n(0) - f_{n-1}(0)| \int_{y_{n-1}}^{\infty} \varphi(s)\, ds \} \end{aligned}$$

Hence

$$(10) \qquad \begin{aligned} \underset{0 \leq x \leq \infty}{\text{Max}} |f_{n+1}(x) - f_n(x)| &\leq a \underset{0 \leq x \leq \infty}{\text{Max}} |f_n(x) - f_{n-1}(x)| \int_0^{\infty} \varphi(s)\, ds \\ &\leq a \underset{0 \leq x \leq \infty}{\text{Max}} |f_n(x) - f_{n-1}(x)|. \end{aligned}$$

Thus the series $\sum_{n=0}^{\infty} (f_{n+1}(x) - f_n(x))$ converges uniformly in a finite interval for all $x \geq 0$, and $f_n(x)$ converges to $f(x)$ for all $x \geq 0$. Since each $f_n(x)$ is continuous, $f(x)$ is also continuous.

To prove uniqueness, let $F(x)$ be another solution which is uniformly bounded for $x \geq 0$. Using the same technique as above for the two equations

$$(11) \qquad \begin{aligned} F(x) &= \underset{y \geq x}{\text{Min}}\, T(y, x, F) \\ f(x) &= \underset{y \geq x}{\text{Min}}\, T(y, x, f), \end{aligned}$$

we readily show that $F(x) - f(x)$ is identically zero. The case where Min is replaced by Inf in (1) is again handled by an approximation process.

Finally, let us note that if we take

$$(12)\qquad f_1(x) = \operatorname*{Min}_{y \geq x} [k(y-x) + a \int_y^\infty p(s-y)\,\varphi(s)\,ds]$$

$$f_2(x) = \operatorname*{Min}_{y \geq x} [T(y, x, f_1)],$$

and so on, we obtain monotone increasing convergence, since $f_2(x) \geq f_1(x)$, and hence, inductively, $f_{n+1}(x) \geq f_n(x)$ for all n.

On the other hand, we may also obtain monotone convergence by approximating in policy space. We may set $y = x$ for all $x \geq 0$ and obtain as our first approximation

$$(13)\qquad f_1(x) = a \int_0^\infty p(s-x)\,\varphi(s)\,ds + a \int_0^x f_1(x-s)\,\varphi\,ds + a f_1(0) \int_x^\infty \varphi(s)\,ds$$

for $x \geq 0$.

This equation is a "renewal equation" whose solution we shall discuss in an appendix to the following chapter.

Determining $f_2(x)$ by means of the equation

$$(14)\ f_2(x) = \operatorname*{Min}_{y \geq x} [k(y-x) + a \int_y^\infty p(s-y)\varphi(s)\,ds + a f_1(0) \int_y^\infty \varphi(s)\,ds + a \int_0^y f_1(y-s)\,\varphi(s)\,ds],$$

it follows that $f_2(x) \leq f_1(x)$. We thus obtain monotone decreasing convergence if we set

$$(15)\qquad f_{n+1}(x) = \operatorname*{Min}_{y \geq x} T(y, x, f_n).$$

Exercises and Research Problems for Chapter IV

1. Determine the structure of the optimal policies associated with the functional equation

$$f(p) = \operatorname*{Max}_{q} [R(p, q) + f(T(p, q))]$$

under the assumption that $R(p, q)$ and $T(p, q)$ are convex functions of p and q, and that $R(p, q)$ and $T(p, q)$ are monotone increasing functions of p for each q.

2. Carry through the details of an existence and uniqueness theorem for the system of equations

$$f_i(p) = \operatorname*{Max}_q \left[g_i(p, q) + \sum_{j=1}^{N} \int_{r \varepsilon D} f_j(r)\, dG_{ij}(p, q, r)\right], \; i = 1, 2, \ldots, N.$$

3. Show that we obtain an equation belonging to this class if we add to Problem 45 of chapter I the further condition that at any stage there is a probability p_1 that the tradein value will be ruled by the function $t_1(x)$ and a probability $p_2 = 1 - p_1$ that it will be ruled by the function $t_2(x)$.

4. Consider the multi-dimensional process where the resources at any stage are measured by the non-negative vector p. At each stage p is divided into r non-negative vectors q_j, $p = \sum_{j=1}^{r} q_j$. As a result of this allocation, we obtain a return $R(q) = R(q_j)$ and assume a cost of $\sum_{j=1}^{r} (c_j, q_j)$. Here (c, q) denotes the inner product of the two vectors.

Let $F_N(z)$ denote the cost incurred obtaining a total return of z in N stages, employing an optimal policy. Show that

$$F_1(z) = \operatorname*{Min}_{\substack{R(q) = z \\ q \geq 0}} \sum_{j=1}^{N} (c_j, q_j),$$

$$F_{n+1}(z) = \operatorname*{Min}_{q \geq 0} \left[\sum_{j=1}^{N} (c_j, q_j) + F_N(z - R(q))\right].$$

5. Under what conditions does the limiting equation

$$F(z) = \operatorname*{Min}_{q \geq 0} \left[\sum_{j=1}^{N} (c_j, q_j) + F(z - R(p))\right], \; F(0) = 0,$$

have a solution?

6. How can the following problem be formulated mathematically? We are lost in a forest whose shape and dimensions are precisely known to us. How do we get out in the shortest time?

7. Consider the case where the "forest" is the region between two parallel lines. (Gross).

8. Generalize the result of Theorem 5 by considering processes in which we have either a denumerable number of different transformations at each stage, or a continuum of transformations.

9. Consider the still more general process where there are a denumerable number, or continuum, of states.

10. Derive the functional equations corresponding to non-linear criteria and establish the corresponding existence and uniqueness theorems.

11. Consider, in particular, the criterion, for stochastic processes, of maximizing the probability of obtaining at least a return R_o.

12. Consider the equation

$$x^2 + ax = b, \quad a, b > 0.$$

Since for $x \geq o$,

$$x^2 = \operatorname*{Max}_{u \geq 0} [2xu - u^2],$$

the equation may be written

$$\operatorname*{Max}_{u \geq 0} [2xu + ax - u^2] = b,$$

yielding, for the positive root

$$x = \operatorname*{Min}_{u \geq 0} \left(\frac{u^2 + b}{2u + a} \right)$$

On the other hand, setting $x^2 = y$, we may write

$$y + ay^{\frac{1}{2}} = b,$$

$$2y^{\frac{1}{2}} = \operatorname*{Min}_{u \geq 0} \left[\frac{y}{u} + u \right],$$

obtaining

$$y = \operatorname*{Max}_{u \geq 0} \left[\frac{b - au/2}{1 + a/2u} \right].$$

Thus

$$\sqrt{\frac{b - au/2}{1 + a/2u}} \leq x \leq \frac{b + u^2}{a + 2u},$$

for all $0 \leq u \leq 2\, b/a$.

13. Generalize these results, considering the equation $x^n + ax = b$. Show that

$$x^n = \operatorname*{Max}_{u \geq 0} (xu - g(u)), \quad n > 1,$$
$$= \operatorname*{Min}_{u \geq 0} (xu + h(u)), \quad 0 < n < 1,$$

for suitable $g(u)$ and $h(u)$, obtaining, $n > 1$,

$$x = \operatorname*{Min}_{u \geq 0} \left[\frac{b + (n - 1)\,(u/n)^{n/(n-1)}}{a + u} \right],$$

and, for $0 < n < 1$

$$x = \operatorname*{Max}_{u \geq 0} \left[\frac{b - (1 - n)\,(u/n)^{n/(n-1)}}{1 + au} \right].$$

14. Show that if $\varphi(x)$ is strictly convex and differentiable, we have

$$\varphi(x) = \operatorname*{Max}_{u} [\varphi(u) - (u - x)\,\varphi'(u)].$$

and if concave,

$$\varphi(x) = \operatorname*{Min}_{u} [\varphi(u) - (u - x)\,\varphi'(u)].$$

Give both analytic and geometric proofs.

15. Consider the multi-dimensional analogue,

$$\varphi(x_1, x_2) = \operatorname*{Max}_{u_1, u_2} \left(\varphi(u_1, u_2) - (u_1 - x_1)\frac{\partial \varphi}{\partial u_1} - (u_2 - x_2)\frac{\partial \varphi}{\partial u_2} \right),$$

for convex functions, and the corresponding result for concave functions.

16. Discuss the possibility of using these results to obtain explicit solutions for non-linear systems of the form

$$\varphi_1(x_1, x_2) = x_1, \quad \varphi_2(x_1, x_2) = x_2,$$

where φ_1 and φ_2 are both concave or both convex.

17. Newton's method furnishes a sequence of successive approximations

$$x_{n+1} = x_n - f(x_n)/f'(x_n)$$

to the solution of $f(x) = 0$. Show that if $f'(x) > 0$ in $[a, b]$ and also $f''(x) > 0$ in this interval, we have

$$x = \operatorname*{Min}_{a \leq y \leq b} [y - f(y)/f'(y)],$$

for a root in this interval.

Obtain corresponding expressions for the multi-dimensional case.

18. Consider the two equations

(a) $v(p) = L(v, p, q) + a(p, q)$,

(b) $u(p) = \operatorname*{Max}_{q} [L(u, p, q) + a(p, q)]$.

where $u(p)$ is a scalar function of a vector p, belonging to a region R, and q a vector variable, belonging to a set S which may or may not depend upon p.

Assume that

(1) There is a unique solution of (a) for any fixed $q = q(p)$, denoted by $v(p, q)$, for p in R.

(2) There is a unique solution of (b) for p in R.

(3) If $w(p) \geq L(w, p, q) + a(p, q)$ for a fixed $q = q(p)$, then $w(p) \geq v(p, q)$.

Prove that under these assumptions we have

$$u(p) = \operatorname*{Max}_q v(p, q)$$

19. Under what assumptions concerning the matrix $A(p, q) = (a_{ij}(p, q))$, can we determine the solutions of the systems

$$u_i(p) = \operatorname*{Max}_q \left[a_i(p, q) + \sum_{j=1}^{N} a_{ij}(p, q)\, u_j(p)\right], \quad i = 1, 2, \ldots, N,$$

or

$$\operatorname*{Max}_q \left[a_i(p, q)\, u_i(p) + \sum_{j=1}^{N} a_{ij}(p, q)\, u_j(p)\right] = c_i, \quad i = 1, 2, \ldots, N,$$

in the above fashion?

20. Let $F_1(x) = G_1(x) = x$, and

$$F_n(x_1, x_2, \ldots, x_n) = \operatorname{Max}(x_1, G_{n-1}(x_2, \ldots, x_n)),$$

$$G_n(x_1, x_2, \ldots, x_n) = \operatorname{Min}(x_1, F_{n-1}(x_2, \ldots, x_n)).$$

Prove that

$$\lim_{n \to \infty} \int_0^1 \cdots \int_0^1 F_n(x_1, x_2, \ldots, x_n)\, dx_1\, dx_2 \ldots dx_n = \pi\sqrt{3}/q,$$

$$\lim_{n \to \infty} \int_0^1 \cdots \int_0^1 G_n(x_1, x_2, \ldots, x_n)\, dx_1\, dx_2 \ldots dx_n = 1 - \pi\sqrt{3}/q.$$

(Gross–Wang, *Amer. Math. Monthly*, Vol 63 (1956), p. 589).

21. Let the y_i be independent random variables assuming the values 1 with probability p and the value 0 with probability $1 - p$. Let the x_i be a set of positive quantities. Set

$$g_N(x; x_i) = \operatorname{Prob}\left\{ \sum_{i=1}^{N} x_i y_i \Big/ \sum_{i=1}^{N} x_i \geq x \right\},$$

and $f_N(x) = \operatorname*{Inf}_{x_i} g_N(x; x_i)$

Show that

$$f_N(x) = \operatorname*{Inf}_{0 \leq x_N \leq 1} \left[p f_{N-1}\left(\frac{x - x_N}{1 - x_N}\right) + (1 - p) f_{N-1}\left(\frac{x}{1 - x_N}\right)\right],$$

and thus obtain a non-trivial uniform lower bound for $g_N(x; x_i)$. (Harris)

22. Under what conditions does there exist a unique solution of the equation

$$u(x) = \operatorname*{Min}_{i} \sum_{j=1}^{N} p_j(x)\, u(x + a_{ij}),\ 0 < x < C,$$

$$u(x) = 0,\ x \leq 0,$$

$$u(x) = 1,\ x \geq C,$$

where, for $0 < x < C$,

(a) $p_j(x) \geq 0$

(b) $\sum_{j=1}^{N} p_j(x) = 1.$

Consider the case where x assumes only a discrete set of values, $\{k\Delta\}$, and $a_{ij} = m_{ij}\Delta$, where m_{ij} is a positive or negative integer.

23. Consider problem 15 in the exercises at the end of Chapter 3. Show that the problem of determining minimum cost is equivalent to the problem of determining the minimum of $L_N(x) = \sum_{i=1}^{N} x_i$ subject to the constraints

a. $x_i \geq 0$,

b. $x_k + x_{k+1} + \dots + x_{k+R} \geq a_k,\ k = 1, 2, \dots, N,$

where $x_{N+k} = x_k$.

(*Management Science*, 1957).

24. Consider the more general problem of determining the minimum of $L_N(y) = y_1 + y_2 + \dots + y_N$ subject to the constraints

(a) $y_i \geq 0$

(b) $y_1 \geq r,\ y_N \geq s,$

(c) $y_1 + y_2 \geq b_1,$

$y_2 + y_3 \geq b_2,$

$\vdots$

$y_{N-1} + y_N \geq b_{N-1}.$

Write, for fixed r, $f_N(s) = \text{Min } L(y)$, $N \geq 2$. Show that

$$f_2(s) = \text{Max}(s + r, b_1),$$
$$f_N(s) = \underset{z \geq s^*}{\text{Min}} [z + f_{N-1}(b_{N-1} - z)],$$

where $s^* = \text{Max}(s, 0)$.

Show that

$$f_k(s) = \text{Max}(s + u_k, v_k), \ k = 1, 2, \ldots,$$

where u_k and v_k are functions of r.

25. Show that

$$u_k = \text{Max}(r + \alpha_k, \beta_k),$$
$$v_k = \text{Max}(r + \gamma_k, \delta_k),$$

for $k \geq 3$, where

$$\alpha_{k+1} = \gamma_k,$$
$$\beta_{k+1} = \delta_k,$$
$$\gamma_{k+1} = \text{Max}(\alpha_k + b_k, \gamma_k),$$
$$\delta_{k+1} = \text{Max}(\beta_k + b_k, \delta_k).$$

26. Consider, in like fashion, the problem of minimizing $L_N(x) = \sum_{i=1}^{N} x_i$ subject to the constraints

a. $x_i \geq 0$,

b. $x_1 \geq x$,

c. $x_1 + x_2 \geq y$,

d. $x_1 + x_2 + x_3 \geq b_1$,

$x_2 + x_3 + x_4/ \geq b_2$,

$\vdots$

$x_{N-2} + x_{N-1} + x_N \geq b_{N-2}$,

$x_{N-1} + x_N \geq s$,

$x_N \geq r$.

27. Consider the problem of minimizing $L_N(x) = \sum_{i=1}^{N} c_i x_i$ subject to the constraints

a. $x_i \geq 0$,

b. $b_{11} x_1 + b_{12} x_2 \geq b_1$,

$b_{22} x_2 + b_{23} x_3 \geq b_2$,

$\vdots$

$b_{N-1, N-1} x_{N-1} + b_{N-1, N} x_N \geq b_{N-1}$,

c. $x_1 \geq x$, $x_N \geq r$.

Obtain the corresponding functional equation and the analogues of the above results under suitable assumptions concerning the coefficients b_{ij}.

28. Let us suppose that we are given a map containing N distinct locations numbered in some fashion $i = 1, 2, \ldots, N$, and a matrix $T = (t_{ij})$ telling us the time required to travel from i to j, with $t_{ii} = 0$. Starting at the first location, we wish to pursue a route which minimizes the total time required to travel to the Nth point, using any of the other locations, and only these, as intermediary stops.

Let f_i denote the time required to go from i to N, $i = 1, 2, \ldots, N - 1$, $f_N = 0$, using an optimal policy. Show that

$$f_i = \operatorname*{Min}_j \, [t_{ij} + f_j], \; i = 1, 2, \ldots, N - 1.$$

29. Show this equation has a solution $\{f_i\}$ unique up to an additive constant.

30. Show that any one of these solutions suffices to determine the optimal policy.

31. Consider the following approximation in policy space,

$$f_i^{(1)} = t_{i, i+1} + t_{i+1, i+2} + \ldots + t_{N-1, N},$$

for $i = 1, 2, \ldots, N - 1$, and let the sequence $\{f_j^{(k)}\}$ be defined by

$$f_i^{(k+1)} = \operatorname*{Min}_j \, [t_{ij} + f_j^{(k)}], \; i = 1, 2, \ldots, N - 1,$$

$k = 1, 2, \ldots$.

Show that the vectors $\{f_{i.}^{(k)}\}$ converge to a solution of the above functional equation, and thus may be used to determine optimal policies.*

32. Consider the problem of maximizing $\sum_{i=1}^{2N} g_i(x_i)$ subject to $\sum_{i=1}^{2N} x_i \leq c$, $x_i \geq 0$. Show that this is equivalent to maximizing $f_N(y_1) + h_N(y_2)$ subject to $y_1, y_2 \geq 0$, $y_1 + y_2 = c$, where

* R. Bellman, A Routing Problem, *Quarterly of Applied Mathematics*, 1957.

$$f_N(y_1) = \underset{R_1}{\text{Max}} \sum_{i=1}^{N} g_i(x_i),$$

$$h_N(y_2) = \underset{R_2}{\text{Max}} \sum_{i=N+1}^{2N} g_i(x_i),$$

and R_1, R_2 are defined by

$$R_1: \; x_i \geq 0, \; \sum_{i=1}^{N} x_i \leq y_1,$$

$$R_2: \; x_i \geq 0, \; \sum_{i=N+1}^{2N} x_i \leq y_2.$$

What computational advantages are there in employing this technique and its natural extension? Discuss the multi-dimensional case.

33. A gambler receives advance information concerning the outcomes of a sequence of independent sporting events over a noisy communication channel. We assume that the outcome of each event is the result of play between two evenly matched teams, and that p is the probability of a correct transmission, and $q = 1 - p$, the probability of incorrect transmission.

Assuming that the gambler starts with an initial amount x, and bets on the outcome of each event so as to maximize his expected capital at the end of N stages of play, show that he wagers his entire capital at each stage, provided that $p > 1/2$, and nothing if $p < 1/2$.

34. Let us assume that the gambler plays so as to maximize the expected value of the logarithm of his capital after N stages. Assuming that he uses the same betting policy at each stage, determine this ratio of the amount bet to the total capital.

(J. Kelly, "A New Interpretation of Information Rate," 1956, Symposium on Information Theory, *Transactions I. R. E.* 1956, pp. 185–189).

35. Let us assume that the gambler plays so as to maximize the expected value of the logarithm of his capital after N stages. Let $f_N(x)$ denote the expected value obtained using an optimal policy. Show that

$$f_{N+1}(x) = \underset{0 \leq y \leq x}{\text{Max}} \left[p f_N(x + y) + q f_N(x - y) \right], \; N = 1, 2, \ldots,$$

assuming that there are equal odds, with

$$f_1(x) = \underset{0 \leq y \leq x}{\text{Max}} \left[p \log(x + y) + q \log(x - y) \right].$$

(For this and the following results, see R. Bellman and R. Kalaba, "On the Role of Dynamic Programming in Statistical Communication Theory", *Transactions I. R. E.*, 1957.

36. Show inductively that

$$f_N(x) = \log x + N k,$$

where

$$k = \underset{0 \leq r \leq 1}{\text{Max}} \; [p \log (1 + r) + q \log (1 - r)],$$

and hence that there is a number r_0 such that the optimal policy at each stage is determined by the relation $y = r_0 x$.

37. Consider the time-dependent case where the probability of correct transmission depends on the stage. Establish the corresponding functional equation and deduce the structure of the optimal policy.

38. For the case where the purpose of the process is to maximize the expected value of the return, or the logarithm of the return after N stages, the above analysis shows that the optimal policy is independent of the quantity of resources available at each stage.

Consider the problem of determining the class of criterion functions possessing this property. Let $\varphi(x)$ be a monotone increasing concave function defined over $0 \leq x < \infty$, normalized by the condition $\varphi'(1) = 1$ and consider the one-stage process where we wish to maximize

$$E(y) = p \, \varphi(x + y) + (1 - p) \, \varphi(x - y)$$

for $0 \leq y \leq x$, where $1 \geq p > 1/2$. Show that if for all $x > 0$, there is a maximum of the form $y = r(p) \, x$, then we must have

$$\varphi(y) = \frac{y^{k+1}}{k+1} + c_1, \; k > -1,$$

or, as an extreme case,

$$\varphi(y) = \log y + c_1.$$

39. Consider the case where successive signals are not independent. Let the probability of a correct transmission at the kth stage depend upon the transmission of the signal at the $(k - 1)^{st}$ stage. Define, for $x > 0$, $k = 1, 2, \ldots, N$,

$f_k(x)$ = expected value of the logarithm of the final capital obtained from the remaining k stages of the original N-stage process, starting with an initial capital x, the information that the $(k - 1)^{st}$ signal was transmitted correctly, and using an optimal policy.

$g_k(x)$ = the corresponding function in the case where the $(k - 1)^{st}$ signal was transmitted incorrectly.

Then

$$f_k(x) = \underset{0 \leq r \leq 1}{\text{Max}} \; [p_{N-k+1} f_{k-1}(x + y) + (1 - p_{N-k+1}) \, g_{k-1}(x - y)],$$

$$g_k(x) = \underset{0 \leq y \leq x}{\text{Max}} \left[r_{N-k+1} f_{k-1}(x+y) + (1 - r_{N-k+1}) g_{k-1}(x-y) \right],$$

where

p_k = probability of correct transmission of the kth signal if the $(k-1)^{st}$ signal was transmitted correctly.

q_k = probability of correct transmission of the kth signal if the $(k-1)^{st}$ signal was transmitted incorrectly.

Show that $f_k(x) = \log x + a_k$, $g_k(x) = \log x + b_k$. Determine a_k and b_k and the structure of the optimal policy.

40. Consider the situation in which the channel transmits any of M different symbols. Upon receiving a symbol the gambler must make bets on what he believes the transmitted signal actually was. Assume that the gambler possesses the following information:

p_{ij} = the conditional probability that the j-signal was sent if the i-signal is received.

q_i = the probability of receiving the i-signal.

r_j = the return from a unit winning bet on signal j.

Assume that the gambler is free to bet an amount z_i on the ith signal, subject to the restriction that $\sum_i z_i \leq x$. Defining the sequence $\{f_N(x)\}$ as above, show that

$$f_N(x) = \sum_{i=1}^{M} q_i \underset{\substack{\Sigma z_i \leq x \\ z_i \geq 0}}{\text{Max}} \left[\sum_{j=1}^{M} p_{ij} f_{N-1}\left(r_j z_j + x - \sum_{s=1}^{M} z_s\right) \right], \quad N \geq 2,$$

$$f_1(x) = \sum_{i=1}^{M} q_i \underset{\substack{\Sigma z_i \leq x \\ z_i \geq 0}}{\text{Max}} \left[\sum_{j=1}^{M} p_{ij} \log\left(r_j z_j + x - \sum_{s=1}^{M} z_s\right) \right].$$

Prove, as before, that $f_N(x) = \log x + Na_k$, determine a_k and the structure of the optimal policy. Show that the optimal policy is independent of the q_i.

41. Consider the case in which there are a continuum of different signals. Let $dG(u, v)$ = conditional probability that a signal with label between v and $v + dv$ is sent if the u-signal is received.

$dH(u)$ = probability that a signal with label between u and $u + du$ is received at any stage.

Show that the corresponding functional equations are

$$f_N(x) = \int_{-\infty}^{\infty} \left[\underset{z(v)}{\text{Max}} \int_{-\infty}^{\infty} f_{N-1}(2z(v))\, dG(u, v) \right] dH(u),$$

$$f_1(x) = \int_{-\infty}^{\infty} \left[\underset{z(v)}{\text{Max}} \int_{-\infty}^{\infty} \log (2z(v))\, dG(u, v) \right] dH(u),$$

assuming for the sake of simplicity that the odds are even, and that all money must be bet. The maximization is over all functions satisfying

$$\text{a. } z(v) \geq 0, \qquad \text{b. } \int_{-\infty}^{\infty} z(v)\, dv = x.$$

Obtain the form of $f_N(x)$ and the structure of the optimal policy.

42. Consider the case where p itself is a random variable, subject to a known probability distribution.

43. Consider the case in which the probability distrinution is unknown. We do, however, have an *a priori* estimate $dG(p)$, and agree, after k successful transmissions and l unsuccessful transmissions, that the new *a priori* estimate is to be

$$dG_{kl}(p) = \frac{p^k (1-p) dG(p)}{\int_0^1 p^k (1-p) dG(p)}$$

44. Several industrial plants are located along a river, numbered from north to south, 1, 2, . . ., N. A certain quantity of water flows down this river, to be allocated along the way to these plants. Assume to begin with that water allocated to a plant cannot be used by any other plants, and determine the allocation policy which maximizes the return to the community. (W. Hall)

45. Consider the same problem under the assumption that a certain quantity of the water allocated to each industry returns to the river, sometimes immediately, and sometimes several stages further down. (W. Hall)

46. Suppose that the waste products of each industry pollute the water, and the cost of using this water depends on the pollution level. Determine the optimal allocation policy in this case. (W. Hall)

47. Suppose that the quantity of water available is seasonal, and that the demand is seasonal. Dams exist at various places along the river where water can be stored. Determine the optimal allocation policy. (W. Hall)

48. There are n different industrial plants whose construction along a river is being considered. The ith plant has production value v_i, discharges waste products in quantity w_i into the river, and has a tolerance level t_i,

which if the plant is to be utilized, must exceed the sum of the wastes from the upstream plants. We wish to choose a subset of the n plants to build along the river so as to maximize the economic value of the plants. (L. M. K. Boelter)

Show that this is a maximization problem over $2^n n!$. choices, which can be reduced to $[n!\cdot e] - 1$ choices. (Gross–Johnson)

49. Show that any optimal solution can be reordered by increasing values of $t_i + w_i$ without loss of optimality, and thus that there are fewer than 2^n cases to consider. (O. Gross–S. Johnson)

50. If $v_i = 1$ for $i = 1, 2, \ldots, n$, show that an optimal solution may be found using the following procedure:

a. Order and renumber the items according to the magnitude of $t_i + w_i$.

b. Compute $s_i = \sum_{j=1}^{i} w_i$, and $d_i = t_i - s_{i-1}$.

c. If $d_k < 0$ is the first violation, delete an item in the set $i \leq k$ whose w_i is largest.

d. Recompute as in step (b) for the new set, and repeat steps (b) and (c) until all violations are removed. (O. Gross–S. Johnson)

51. Show that in the general case an optimal solution has no greater number of items than there are in the optimal solution of the same problem with all the v_i equal. (O. Gross–S. Johnson)

52. Consider the problem of finding an approximate solution of the equations $f(x, y) = a$, $g(x, y) = b$. Let $\{x_k, y_k\}$, $k = 0, 1, 2, \ldots$, be a sequence of guesses, and

$$d_N = (f(x_N, y_N) - a)^2 + (g(x_N, y_N) - b)^2.$$

Assuming that $x_0 = c_1$, $y_0 = c_2$, and that $(x_{i+1} - x_i)^2 + (y_{i+1} - y_i)^2 \leq r^2$, for $i = 0, 1, 2, \ldots$, let for $N = 0, 1, 2, \ldots$

$$f_N(c_1, c_2) = \min_{\{x_i, y_i\}} d_N.$$

Show that

$$f_{N+1}(c_1, c_2) = \min_R [f_N(x_1, y_1)],$$

where R is the region determined by $(x_1 - c_1)^2 + (y_1 - c_2)^2 \leq r^2$.

53. Set $x_1 = c_1 + r \cos \theta$, $y_1 = c_2 + r \sin \theta$ and assume that r is a small quantity. Then

$$\begin{aligned} f_{N+1}(c_1, c_2) &= \operatorname*{Min}_{\theta} [f_N (c_1 + r \cos \theta, c_2 + r \sin \theta)] \\ &= \operatorname*{Min}_{\theta} [f_N (c_1, c_2) + r [\cos \theta \, \partial f_N / \partial c_1 + \sin \theta \, \partial f_N / \partial c_2]]. \end{aligned}$$

From this determine approximate values for $\cos \theta$ and $\sin \theta$. What is the connection with the classical gradient method?

54. Consider the problem of determining the Cebycev norm

$$d_N = \operatorname*{Min}_{c_i} \operatorname*{Max}_{0 \leq x \leq 1} \left| f(x) - \sum_{x=0}^{N} c_k x^k \right| .$$

Discuss the convergence of the following scheme. Let $\{c^0{}_k\}$ be an initial approximation, and c_0' determined as the minimum of

$$\operatorname*{Max}_{0 \leq x \leq 1} \left| f(x) - c_0 - \sum_{k=1}^{N} c^0{}_k x^k \right| .$$

The let c_1' be determined as the minimum of

$$\operatorname*{Max}_{0 \leq x \leq 1} \left| f(x) - c_0' - c_1 x - \sum_{k=2}^{N} c^0{}_k x^k \right| ,$$

and so on.

55. Suppose that we wish to send a rocket to the moon. Since there are questions of cost and engineering involved in carrying large quantities of fuel, and the containers for large quantities of fuel, we attempt to cut down on the quantity of fuel required and the size of the rocket by building a multi-stage rocket of the following type:

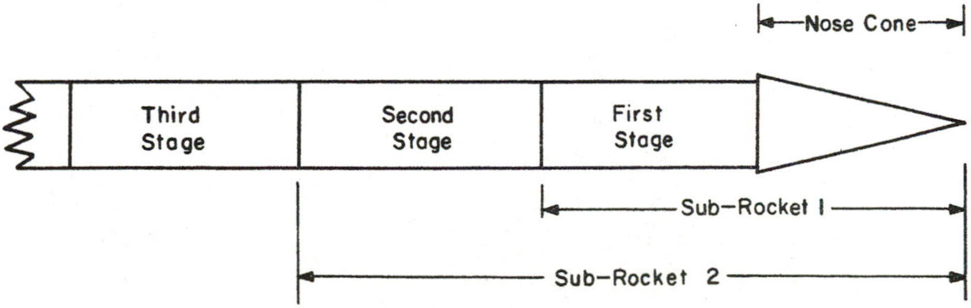

After the fuel carried in the last stage, the kth stage is consumed, this stage drops off, leaving a $(k - 1)$ stage sub-rocket, and so on.

The problem is to build a k-stage rocket of minimum weight which will attain a final velocity of v. Let

W_k = initial gross weight of sub-rocket k.
w_k = initial gross weight of stage k.
p_k = initial propellant weight of stage k.
v_k = change in rocket velocity during burning of stage k.

Assume that the change in velocity v_k is a known function of W_k and p_k, so that $v_k = v(W_k, p_k)$ and thus $p_k = p(W_k, v_k)$. Since $W_k = W_{k-1} + w_k$, and the weight of the kth stage is a known function, $g(p_k)$, of the propellant carried in the stage, we have

$$w_k = g(p(W_{k-1} + w_k, v_k)),$$

whence, solving for w_k, we have

$$w_k = w(W_{k-1}, v_k).$$

Let $f_k(v)$ denote the minimum weight of sub-rocket k achieving a terminal velocity of v. Then

$$f_k(v) = \min_{0 \leq v_k \leq v} [w(f_{k-1}(v - v_k), v_k) + f_{k-1}(v - v_k)],$$

for $k \geq 2$, with

$$f_0(v) = W_0 = \text{weight of nose cone}$$
$$f_1(v) = \min_{0 \leq v_1 \leq v} (w(W_0, v_0) + W_0).$$

(R. P. Ten Dyke)

56. Consider the problem of maximizing the linear form

$L_N(x) = \sum_{i=1}^{3N} x_i$ over all non-negative x_i satisfying the constraints

$$a_{11}x_1 + a_{12}x_2 + a_{13}x_3 \leq c_1,$$
$$a_{21}x_1 + a_{22}x_2 + a_{23}x_3 \leq c_2,$$
$$a_{31}x_1 + a_{32}x_2 + a_{33}x_3 + b_1x_4 \leq c_3,$$
$$a_{44}x_4 + a_{45}x_5 + a_{46}x_6 \leq c_4,$$
$$a_{54}x_4 + a_{55}x_5 + a_{56}x_6 \leq c_5,$$
$$a_{64}x_4 + a_{65}x_5 + a_{66}x_6 + b_2x_7 \leq c_6,$$
$$\vdots$$
$$a_{3N-2,3N-2}x_{3N-2} + a_{3N-2,3N-1}x_{3N-1} + a_{3N-2,3N}x_{3N} \leq c_{3N-2},$$
$$a_{3N-1,3N-2}x_{3N-2} + a_{3N-1,3N-1}x_{3N-1} + a_{3N-1,3N}x_{3N} \leq c_{3N-1},$$
$$a_{3N,3N-2}x_{3N-2} + a_{3N,3N-1}x_{3N-1} + a_{3N,3N}x_{3N} \leq c_{3N},$$

and $x_i \geq 0$, where $a_{ij} \geq 0$, $b_i > 0$.

Define the sequence of functions

$$f_N(z) = \operatorname*{Max}_{x_i} L_N(x),$$

where the x_i are subject to the constraints given above with the exception that the last constraint is now

$$a_{3N,\,3N-2}\,x_{3N-2} + a_{3N,\,3N-1}\,x_{3N-1} + a_{3N,\,3N}\,x_{3N} \leq z.$$

Show that

$$f_N(z) = \operatorname*{Max}_{[x_{3N-2},\,x_{3N-1},\,x_{3N}]} [x_{3N-2} + x_{3N-1} + x_{3N} + f_{N-1}(c_{3N-3} - b_{N-1}\,x_{3N-2})],$$
$$N \geq 1,$$

where $x_{3N-2}, x_{3N-1}, x_{3N}$ are subject to

$$a_{3N-2,\,3N-2}\,x_{3N-2} + a_{3N-2,\,3N-1}\,x_{3N-1} + a_{3N-2,\,3N}\,x_{3N} \leq c_{3N-2},$$
$$a_{3N-1,\,3N-2}\,x_{3N-2} + a_{3N-1,\,3N-1}\,x_{3N-1} + a_{3N-1,\,3N}\,x_{3N} \leq c_{3N-1},$$
$$a_{3N,\,3N-2}\,x_{3N-2} + a_{3N,\,3N-1}\,x_{3N-1} + a_{3N,\,3N}\,x_{3N} \leq z,$$
$$b_{N-1}\,x_{3N-2} \leq c_{3N-3},\ x_i \geq 0.$$

The function $f_0(z)$ is taken to be identically zero.

57. Obtain corresponding results for the case where the matrices are of different order.

58. Consider the case where the $3k^{th}$ equation, $k = 1, 2, \ldots,$ has the form

$$a_{3k-2,\,3k-2}\,x_{3k-2} + a_{3k-2,\,3k-1}\,x_{3k-1} + a_{3k-2,\,3k}\,x_{3k} + b_2\,x_{3k+1}$$
$$+ c_2\,x_{3k+2} + d_2\,x_{3k+3} \leq c_{3k}.$$

59. Show that the above functional equation can be reduced to the form

$$f_N(z) = \operatorname*{Max}_{x_{3N-2}} [g_N(x_{3N-2}, z) + f_{N-1}(c_{3N-3} - b_{N-1}\,x_{3N-2})],$$

where x_{3N-2} satisfies an inequality

$$0 \leq x_{3N-2} \leq \operatorname{Min}[a_N, z/a_{3N,\,3N-2}].$$

60. Consider the problem of resolving a set of linear equations of the form

$$a_{11}\,x_1 + a_{12}\,x_2 + a_{13}\,x_3 = c_1,$$
$$a_{21}\,x_1 + a_{22}\,x_2 + a_{23}\,x_3 = c_2,$$
$$a_{31}\,x_1 + a_{32}\,x_2 + a_{33}\,x_3 + b_1\,x_4 = c_3,$$
$$b_1\,x_3 + a_{44}\,x_4 + a_{45}\,x_5 + a_{46}\,x_6 = c_4,$$

$$a_{54} x_4 + a_{55} x_5 + a_{56} x_6 \qquad = c_5,$$
$$a_{64} x_4 + a_{65} x_5 + a_{66} x_6 + b_2 x_7 = c_6$$
$$\vdots$$
$$b_{N-1} x_{3N-3} + a_{1+3N,\,1+3N} x_{1+3N} + a_{1+3N,\,2-3N} x_{2+3N}$$
$$+ a_{1+3N,\,3+3N} x_{3+3N} = c_{1+3N}$$
$$a_{2+3N,\,1+3N} x_{1+3N} + a_{2+3N,\,2+3N} x_{2+3N}$$
$$+ a_{2+3N,\,3+3N} x_{3+3N} = c_{2+3N}$$
$$a_{3+3N,\,1+3N} x_{1+3N} + a_{3+3N,\,2+3N} x_{2+3N}$$
$$+ a_{3+3N,\,3+3N} x_{3+3N} = c_{3+3N},$$

where (a_{ij}) is a symmetric matrix, and, in addition, positive definite.

Linear systems of this type arise in the study of multicomponent systems where there is weak coupling between stages.

The problem of solving this system is equivalent to that of determining the minimum of the inhomogeneous quadratic form

$$(x^1, A_1 x^1) + (x^2, A_2 x^2) + \ldots + (x^N, A_N x^N)$$
$$- 2 (c^1, x^1) - 2 (c^2, x^2) + \ldots - 2 (c^N, x^N)$$
$$+ 2 b_1 x_3 x_4 + 2 b_2 x_6 x_7 + \ldots + 2 b_{N-1} x_{3N-1} x_{3N-2}$$

where the vectors x^k and c^k are defined by

$$x^k = (x_{3k-2}, x_{3k-1}, x_{3k}),\ c^k = (c_{3k-2}, c_{3k-1}, c_{3k}),$$

and $A_k = (a_{i+3k,\, j+3k})$, $i, j = 1, 2, 3$.

Show that the problem can be reduced to that of determining the sequence $\{f_N(z)\}$ defined by the recurrence relation

$$f_N(z) = \operatorname*{Min}_{(x_{3N},\, x_{3N-1},\, x_{3N-2})} [(x^N, A_N x^N) - 2 z x_{3N} - 2 (c^N, x^N) + $$
$$+ f_{N-1}(b_{N-1} x_{3N-2})].$$

(*Illinois Journal of Mathematics*, 1957).

61. Show that this may be reduced to the form

$$f_N(z) = \operatorname*{Min}_{y} [g_N(z, y) + f_{N-1}(b_{N-1} y)],$$

where

$$g_N(z, y) = \operatorname*{Min}_{(x_{3N},\, x_{3N-1})} [(x^N, A_N x^N) - 2 z x_{3N} - 2 (c^N, x^N)].$$

62. Show that $f_N(z) = u_N + v_N z + w_N z^2$, where u_N, v_N, w_N are independent of z, determine the recurrence relations connecting (u_N, v_N, w_N) and $(u_{N-1}, v_{N-1}, w_{N-1})$, and thus determine the solution of the linear system.

63. Consider the problem of determining the maximum of

$$Q_N(x) = \sum_{i=1}^{N} (x^i, A_i x^i) + 2 \sum_{i=1}^{N-1} b_i x_{3i} x_{1+3i}$$

over the sphere S_N, $\sum_{i=1}^{N} (x^i, x^i) = 1$.

Consider the associated functions of z defined by

$$f_N(z) = \operatorname*{Max}_{S_N} [q_N(x) + 2 z x_{3N}],$$

and obtain the recurrence relation connecting $f_N(z)$ and $f_{N-1}(z)$.

64. Generalize the foregoing results to the case where the matrices A_k are not necessarily of the same dimension.

65. Obtain existence and uniqueness theorems under appropriate assumptions for the following functional equations

a. $f(p) = \operatorname*{Min}_{q} \operatorname{Max} [g(p, q), f(T(p, q))]$

b. $f(p) = \operatorname*{Min}_{q} \operatorname{Max} [g(p, q), h(p, q) f(T(p, q))]$

c. $f(p) = \operatorname*{Min}_{q} \operatorname{Max} [g(p, q), r(p, q) + \int_R f(z)\, dG(z, p, q)]$.

66. Consider the problem of assigning m different types of machines to n different tasks. Let $A_{ij} \geq 0$ be the amount of task j performed by a unit input of machine i, and assume that

a. If $A_{ij} > 0$, and $i' < i$, then $A_{i'j} > 0$.

b. If $A_{ij} > 0$, and $j' > j$, then $A_{ij'} > 0$.

c. If $i < i'$, $j < j'$, $A_{i'j} > 0$, then
$(A_{ij'}/A_{ij}) < (A_{i'j'}/A_{i'j})$.

Let x_{ij} be the quantity of machines of type i to be used for task j. The matrix $x = (x_{ij})$, $i = 1, 2, \ldots, m$, $j = 1, 2, \ldots, n$, is said to be *feasible* if $x_{ij} \geq 0$, $\sum_{i=1}^{n} A_{ij} x_{ij} = T_j$, $j = 1, 2, \ldots, n$, and $\sum_{j=1}^{n} x_{ij} \leq M_i$, $i = 1, 2, \ldots, m$.

Consider the following policy. Assign x_{11} up to the minimum of T_1 and M_1. If $x_{11} = T_1$, then assign $x_{12} = \min(T_2, M_1 - x_{11})$, and so on. When M_1 is used up in this way, on the jth task for some j, assign x_{2j} in such a way that either task j is finished or all machines of type 2 are assigned. Complete the assignment of machines in this way.

Show that if this policy does not lead to a feasible allocation, then there exists no feasible policy. (Arrow–Markowitz–Johnson)

67. Show that the above policy yields the solution of the problem of maximizing $T_n = \sum_{i=1}^{m} A_{in} x_{in}$ subject to

a. $\sum_{j=1}^{n} x_{ij} = M_1, i = 1, 2, \ldots, m, x_{ij} \geq 0,$

b. $\sum_{i=1}^{m} A_{ij} x_{ij} = T_j, j = 1, 2, \ldots, n-1,$

provided that the A_{ij} satisfy the above conditions. (Johnson)

68. Show that the problem of maximizing the sum $\sum_{i=1}^{N} g_i(x_i, y_i)$ subject to the constraints

a. $x_i \geq 0, \sum_{i=1}^{N} x_i = x,$

b. $y_i \leq 0, \sum_{i=1}^{N} y_i = y,$

can, under appropriate assumptions concerning the functions $g_i(x, y)$, be reduced to the problem of maximizing

$$S_N = \sum_{i=1}^{N} g_i(x_i, y_i) - \lambda \sum_{i=1}^{N} y_i,$$

subject to the constraints

a. $x_i \geq 0, \sum_{i=1}^{N} x_i = x,$

b. $y_i \geq 0.$

This last problem leads to the recurrence relations

$$f_n(x) = \underset{0 \leq x_n \leq x}{\text{Max}} \left[\underset{y \geq 0}{\text{Max}} [g_n(x_n, y) - \lambda y] + f_{n-1}(x - x_n) \right],$$

involving a one-dimensional sequence, for each fixed λ.

How does one use the solution of this second problem to solve the original? (*Proc. Nat. Acad. Sci.*, 1956).

69. Each year the walnut crop consists of walnuts of different grades, say $G_1, G_2, \ldots, G_k$, in quantities $q_1, q_2, \ldots, q_k$. Using various quantities of each grade, assortments of walnuts are put together for commercial sale at different prices. Assume that there are fixed demands d_i for the ith assortment, and that each assortment mixes walnuts of different grades in its own fixed ratios. How many packets of each assortment should be made in order to maximize total profit?

70. Consider the case where the demand is stochastic with known distributions for each type of packet.

Bibliography and Comments for Chapter IV

§ 1. This chapter follows R. Bellman, "Functional equations in the Theory of Dynamic Programming — I, Functions of Points and Point Transformations," *Trans. Amer. Math. Soc.*, vol., Vol. 80 (1955), pp. 51–71. An entirely different treatment of a more abstract type, making use of Tychonoff's Theorem, is contained in an unpublished paper by S. Karlin and H. N. Shapiro, "Decision Processes and Functional Equations." The RAND Corporation, RM–933, Sept. 1952.

See also, S. Karlin, "The Structure of Dynamic Programming Models," *Naval Research Logistics Quarterly*, Vol 2 (1955), pp. 285–294.

§ 6. A discussion of the importance of stability theory in the domain of differential equations may be found in R. Bellman, *Stability Theory of Differential Equations*, McGraw-Hill, 1952.

§ 8. The choice of $f_0(p)$ in (8.6) is due to a suggestion of H. N. Shapiro.

§ 9. This equation will be discussed in extenso in the following chapter.

CHAPTER V

The Optimal Inventory Equation

§ 1. Introduction

In this chapter we wish to study a class of analytic problems arising from an interesting stochastic allocation process occurring in the study of inventory and stock control.

Although the general equation seems to be quite difficult to treat, we can obtain an explicit solution of a particular case where certain simple, but not too far from realistic, assumptions are made, and we can determine the structure of the optimal policy in some other cases.

These explicit solutions are useful since they lay bare certain meaningful combinations of essential parameters. Since the inverse problem of the estimation of parameters from observed data plays a critical role in this theory, this is a feature which can be of importance.

Furthermore, and this is a remark pertinent to all decision processes, the analytic form of the solution will occasionally possess a simple economic interpretation, which when verbalized, opens the way to the approximation of optimal policies for more complicated processes.[1]

Apart from the results we obtain, the methods we employ to investigate the structure of optimal policies possess an independent interest. The reader has already encountered them, in part, in § 12 of Chapter I, and will encounter them again in a later chapter devoted to the calculus of variations. What stands out quite vividly is the fact that the method of successive approximations is not only useful in the production of existence and uniqueness theorems, to which relatively dull task it is usually relegated, but is, in addition, a powerful analytic tool for the discovery and proof of properties of the solution of a functional equation, and in our case, for the determination of the behavior of optimal policies.

We shall begin with the formulation of a class of related problems occurring in the study of "optimal inventory." Following this, we devote a section to the simple formal observation upon which all the analysis in this chapter hinges.

We then consider a number of cases in which the optimal policy is

[1] This idea has, of course, been used extensively in the physical and engineering world.

characterized in an especially simple and intuitive way, namely, by the maintenance of a constant "stock level". In particular, this is the case, in both the multi-dimensional as well as the one-dimensional case, if all the ordering costs are directly proportional to the amounts ordered.

If the initial ordering cost includes a fixed cost which is independent of the amount ordered, the problem seems to become very much more difficult. This fixed cost may represent a "red tape" cost, or a "set-up" cost, in the case of manufacturing processes. We shall not treat any problems of this type here, since at the present time practically no solutions of the corresponding functional equations exist, and very little seems to be known concerning the character of the optimal policies arising from processes of this more realistic type.

To illustrate further the method of successive approximations, we shall consider two processes, each variants of the relatively simple process discussed above. In the first, linearity is discarded, in that the cost is taken to be a convex function of the amount ordered; in the second, simultaneity is voided, in that there is assumed to be a time-lag in satisfying an order. Although the optimal policies cannot be described in simple terms, we can determine their general structure.

From the mathematical point of view, we have to deal with a very interesting class of quasi-linear integral equations, nonlinear versions of the renewal equation which we shall discuss in an appendix. As usual, these nonlinear equations possess certain quasi-linear properties which we can occasionally use as handholds and footholds in making our way through this tortuous terrain.

§ 2. Formulation of the general problem

The problem we shall discuss here, in various masquerades, is one very particular case of the general problem of decision-making in the face of an uncertain future. The version we shall consider is concerned with the problem of stocking a supply of items to meet an uncertain demand, under the assumptions that there are various costs associated with oversupply and undersupply.

The situation may be described as follows: At various specified times, determined in advance or dependent upon the process itself, we have an opportunity to order supplies of a certain set of items, where the cost of ordering depends naturally upon the number ordered of each item, and where there may or may not be, in addition, some fixed costs, administrative or otherwise, which are independent of the number ordered. At various other times, demands are made upon the stocks of these items. The interesting case is that where these demands are not known in advance, but where we do know the joint distribution of the demands which can be made at any particular time. The incentive for ordering lies in a penalty which is assessed whenever the demand for an item exceeds the supply. Different penalties may be levied in different fields of activity. A case which we shall treat in great detail is that where the penalty is directly proportional to the excess of demand over supply. Its importance lies in the fact that we can solve the functional equations arising from the process explicitly under the crucial assumption that the cost of initial ordering depends only upon the amount ordered, and is either a linear function, or, more generally, convex.

Speaking loosely, we wish to determine the ordering policy at each stage which will minimize some average function of the overall cost of the process. In practical applications, an important aspect of the problem, which we shall not discuss here, is that of determining suitable criteria for the various costs, which are both realistic and analytically malleable.

In the following subsections we shall consider various sets of assumptions which yield various functional equations, all of which belong to a common family. Additional processes will be discussed in the exercises.

A. Finite total time period

The first process we shall consider involves the stocking of only one item. We shall assume that orders are made at each of a finite number of equally-spaced times, and immediately fulfilled. After the order has been made and filled, a demand is made. This demand is satisfied as far as possible, with excess demand leading to a penalty cost.

Let us assume that we know completely the following functions:

(1) a. $\varphi(s)ds$ = probability that the demand will lie between s and $s + ds$.[2]

b. $k(z)$ = the cost of ordering z items initially to increase the stock level.

c. $p(z)$ = the cost of ordering z items to meet an excess, z, of demand over supply, the *penalty cost*.

Observe that we assume that these functions are independent of time. Furthermore, we suppose that these orders can be filled immediately.

Let x denote the stock level at the initiation of the process. Assuming that there are n stages, we will order a quantity y_1 at the first stage, y_2 at the second stage, and so on.

[2] We shall avoid Stieltjes integrals throughout to simplify the discussion. It will readily be seen that most of our results carry over to the more general situation when suitable attention is paid to possible nonuniqueness of roots of equations. This is left as a set of exercises, of nontrivial nature, for the reader.

A set of functions $(y_1, y_2, \ldots, y_n)$, $y_k = y_k(x)$, specifying for each k the quantity y_k to be ordered at the kth stage when the stock level is x will be called a *policy*. Corresponding to each policy, there will be a certain expected total cost for this n-stage process, composed of initial ordering and penalty costs.

The problem we set ourselves is that of determining the policy, or policies, which minimize the expected total cost. A policy which yields this minimum expected cost is called optimal. All this is in accordance with our previous notation.

We obtain an equally interesting but more difficult class of problems if we attempt to minimize the probability that the cost exceeds a fixed level.

At any stage, the problem is characterized completely by two state variables, x, the supply of stock, and n, the number of remaining stages. Let us then define

(2) $f_n(x) =$ expected total cost for an n-stage process starting with an initial supply x, and using an optimal ordering policy.

Let us now proceed to obtain a functional equation for $f_n(x)$. We have

$$f_1(x) = k(y - x) + \int_y^\infty p(s - y)\,\varphi(s)\,ds, \tag{3}$$

if a quantity $y - x \geq 0$ is ordered.

Although it may seem odd to order a quantity $y - x$, instead of say y, it turns out that it is simpler to think of ordering up to a certain *level*, y. The optimal stock level turns out to be a more basic quantity than the amount ordered.

Since y is to be chosen to minimize the expected cost, we see that $f_1(x)$ is given by

$$f_1(x) = \operatorname*{Min}_{y \geq x}\left[k(y - x) + \int_y^\infty p(s - y)\,\varphi(s)\,ds\right]. \tag{4}$$

In general, for $n \geq 2$ we have

$$\begin{aligned} f_n(x) = \operatorname*{Min}_{y \geq x}\Big[k(y - x) &+ \int_y^\infty p(s - y)\,\varphi(s)\,ds\ + \\ f_{n-1}(0)\int_y^\infty \varphi(s)\,ds &+ \int_0^y f_{n-1}(y - s)\,\varphi(s)\,ds\Big], \end{aligned} \tag{5}$$

upon enumerating the various cases corresponding to the possibility of an excess of demand over supply, and corresponding to the possibility of being able to fulfill the demand.

B. Unbounded time period—discounted cost

If we wish to consider an unbounded period of time over which this process operates, we must introduce some device to prevent infinite costs from entering.

The most natural such device is that of discounting the future costs, using a fixed discount ratio, a, $0 < a < 1$, for each period. This possesses a certain amount of economic justification and a great deal of mathematical virtue, particularly in its invariant aspect.

If we set

(6) $f(x) =$ expected total discounted cost starting with an initial supply x and using an optimal policy,

we obtain, by the same enumeration of possibilities, in place of (5) the functional equation

$$(7)\quad f(x) = \operatorname*{Min}_{y \geq x} \Big[k(y-x) + a\int_y^\infty p(s-y)\,\varphi(s)\,ds + af(0)\int_y^\infty \varphi(s)\,ds + a\int_0^y f(y-s)\,\varphi(s)\,ds\Big].$$

The advantage of (7) over (5) is the usual one that it contains $f(x)$, one function of one variable, in place of a sequence of functions, $\{f_n(x)\}$.

C. Unbounded time period—partially expendable items

If we assume that some of the items supplied upon demand may be partially recovered, so that a demand of s items results in a return of bs items, $0 \leq b \leq 1$, which may be used again, the analogue of (7) is

$$(8)\quad f(x) = \operatorname*{Min}_{y \geq x} \Big[k(y-x) + a\int_y^\infty p(s-y)\,\varphi(s)\,ds + a\int_y^\infty f(bs)\varphi(s)ds + a\int_0^y f(y-s+bs)\,\varphi(s)\,ds\Big].$$

D. Unbounded time period—one period lag in supply

Let us now assume that when we order a quantity z it does not become available until one period later. If the current supply is x and y was on order from the period before, $x + y$ will be available to meet the next demand. The functional equation corresponding to (7) is now of more complicated form

$$(9)\quad f(x) = \operatorname*{Min}_{z \geq 0} \Big[k(z) + a\int_x^\infty p(s-x)\,\varphi(s)\,ds + af(z)\int_x^\infty \varphi(s)\,ds + a\int_0^x f(x-s+z)\,\varphi(s)\,ds\Big].$$

The quantity x now represents the total quantity available at any stage to meet the demand.

E. Unbounded time period—two period lag

If we have a two period lag, we require two state variables to describe the state of the process, namely,

$$\begin{aligned} &x = \text{quantity of stock available to meet next demand,} \\ &y = \text{quantity to be delivered one period hence.} \end{aligned} \tag{10}$$

Hence we define

(11) $f(x, y) =$ expected total cost with x and y as above, using an optimal policy.

Then $f(x, y)$ satisfies the equation

$$f(x, y) = \operatorname*{Min}_{z \geq 0} \Big[k(z) + a \int_x^\infty p(s - x)\, \varphi(s)\, ds + af(y, z) \int_x^\infty \varphi(s)\, ds + a \int_0^x f(x - s + y, z)\, \varphi(s)\, ds\Big]. \tag{12}$$

We shall not consider the equations in (8), (9), or (12) here, although they are amenable to the same techniques of successive approximation we shall apply to the others. There does not seem to exist any explicit solution comparable in simplicity to that obtainable for (7).

§ 3. A simple observation

In this section, we wish to present, in as simple a form as possible, the fundamental analytic property of functional equations of the form

$$u(x) = \operatorname*{Min}_{y} v(x, y),\ y \in R(x), \tag{1}$$

upon which all the subsequent work in this chapter depends.[3]

In general, the variation will be over some region, $R(x)$, in this case, a set of intervals, dependent upon x. Let us assume that over some interval of x-values, $a \leq x \leq b$, the minimum is attained inside the region $R(x)$, and that v is differentiable. Then at the minimizing value of y we have

$$0 = v_y\,. \tag{2}$$

This determines a function $y(x)$, which need not be single-valued but which we do assume differentiable.

[3] This property has already been used, without explicit remark in § 11 of Chapter I.

On any one particular branch of this function $y(x)$, we have

$$u(x) = v(x, y). \tag{3}$$

The crucial observation is now that for $a \leq x \leq b$, we have

$$u'(x) = v_x + v_y \, dy/dx = v_x, \tag{4}$$

since $v_y = 0$, by (2).

Similarly, if

$$u(x_1, x_2) = \operatorname*{Min}_{y_1, y_2} [v(x_1, x_2, y_1, y_2)], \quad (y_1, y_2) \,\varepsilon\, R(x_1, x_2), \tag{5}$$

and we assume that the minimum is always attained inside the region, we have

$$\begin{aligned} u_{x_1} &= v_{x_1}, \\ u_{x_2} &= v_{x_2}, \end{aligned} \tag{6}$$

at the minimizing points.

Let us now apply these remarks to the functional equation ot (2.7), under the assumption that $k(z) = kz$, $k > 0$ and $p(z) = pz$, linear functions of z. We have

$$f(x) = \operatorname*{Min}_{y \geq x} \left[ky - kx + a \int_y^\infty p(s - y) \varphi(s) \, ds + af(0) \int_y^\infty \varphi(s) \, ds + a \int_0^y f(y - s) \varphi(s) \, ds\right]. \tag{7}$$

If the minimum is attained at a point $y > x$, we have at this point

$$k - ap \int_y^\infty \varphi(s) \, ds + a \int_0^y f'(y - s) \varphi(s) \, ds = 0, \tag{8}$$

an equation independent of x!

Furthermore, for this value of y, we have

$$f'(x) = -k. \tag{9}$$

These two results, correctly combined and interpreted, furnish the clues to the solutions of the problems involving proportional costs. We shall discuss them in more detail in later sections, and we shall also utilize their multi-dimensional analogues.

§ 4. Constant stock level—preliminary discussion

In this and the next few sections we shall consider several processes characterized by the principle of "constant stock level." The common feature of these models is the assumption that the cost of initial ordering is directly proportional to the amount ordered, and that the distribution of demand remains the same from stage to stage. The addition of an administrative fixed cost, "red-tape" cost, changes the nature of the optimal policy in an essential manner.[4] This cost may also represent "set-up" cost in manufacturing processes.

In § 5, we shall obtain the complete solution, for an arbitrary distribution function $\varphi(s)$, for the case where the penalty cost is also directly proportional to the number ordered. In § 6 we extend this result to the multi-dimensional case, and show that the solution for the case where there are many items subject to a joint distribution of demand possesses the very important property of sub-optimality.

Turning from the consideration of these processes involving unbounded time intervals, we consider the finite process described in § 2 and show that again the assumption of direct proportionality entails a principle of constant stock level at each stage. This level, of course, changes with the stage.

This section serves as an excellent introduction to the use of successive approximations as an analytic tool in the study of these functional equations.

We enter territory where the going is much rougher when we consider the case where the penalty cost includes a "red-tape" term which is independent of the amount ordered. The form of the solution now seems in the general case to depend upon the form of the demand function. Nevertheless, several important classes of distribution functions fall within catagories which we can handle precisely.

Finally, we indicate briefly the form of the general solution without, however, being able to make any constructive use of it.

§ 5. Proportional cost—one-dimensional case

In this section we present the solution of the case where both cost functions, direct ordering and penalty ordering, are directly proportional to the amounts ordered.

THEOREM 1. *Consider the equation*

$$(1)\quad f(x) = \underset{y \geq x}{\text{Min}} \left[k(y-x) + a\int_y^\infty p(s-y)\varphi(s)\,ds + af(0)\int_y^\infty \varphi(s)\,ds + a\int_0^y f(y-s)\varphi(s)\,ds \right].$$

where we impose the conditions

[4] In the sense that it changes the policy from one of known form to one of unknown form.

(2) a. *k and p are positive constants,*

b. $\varphi(s) > 0, \int_0^\infty \varphi(s)\, ds = 1, \int_0^\infty s\, \varphi(s)\, ds < \infty,$

c. $0 < a < 1,$

d. $ap > k.$

Let $\bar{x}$ be the unique root of

$$k = ap \int_y^\infty \varphi(s)\, ds + ak \int_0^y \varphi(s)\, ds \text{ [5]}. \tag{3}$$

Then the optimal policy has the form

(4) a. *for* $0 \leq x \leq \bar{x}, y = \bar{x},$

b. *for* $x \geq \bar{x}, y = \bar{x}.$

In other words, the optimal stock level is $\bar{x}$.

If $ap \leq k$, the solution is given by $y = x$ for $x \geq 0$, i.e. *never order.*

Proof. In order to understand the genesis of this solution, let us proceed heuristically. If we can obtain a plausible solution by some formal means and then verify directly that it satisfies the equation in (1) above, the uniqueness theorem established in § 9 of Chapter IV tells us that it is *the* solution. Let us point out, however, that the method of successive approximations would have led us to this solution in a systematic fashion.

As pointed out in § 3, if the minimum occurs at $y > x$, the minimizing values of y must be roots of the equation

$$k + a\left[-p \int_y^\infty \varphi(s)\, ds + \int_0^y f'(y-s)\, \varphi(s)\, ds\right] = 0, \tag{5}$$

and at this value of y we have

$$f'(x) = -k. \tag{6}$$

Now let us pull ourselves up by our bootstraps. If the solution has the conjectured form, the complicated term, $\int_0^y f'(y-s)\, \varphi(s)\, ds$ may be replaced by the simpler term $-k \int_0^y \varphi(s)\, ds$, so that equation (5) may be replaced by

[5] The interpretation of this equation is that the run-out probability must be set at the level where the marginal cost for holding inventory is just balanced by the marginal penalty for run-out.

$$k - ap \int_y^\infty \varphi(s)\, ds - ak \int_0^y \varphi(s)\, ds = 0, \tag{7}$$

precisely the equation of (3).

Since $\int_0^\infty \varphi(s)\, ds = 1$, this equation reduces to

$$\int_0^y \varphi(s)\, ds = (ap - k)/a(p - k), \tag{8}$$

which possesses exactly one root under the assumption that $\varphi(s) > 0$. Observe that the limiting cases behave properly. If $ap - k = 0$, $y = 0$, if $a = 1$, $y = \infty$; if $p = \infty$, $y = \infty$.

Having determined $\bar{x}$, we proceed to determine $f(x)$ as follows. For $0 \leq x \leq \bar{x}$ we have

$$f(x) = k(\bar{x} - x) + a\Big[\int_{\bar{x}}^\infty p(s - \bar{x})\varphi(s)\, ds + f(0)\int_{\bar{x}}^\infty \varphi(s)\, ds + \int_0^{\bar{x}} f(\bar{x} - s)\varphi(s)\, ds\Big], \tag{9}$$

and $f'(x) = -k$, or,

$$f(x) = f(0) - kx. \tag{10}$$

Substituting (10) in (9), and setting $x = 0$, we obtain the following result for $f(0)$[6],

$$f(0) = \frac{k\bar{x} + pa\int_{\bar{x}}^\infty (s - \bar{x})\varphi(s)\, ds - ak\int_0^{\bar{x}} (\bar{x} - s)\varphi(s)\, ds}{(1 - a)} \tag{11}$$

To determine $f(x)$ for $x \geq \bar{x}$[7] we have the equation

$$f(x) = a\Big[\int_x^\infty p(s - x)\varphi(s)\, ds + f(0)\int_x^\infty \varphi(s)\, ds + \int_0^x f(x - s)\varphi(s)\, ds\Big] \tag{12}$$

which we write in the form

$$f(x) = u(x) + a\int_0^x f(x - s)\varphi(s)\, ds, \tag{13}$$

[6] Note that the $\bar{x}$ we obtain from (7) is the value of $\bar{x}$ which minimizes this expression for $f(0)$.

[7] Observe that as far as applications are concerned, this part of the solution is of very little interest, since for only one initial interval, if at all, will x ever exceed $\bar{x}$.

where $u(x)$ is a known function of x. This, in turn, we write

$$(14)\ f(x) = u(x) + a \int_0^{x-\bar{x}} f(x-s)\,\varphi(s)\,ds + a \int_{x-\bar{x}}^{x} f(x-s)\,\varphi(s)\,ds.$$

In the interval $[x-\bar{x}, x]$, $f(x-s)$ is known, hence we may write, combining the $u(x)$ term and the second integral

$$(15) \qquad f(x) = v(x) + a \int_0^{x-\bar{x}} f(x-s)\,\varphi(s)\,ds, \quad x \geq \bar{x}.$$

If we now set $x - \bar{x} = z$ and $f(\bar{x} + z) = g(z)$, we see that $g(z)$ satisfies the equation

$$(16) \qquad g(z) = v(\bar{x} + z) + a \int_0^{z} g(z-s)\,\varphi(s)\,ds, \quad z \geq 0,$$

a simple renewal equation whose properties are discussed in the appendix.

Actually, it is much simpler to differentiate (12) first and then proceed as above. Let us observe, parenthetically, that it seems to be a general characteristic of functional equations in the theory of dynamic programming that the derivatives satisfy simpler equations, and are the more basic quantities. This is due to the fact that they represent "marginal returns", or "prices", which in purely mathematical language means that they represent Lagrange multipliers. This, in turn, is connected with the general problem of constructing dual processes, a subject we shall not pursue here.

Let us now turn to a proof that the conjectured solution is actually a solution. Call the function obtained above $F(x)$ and denote the constant $f(0)$ determined in (11) by C. Then $F(x)$ is completely determined by the following equations.

$$(17) \quad \text{a.} \quad F(x) = C - kx, \; 0 \leq x \leq \bar{x}$$

$$\text{b.} \quad F(x) = a\Big[\int_x^{\infty} p(s-x)\,\varphi(s)\,ds + F(0)\int_x^{\infty} \varphi(s)\,ds + \int_0^{x} F(x-s)\,\varphi(s)\,ds\Big], \; x \geq \bar{x},$$

An essential point in our verification of the solution is the fact that $F(x) + kx$ is strictly increasing for $x \geq 0$. This we establish as follows. From (17b), we see that

$$(18) \qquad F'(x) = -ap \int_x^{\infty} \varphi(s)\,ds + a \int_0^{x} F'(x-s)\,\varphi(s)\,ds,$$

for $x > \bar{x}$. In $[x - \bar{x}, x]$, we have $0 \leq x - s \leq \bar{x}$ and hence $F'(x - s) = -k$, as we see from (17a). Thus for $x > \bar{x}$.

(19) $$F'(x) = -ap \int_x^\infty \varphi(s)\, ds - ka \int_{x-\bar{x}}^{x} \varphi(s)\, ds + a \int_0^{x-\bar{x}} F'(x-s)\, \varphi(s)\, ds,$$

or

(20) $$F'(x) + k = [k - ap \int_x^\infty \varphi(s)\, ds - ak \int_0^x \varphi(s)\, ds] + a \int_0^{x-\bar{x}} [F'(x - s) + k]\, \varphi(s)\, ds.$$

The expression

(21) $$u(x) = k - ap \int_x^\infty \varphi(s)\, ds - ak \int_0^x \varphi(s)\, ds$$

is zero at $x = \bar{x}$ and positive thereafter. Setting $x - \bar{x} = z$ and $F'(x + z) + k = g(z)$, we see that $g(z)$ satisfies the equation

(22) $$g(z) = u(x + z) + \int_0^z g(z - s)\, \varphi(s)\, ds,\ z \geq 0.$$

It follows, *cf.* p. 177f, that $g(z) > 0$ for $z > 0$.

Hence, $F'(x) + k > 0$ for $x > \bar{x}$, and $F(x) + kx$ is strictly increasing for $x > \bar{x}$.

Let us now return to the problem of demonstrating that $F(x)$ satisfies the equation in (1). Consider first the case where $x > \bar{x}$. Then

(23) $$F(x) = \operatorname*{Min}_{y \geq x} [\quad \ldots \quad] = \operatorname*{Min}_{y \geq x} [k(y - x) + F(y)],$$

using the representation in (17b). Since $ky + F(y) \geq kx + F(x)$ for $y \geq x$, we see that the minimum occurs at $y = x$, yielding $F(x)$, as desired.

Now consider the interval $0 \leq x < \bar{x}$. Write

(24) $$\operatorname*{Min}_{y \geq x} = \operatorname*{Min} \begin{bmatrix} \operatorname*{Min}_{y \geq \bar{x}} [\quad\quad] \\ \operatorname*{Min}_{\bar{x} \geq y \geq x} [\quad\quad] \end{bmatrix}$$

As above, the minimum over $y \geq \bar{x}$ reduces to the value at $y = \bar{x}$. Hence

(25) $$\operatorname*{Min}_{y \geq x} [\ .\ .\ .\ ] = \operatorname*{Min}_{\bar{x} \geq y \geq x} [\ .\ .\ .\ ]$$

Since $F(x) = C - kx$ for $0 \leq x \leq \bar{x}$, it follows that the minimum is assumed at $y = \bar{x}$, as in the original derivation of the value of x.

In the case $ap \leq k$, taking $\bar{x} = 0$ in (17) yields an F which is easily seen to satisfy (23), since, as above, $F(y) + ky$ is non-decreasing.

This completes the proof. It is interesting to note that the solution for $0 \leq x \leq \bar{x}$, the most important part of the solution, can be found without reference to the form of the solution for $x > \bar{x}$.

This completes the verification of the fact that $F(x)$ is *a* solution, and consequently *the* solution, within the class of uniformly bounded functions over $x \geq 0$.

§ 6. Proportional cost—multi-dimensional case

Let us now consider the multi-dimensional version of the problem. Here we have N items whose stock levels will be denoted by $x_1, x_2, \ldots, x_n$, and whose demand $(s_1 s_2, \ldots, s_n)$ at any time is subject to a joint distribution function whose density is $\varphi(s_1, s_2, \ldots, s_n)$.

In formulating the functional equation for the function $f(x_1, x_2, \ldots, x_n)$, the minimum expected over-all discounted cost, let us, for the sake of simplicity, consider only the two-dimensional case.

The remarkable fact that emerges is that the form of the solution is precisely the same as if $\varphi(s_1, s_2, \ldots, s_n)$ had the form $\varphi_1(s_1)\varphi_2(s_2) \ldots \varphi_n(s_n)$, i.e. uncorrelated demands. It is this which yields the important sub-optimalization property of the solution which we discuss below. An enumeration of cases yields the following functional equation for $f(x_1, x_2)$:

$$(1) \quad f(x_1, x_2) = \operatorname*{Min}_{y_i \geq x_i} [k_1(y_1 - x_1) + k_2(y_2 - x_2) + a[\int_{y_1}^{\infty}\int_{y_2}^{\infty} [p_1(s_1 - y_1) + p_2(s_2 - y_2)]\, \varphi(s_1, s_2)\, ds_1\, ds_2$$

$$+ f(0, 0)\int_{y_1}^{\infty}\int_{y_2}^{\infty} \varphi(s_1, s_2)\, ds_1\, ds_2$$

$$+ \int_{y_1}^{\infty}\int_{0}^{y_2} [p_1(s_1 - y_1) + f(0, y_2 - s_2)]\, \varphi(s_1, s_2)\, ds_1\, ds_2$$

$$+ \int_{0}^{y_1}\int_{y_2}^{\infty} [f(y_1 - s_1, 0) + p_2(s_2 - y_2)]\, \varphi(s_1, s_2)\, ds_1\, ds_2$$

$$+ \int_{0}^{y_1}\int_{0}^{y_2} f(y_1 - s_1, y_2 - s_2)\, \varphi(s_1, s_2)\, ds_1\, ds_2]]$$

Let us simplify our notation a bit by setting $\varphi(s_1, s_2)\, ds_1\, ds_2 = dG(s_1, s_2)$ and call the quantity within the brackets $K(y_1, y_2)$. We then have

$$(2) \qquad \frac{\partial K}{\partial y_1} = k_1 + a[-p_1 \int_{y_1}^{\infty} (\int_{s_2 = 0}^{\infty} dG(s_1, s_2))$$

$$+ \int_0^{y_1} \frac{\partial f}{\partial y_1} (y_1 - s_1, 0) \quad (\int_{s_2 = y_2}^{\infty} dG\,(s_1, s_2))$$

$$+ \int_0^{y_1} \int_0^{y_2} \frac{\partial f}{\partial y_1} (y_1 - s_1, y_2 - s_2)\, dG\,(s_1, s_2)],$$

$$\frac{\partial K}{\partial y_2} = k_2 + a\,[- p_2 \int_{y_2}^{\infty} (\int_{s_1 = 0}^{\infty} dG\,(s_1, s_2)$$

$$+ \int_0^{y_2} \frac{\partial f}{\partial y_2} (0, y_2 - s_2) \quad (\int_{s_1 = y_1}^{\infty} dG\,(s_1, s_2))$$

$$+ \int_0^{y_1} \int_0^{y_2} \frac{\partial f}{\partial y_2} (y_1 - s_1, y_2 - s_2)\, dG\,(s_1, s_2)],$$

Furthermore, as above, if $y_1 > x_1, y_2 > x_2$, we have

$$\frac{\partial f}{\partial x_1} = - k_1, \frac{\partial f}{\partial x_2} = - k_2 \tag{3}$$

Consequently, if we assume that the solution here has the same form as in the one-dimensional case, the critical levels $\bar{x}_1$ and $\bar{x}_2$ are given as roots of the equations

(4) a. $k_1 + a\,[- p_1 \int_{\bar{x}_1}^{\infty} (\int_{s_2 = 0}^{\infty} dG\,(s_1, s_2)) - k_1 \int_0^{\bar{x}_1} (\int_{s_2 = 0}^{\infty} dG(s_1, s_2))] = 0$

b. $k_2 + a\,[- p_2 \int_{\bar{x}_2}^{\infty} (\int_{s_1 = 0}^{\infty} dG\,(s_1, s_2)) - k_2 \int_0^{\bar{x}_2} (\int_{s_1 = 0}^{\infty} dG(s_1, s_2))] = 0$

These roots exist and are unique provided we make the same assumptions as above, namely, $ap_1 > k_1$, $ap_2 > k_2$, and $dG > 0$.

We see that $\bar{x}_1$ depends for its determination only upon the conditional distribution $\int_{s_2 = 0}^{\infty} dG\,(s_1, s_2)$, and similarly to determine $\bar{x}_2$ we require only $\int_{s_1 = 0}^{\infty} dG\,(s_1, s_2)$.

This is the important property of *suboptimalization* mentioned above.

The verification of the solution follows precisely the same lines as that for the one-dimensional case, and hence will be omitted, since the details are, of course, much more tedious.

Let us state our conclusion as

THEOREM 2. *Let us impose the following conditions upon the equation in* (1):

(5) a. *k_i and p_i are positive constants,*

b. $\varphi > 0, \int_0^{\infty} \int_0^{\infty} \varphi\, ds_1\, ds_2 = 1, \int_0^{\infty} \int_0^{\infty} s_i\, \varphi\, ds_1\, ds_2 < \infty$

c. $0 < a < 1$,

d. $ap_i > k_i$,

Let $\bar{x}_i$ be the unique root of

$$(6) \quad k_i = ap_i \int_y^\infty \left(\int_{s_i=0}^\infty \varphi(s_1, s_2)\, ds_2 \right) ds_1 - ak_i \int_0^y \left(\int_{s_i=0}^\infty \varphi(s_1, s_2)\, ds_2 \right) ds$$

Then the optimal policy has the form

$$(7) \quad \begin{aligned} &\text{a. } \textit{for } 0 \leq x_i \leq \bar{x}_i,\ y_i = \bar{x}_i \\ &\text{b. } \textit{for } x_i \geq \bar{x}_i,\ y_i = x_i \end{aligned}$$

In other words, the optimal stock level for the ith item is $\bar{x}_i$.

If $ap_i \leq k_i$ for any i, we set $x_i = 0$ $(i = 1, 2)$.

It is clear that this form of the solution extends immediately to the N-dimensional case.

§ 7. Finite time period

Let us now consider the corresponding problem for a finite process where we do not discount future costs. We now wish to minimize the total expected cost.

We define

(1) $f_N(x)$ = expected cost over an N-stage period starting with an initial quantity x and using an optimal N-stage policy.

Then

$$(2) \quad f_1(x) = \operatorname*{Min}_{y \geq x} \left[k(y-x) + p \int_y^\infty (s-y)\, \varphi(s)\, ds \right]$$

$$f_{n+1}(x) = \operatorname*{Min}_{y \geq x} \left[k(y-x) + p \int_y^\infty (s-y)\, \varphi(s)\, ds + f_n(0) \int_y^\infty \varphi(s)\, ds + \int_0^y f_n(y-s)\, \varphi(s)\, ds \right],\ n = 1, 2, \ldots$$

We wish to prove, under the natural assumption $p > k$,

THEOREM 3. *For each n, the optimal policy has the form*

$$(3) \quad \begin{aligned} &\text{a. } \textit{for } x \leq \bar{x}_n,\ y = \bar{x}_n, \\ &\text{b. } \textit{for } x \geq \bar{x}_n,\ y = x \end{aligned}$$

where the sequence $\bar{x}_n$ is monotone increasing in n.

PROOF. The proof will be inductive. We have, with $f_1(x)$ defined as in (2), as our critical stock level the solution of

$$(4) \quad k = p \int_y^\infty \varphi(s)\, ds,$$

which, if it exists, is unique, and which does exist if $p > k$, as is reasonable. Call this value $\bar{x}_1$. It is clear then that for $n = 1$, the optimal policy is $y = \bar{x}_1$ for $x \leq \bar{x}_1$; $y = x$ for $x \geq \bar{x}_1$. When $y = \bar{x}_1$ we have $f'_1(x) = -k$, and for $x \geq \bar{x}_1$, we have

$$f_1(x) = p \int_x^\infty (s - x) \varphi(s)\, ds, \tag{5}$$

$$f'_1(x) = -p \int_x^\infty \varphi(s)\, ds \geq -k,$$

$$f''_1(x) = p\,\varphi(x) > 0.$$

Hence $f'_1(x) + k \geq 0$ for all $x \geq 0$.

Consider the case $n = 2$. We have

$$f_2(x) = \operatorname*{Min}_{y \geq x} \Big[k(y - x) + p \int_y^\infty (s - y) \varphi(s)\, ds + f_1(0) \int_y^\infty \varphi(s)\, ds + \int_0^y f_1(y - s) \varphi(s)\, ds\Big]. \tag{6}$$

The critical value of y is attained by setting the partial derivative with respect to y equal to zero, or

$$k = p \int_y^\infty \varphi(s)\, ds - \int_0^y f_1'(y - s) \varphi(s)\, ds = F_1(y). \tag{7}$$

The function $F_1(y)$ has the derivative

$$F'_1(y) = -p\,\varphi(y) - f_1'(0)\,\varphi(y) - \int_0^y f_1''(y - s) \varphi(s)\, ds. \tag{8}$$

Since $f_1'' > 0$, $p + f_1'(0) > k + f_1'(0) = 0$, we see that $F_1(y)$ is monotone decreasing, and there can be at most one root of (7). However, $F_1(0) = p > k$, $F_1(\infty) = 0$. Hence there is precisely one root. Call this root $\bar{x}_2$.

The policy is then

$$\begin{aligned} y &= \bar{x}_2, & 0 \leq x \leq \bar{x}_2, \\ y &= x, & \bar{x}_2 \leq x. \end{aligned} \tag{9}$$

The geometric picture is illuminating. Write (6) in the form

$$f_2(x) + kx = \operatorname*{Min}_{y \geq x} v(y), \tag{10}$$

where $v(y)$ is a known function. From what we have demonstrated above, $v(y)$ has the following graph

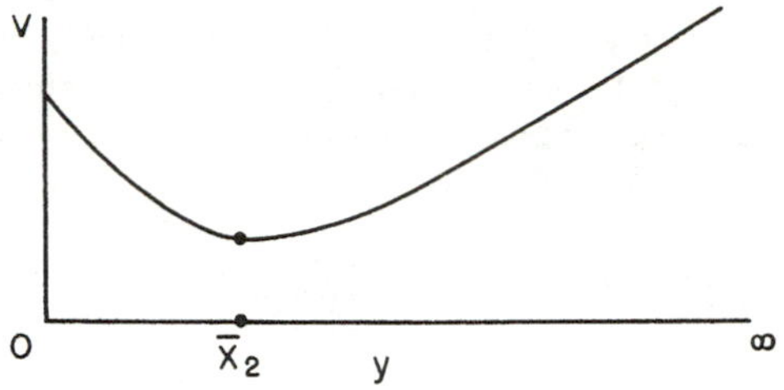

Figure 1

The function $f_2(x) + kx$ is obtained by drawing the tangent to $v(y)$ at $y = \bar{x}_2$ and continuing it to the left until it hits the v-axis. The function $f_2(x) + kx$ is now constant for $0 \leq x \leq \bar{x}_2$ and equal to $v(x)$ for $x \geq \bar{x}_2$.

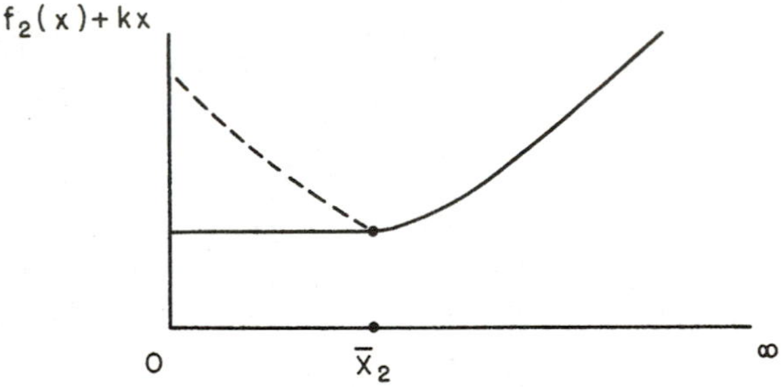

Figure 2

It remains to show that $\bar{x}_2 > \bar{x}_1$. The quantity $\bar{x}_1$ is determined by equation (4), while $\bar{x}_2$ is determined by (7). Since $-f_1' \geq 0$, it follows that the curve

$$\text{(11)} \qquad w = g_2(y) = p \int_y^\infty \varphi(s)\, ds - \int_0^y f_1'(y-s)\, \varphi(s)\, ds$$

always lies above the curve

$$\text{(12)} \qquad w = g_1(y) = p \int_y^\infty \varphi(s)\, ds,$$

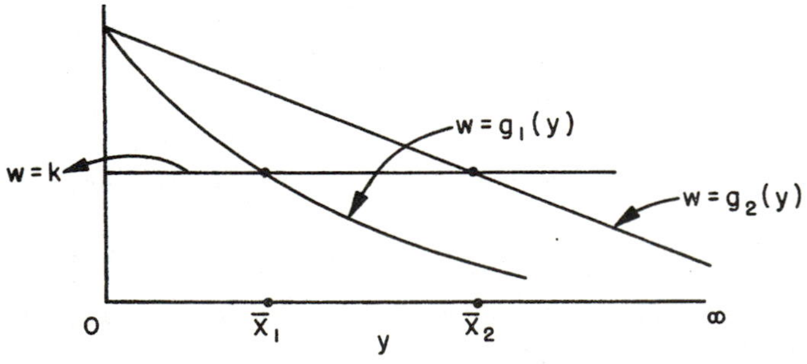

Figure 3

From this it is clear the $\bar{x}_2 > \bar{x}_1$.
In order to continue this proof inductively, we must show that

$$(13) \qquad -f_2'(x) \geq -f_1'(x).$$

We have

$$(14) \qquad -f_1'(x) = k, \; 0 \leq x \leq \bar{x}_1$$

$$-f_1'(x) = p \int_x^\infty \varphi(s)\, ds, \; x \geq \bar{x}_1$$

and

$$(15) \qquad -f_2'(x) = k, \; 0 \leq x \leq \bar{x}_2$$

$$= p \int_x^\infty \varphi(s)\, ds - \int_0^x f_1'(x-s)\, \varphi(s)\, ds, \; x \geq \bar{x}_2.$$

In the intervals $[0, \bar{x}_1]$ and $[\bar{x}_2, \infty]$, the inequality is clear. In $[\bar{x}_1, \bar{x}_2]$, the inequality follows from the monotonicity of $k - p \int_x^\infty \varphi(s)\, ds$, which is zero at $x = \bar{x}_1$.

Finally, we wish to demonstrate the convexity of $f_2(x)$. This is clearly true in $[0, \bar{x}_2]$. In $[\bar{x}_2, \infty]$, we have, using (15)

$$(16) \qquad f_2''(x) = p\,\varphi(x) + f_1'(0)\,\varphi(x) + \int_0^x f_1''(x-s)\,\varphi(s)\, ds.$$

Since $f_1'(0) + p > 0$, $f_1'' \geq 0$, we have $f_2''(x) > 0$.
We now have all the ingredients of an inductive proof.

§ 8. Finite time—multi-dimensional case

The hardy reader may verify that the solution in the multidimensional case has precisely the same general character.

§ 9. Non-proportional penalty cost—red tape

As soon as we consider the case where the penalty is not directly proportional to the excess of demand over supply, we encounter difficulties, and it appears that the simple and elegant solution obtained for the case of proportional cost is no longer valid generally.

There are, however, a number of interesting cases in which we still obtain a solution involving constant stock level. The most interesting of these occur when we take the cost of ordering $(s - y)$ to be $p(s - y) + q$, where q is a fixed administrative cost which appears whenever an excess demand occurs, regardless of the amount of the demand. The initial ordering cost is still assumed to be proportional.

Let us then consider the equation

$$(1)\quad f(x) = \operatorname*{Min}_{y \geq x} [k(y-x) + a[\int_y^\infty [p(s-y)+q]\varphi(s)\,ds + f(0)\int_y^\infty \varphi(s)\,ds + \int_0^y f(y-s)\,\varphi(s)\,ds]],$$

distinguished from the equation we have considered above only by the additional term $aq\int_y^\infty \varphi(s)\,ds$. It is surprising how much complication this innocuous-appearing expression would seem to introduce.

We shall, to begin with, proceed formally on the assumption that there is a constant stock level solution. The critical level is then determined by the solution of

$$(2)\quad 0 = k + a[-p\int_y^\infty \varphi(s)\,ds - q\,\varphi(y) + \int_0^y f'(y-s)\,\varphi(s)\,ds],$$

and we have $f'(x) = -k$ when $y > x$.

It follows then that $\bar{x}$ will be a root of

$$(3)\qquad 0 = k + a[-p\int_y^\infty \varphi(s)\,ds - q\,\varphi(y) - k\int_0^y \varphi(s)\,ds]$$

Unfortunately, it is not true that this equation has a unique root for all density functions $\varphi(s)$. This equation may be written in the form

$$(4)\qquad (1-a)\,k = a(p-k)\int_y^\infty \varphi(s)\,ds + aq\,\varphi(y).$$

A simple condition under which this equation has a unique root is $\varphi'(y) \leq 0$.

If we do assume that this equation has a unique root, the proof is almost exactly as before. There is, however, a more general result where the optimal policy is that of constant stock level, which we shall now discuss.

If the equation above, (3) or (4), does not possess a unique root, it may still happen that the largest root of (4) corresponds to an absolute minimum of the function in the brackets in (1), over the interval $[0, \bar{x}]$.

Thus the picture may be

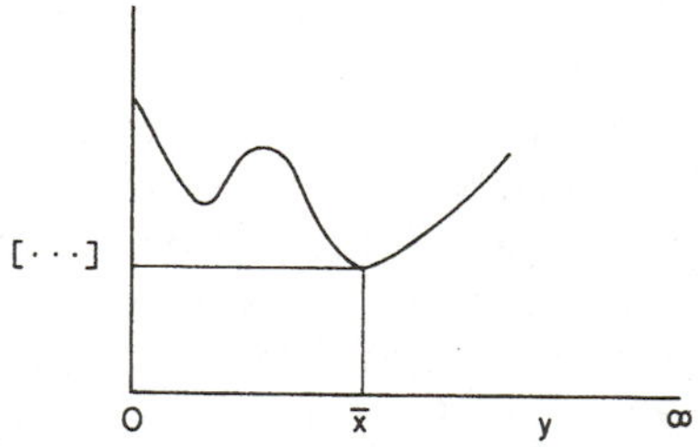

Figure 4

Let us prove

THEOREM 4. *Under the assumptions upon a, k, p, q and $\varphi(s)$, stated in Theorem 1, and the additional assumption that the last minimum of*

$$(6)\quad \psi(y) = ky + a\left[\int_y^\infty [p(s-y)+q]\varphi(s)\,ds - k\int_0^y (y-s)\varphi(s)\,ds\right.$$

is the absolute minimum in $0 \leq y \leq \infty$, the optimal policy in (1) is given by the rule

$$(7)\qquad \text{(a)}\quad y = \bar{x},\ \textit{for}\ 0 \leq x \leq \bar{x}, \qquad y = x,\ \textit{for}\ x \geq \bar{x},$$

where $\bar{x}$ is the value of y where the absolute minimum is attained.

PROOF. Let $\bar{x}$ be the value of y which yields the last minimum, and the absolute minimum in the interval $[0, \infty]$, of the function $\psi(y)$ above. Then, precisely, as in the case where $q = 0$, we have $f(x) = f(0) - kx$ in $0 \leq x \leq \bar{x}$, and $f(0)$ is determined by substituting this result in (1), in the range $0 \leq x \leq \bar{x}$. In the interval $[x, \infty]$, $f(x)$ is determined by setting $y = x$ in (1).

The proof that $f(x)$ actually satisfies the equation now continues in exactly the same way as in the case where $q = 0$.

§ 10. Particular cases

Some particular cases where the above conditions are satisfied are

$$(1)\qquad \text{(a)}\quad \varphi(x) = e^{-(x-a)^2} \Big/ \int_{-a}^\infty e^{-u^2}\,du$$

$$\text{(b)}\quad \varphi(x) = be^{-bx}$$

We leave the verification as exercises for the reader.

§ 11. The form of the general solution

Let $f(x)$ be the solution of (9.1), which is to say

$$(1)\qquad f(x) + kx = \operatorname*{Min}_{y \geq x} F(y),$$

where

$$(2)\quad F(y) = ky + a\left[p\int_y^\infty (s-y)\varphi(s)\,ds + (f(0)+q)\int_y^\infty \varphi(s)\,ds + \int_0^y f(y-s)\varphi(s)\,ds\right]$$

Let $F(y)$ have the graph

(3)

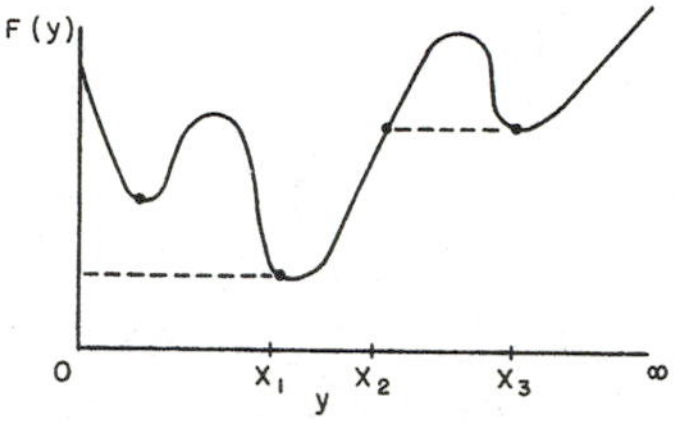

Figure 5

Then, the optimal policy has the following form

(4)
$$\begin{aligned}
&\text{(a)} \quad y = x_1, \ 0 \leq x \leq x_1 \\
&\text{(b)} \quad y = x, \ x_1 \leq x \leq x_2 \\
&\text{(c)} \quad y = x_3, \ x_2 < x < x_3, \\
&\text{(d)} \quad y = x, \ x \geq x_3,
\end{aligned}$$

and

(5)
$$\begin{aligned}
f(x) + kx &= F(x_1), \ 0 \leq x \leq x_1 \\
&= F(x), \ x_1 \leq x \leq x_2 \\
&= F(x_3), \ x_2 \leq x \leq x_3 \\
&= F(x), \ x_3 \leq x < \infty
\end{aligned}$$

However, the problem of determining how many different regions exist, given the cost functions and the demand functions, and how to fit this information together, seems quite difficult, and is unsolved at the present time.

§ 12. Fixed costs

Let us now consider the case where there is a constant red-tape cost in initial ordering. This problem is also unsolved to date.

The equation now has the form

(1)
$$f(x) = \underset{y \geq x}{\text{Min}} \left[k(y - x) + g(y - x) + a\left[\int_y^\infty p(s - y)\,\varphi(s)\,ds \right.\right.$$
$$\left.\left. + f(0)\int_y^\infty \varphi(s)\,ds + \int_0^y f(y - s)\,\varphi(s)\,ds\right]\right],$$

where

(2) $$g(x) = g, \; x > 0$$
$$= 0, \; x = 0.$$

Here g represents the fixed cost.

It is tempting to envisage a solution of the following form

(3) $$y = S \text{ for } 0 \leq x \leq s$$
$$= x \text{ for } s < x$$

where $0 < s < S < \infty$. A policy of this type is called an "sS-policy."

Policies of this type are used in various establishments, and have a fine intuitive flavor. Unfortunately, it is easy to construct relatively simple examples which show that this policy cannot be optimal in all cases, and there the matter rests.

§ 13. Preliminaries to a discussion of more complicated policies

In the previous sections we have considered some processes having solutions of quite simple and intuitive form. We now wish to consider two cases in which the solutions are of a more complicated nature. The first of these will be one involving a time-lag in the fulfilling of orders, the second will treat the case where the initial ordering function is a non-linear convex function of the amount ordered, with no red-tape cost in either case.

In both cases we shall employ the method of successive approximations to determine the properties of the solution.

§ 14. Unbounded process—one period time lag

The functional equation we shall consider is that derived in § 2, namely

(1) $$f(x) = \operatorname*{Min}_{z \geq 0} \Big[kz + a \Big[\int_x^\infty p(s - x)\, \varphi(s)\, ds + f(z) \int_x^\infty \varphi(s)\, ds$$
$$+ \int_0^x f(x - s + z)\, \varphi(s)\, ds \Big]\Big]$$

We shall prove

THEOREM 5. *The optimal policy is given by the rule*

(2) $$z = z(x) \textit{ for } 0 \leq x \leq \bar{x}$$
$$z = 0, \textit{ for } \bar{x} \leq x,$$

where $z(\bar{x}) \geq 0$ *and* $z(\bar{x}) = 0$.

This function $z(x)$ is monotone decreasing in x.

PROOF. The proof will proceed by induction, based upon the following sequence of successive approximations

(3) $$f_0(x) = a\left[\int_x^\infty p(s-x)\varphi(s)\,ds + f_0(0)\int_x^\infty \varphi(s)\,ds + \int_0^x f_0(x-s)\varphi(s)\,ds\right],$$

(a function we have repeatedly encountered before), and for $n = 0, 1, 2, \ldots,$

(4) $$f_{n+1}(x) = \min_{z \geq 0} T(z, x, f_n),$$

where $T(z, x, f_n)$ is the expression contained within the brackets in (1).

Let us now consider $T(z, x, f_0)$ as a function of z, say $M_1(z)$. We have

(5) $$M_1'(z) = k + a f_0'(z)\int_x^\infty \varphi(s)\,ds + a\int_0^x f_0'(x-s+z)\varphi(s)\,ds,$$

and the second derivative is

(6) $$M_1''(z) = af_0''(z)\int_x^\infty \varphi(s)\,ds + a\int_0^x f_0''(x-s+z)\varphi(s)\,ds.$$

Since $f_0'' > 0$, we see that $M_1''(z) > 0$ for all $x \geq 0$. Hence the equation $M_1'(z) = 0$ has at most one root in z for any x. For large x, it is clear that there will be no root, and for small x, say $x = 0$, there will be a root provided that a, p and k are properly related, a point we will check subsequently. Meanwhile, let us show that this root, which we call $z_1(x)$ is monotone decreasing in x.

To show this consider the expression $G_0(x, z) = -af_0'(z)\int_x^\infty \varphi(s)\,ds - a\int_0^x f_0'(x-s+z)\varphi(s)\,ds$, as a function of x for fixed z. Its derivative with respect to x is

(7) $$\frac{\partial G_0}{\partial x} = -a\int_0^x f_0''(x-s+z)\varphi(s)\,ds,$$

which is negative. Hence, the family of curves $w = G_0(x, z)$ looks as follows

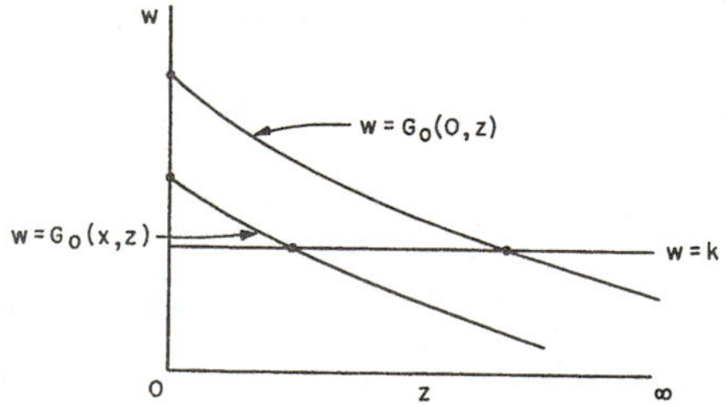

Figure 6

This graph very clearly shows that $z_1(x)$ is monotone decreasing in x, and equal to zero for $x \geq \bar{x}_1$.

In order to obtain similar results for the second approximation, we must show that $f_1''(x) \geq 0$. We have

$$(8) \qquad \begin{aligned} f_1(x) &= T(z_1(x), x, f_0), && 0 \leq x \leq \bar{x}_1, \\ &= T(0, x, f_0), && \bar{x}_1 \leq x. \end{aligned}$$

In $[0, \bar{x}_1]$, we have

$$(9) \qquad f_1'(x) = -ap \int_x^\infty \varphi(s)\,ds + a \int_0^x f_0'(x - s + z)\,\varphi(s)\,ds,$$

and

$$(10) \qquad \begin{aligned} f_1''(x) &= ap\,\varphi(x) + af_0'(z)\,\varphi(x) \\ &+ \left(a \int_0^x f_0''(x - s + z)\,\varphi(s)\,ds\right)(1 + dz_1/dx). \end{aligned}$$

From (9), we see that $f_0'(0) = -ap$. Since $f_0'(x)$ is monotone increasing in x it follows that $ap + af_0'(z) > 0$ for $z \geq 0$. Hence we will have $f_1''(x) > 0$ if we show that $1 + dz_1/dx \geq 0$.

To do this we return to the equation defining z_1, namely $M_1'(z) = 0$. Using the expression in (5), we see that

$$(11) \qquad \begin{aligned} &\left[af_0''(z) \cdot \int_x^\infty \varphi(s)\,ds + a \int_0^x f_0''(x - s + z)\,\varphi(s)\,ds\right] dz/dx \\ &+ a \int_0^x f_0''(x - s + z)\,\varphi(s)\,ds = 0, \end{aligned}$$

which shows anew that $dz_1/dx \leq 0$ and that $1 + dz_1/dx \geq 0$.

We require finally a relation connecting $f_0'(x)$ and $f_1'(x)$. We have

$$(12) \qquad f_0'(x) = -ap \int_x^\infty \varphi(s)\,ds + a \int_0^x f_0'(x - s)\,\varphi(s)\,ds.$$

Hence in $[0, \bar{x}]$ we see that $f_0'(x) \leq f_1'(x)$, since $f_0'(x)$ is monotone increasing in x. Since $f_1(x) = f_0(x)$ for $x \geq \bar{x}$, we have

$$(13) \qquad f_0'(x) \leq f_1'(x)$$

for all $x \geq 0$.

Continuing as in the preceding pages, we see that we obtain a function $z_n(x)$ for each n having the property that

$$(14) \qquad \begin{aligned} &\text{(a)} \quad z_n(x) > 0 \text{ for } 0 \leq x < \bar{x}_n \\ &\text{(b)} \quad z_n(x) = 0 \text{ for } \bar{x}_n \leq x, \end{aligned}$$

and $z_n(x)$ monotone decreasing in x. Furthermore the sequence x_n will be monotone decreasing and possess a limit $\bar{x}$.

It remains to show that $\bar{x} = 0$ if a, p and k are suitably related. This is equivalent to checking as to whether or not $f_0(x)$ is the solution. Returning to (5), we set $x = 0$ and examine the equation

$$(15) \qquad k + af_0'(z) = 0$$

If $k + af_0'(0) < 0$, there will be a solution of this equation. Turning to (12), we see that $f_0'(0) = -ap$. Hence we require

$$(16) \qquad k < a^2 p$$

the intuitive and expected condition for a process involving a one-stage delay.

§ 15. Convex cost function—unbounded process

As another illustration of the power of the method of successive approximations, let us consider the case where the cost of ordering, $g(y-x)$, is a strictly convex function of the amount, $y-x$, ordered. The equation is now

$$(1) \quad f(x) = \operatorname*{Min}_{y \geq x} [g(y-x) + a[\int_y^\infty p(s-y)\varphi(s)\,ds + f(0)\int_y^\infty \varphi(s)\,ds + \int_0^y f(y-s)\varphi(s)\,ds]].$$

As usual, we set

$$(2) \quad f_0(x) = a[\int_x^\infty p(s-x)\varphi(s)\,ds + f_0(0)\int_x^\infty \varphi(s)\,ds + \int_0^x f_0(x-s)\varphi(s)\,ds],$$

and, for $n = 1, 2, \ldots,$

$$(3) \qquad f_{n+1}(x) = \operatorname*{Min}_{y \geq x} T(y, x, f_n).$$

Let us begin with the consideration of $f_1(x)$, assuming that $g(x)$ possesses a continuous derivative for $x \geq 0$.

If $y > x$, y is determined by the equation

$$(4) \qquad g'(y-x) = a[p\int_y^\infty \varphi(s)\,ds - \int_0^y f_0'(y-s)\varphi(s)\,ds].$$

Since we have assumed that $g(x)$ is convex, i.e. $g''(x) > 0$, it follows that this equation can have at most one root, since the left side is monotone increasing and the right-side monotone decreasing.

For $x = 0$, there is a root provided that

$$(5) \qquad g'(0) < ap.$$

For x large, there is no root if $g'(0) > 0$.

If $y > x$, we have

$$f_1'(x) = -g'(y-x), \tag{6}$$

and

$$f_1''(x) = -g''(y-x)(dy/dx - 1). \tag{7}$$

To determine the magnitude of $dy/dx - 1$, we turn to (4). This yields

$$g''(y-x)(dy/dx-1) = [-ap\varphi(y) - af_0'(0)\varphi(y) - (a\int_0^y f_0''(y-s)\varphi(s)\,ds)]\,dy/dx. \tag{8}$$

From this we readily conclude that $dy/dx > 0$ and $dy/dx - 1 < 0$. Hence $f_1''(x) > 0$.

Furthermore, we see that $-f_1' \geq -f_0'$. We now have all the details of an inductive proof of

THEOREM 6. *There is a function $y(x)$ and a number $\bar{x}$ with the properties*

(9) (a) $y(x) \geq x$, *$y(x)$ is monotone increasing*

(b) $y(x) > x$, *for* $x \leq \bar{x}$, $y(x) = x$, $x \geq \bar{x}$

(c) $\bar{x} > 0$ *if* $ap > g'(0)$

This function $y(x)$ is the optimal policy in (1).

Appendix Chapter V—The Renewal Equation

The equation

$$u(x) = f(x) + \int_0^x u(x-s)\varphi(s)\,ds, \tag{1}$$

which occurs in a great many different areas of analysis, is commonly called the *renewal equation.*

There are two important methods available for establishing properties of the solutions, the method of the Laplace transform, and the Liouville-Neumann method of solution — which is successive approximations.

The Laplace transform technique owes its success to the fact that $\int_0^x u(x-s)\varphi(s)\,ds$ is a convolution having the formal property that

$$\int_0^\infty e^{-tx}\left[\int_0^x u(x-s)\varphi(s)\,ds\right]dx = \left(\int_0^\infty e^{-tx}u(x)\,dx\right)\left(\int_0^\infty e^{-st}\varphi(s)\,ds\right). \tag{2}$$

Hence (1) yields, proceeding completely formally,

$$(3) \qquad \int_0^\infty e^{-tx} u(x)\, dx = \int_0^\infty e^{-tx} f(x)\, dx \,/\, (1 - \int_0^\infty e^{-tx} \varphi(x)\, dx),$$

from which a great deal can be deduced concerning the asymptotic behavior of $u(x)$ as $x \to \infty$, using either Tauberian theorems or complex variable theory, under appropriate assumptions concerning f and φ.

However, the properties of most interest to us here, positivity, convexity, et al, can most readily be deduced by considering the sequence of approximants

$$(4) \qquad u_0 = f(x)$$

$$u_{n+1} = f(x) + \int_0^x u_n(x-s)\, \varphi(s)\, ds,$$

and showing that each function $u_n(x)$ has the required property.

This approach is justified by the following result.

THEOREM 9. *Let us assume that*

(5) a. $f(x)$ *is bounded in every finite interval* $[0, x_0]$

b. $\int_0^\infty |\varphi(s)|\, ds < 1.$

Then there is a unique solution to (1) *which is bounded in any interval* $[0, x_0]$.

This solution may be obtained as the limit of the sequence given by (4). *If* $f(x)$ *is differentiable and* $\varphi(x)$ *is continuous, we have*

$$(6) \qquad u'(x) = f'(x) + u(0)\, \varphi(x) + \int_0^x u'(x-s)\, \varphi(s)\, ds.$$

If $f(x) \geq 0$, $\varphi(x) \geq 0$, *then* $u(x) \geq 0$.

There are a number of other combinations of conditions corresponding to those given in (5a) and (5b) which also yield existence and uniqueness.

The proof of Theorem 9 is readily obtained following the techniques we have by now applied many times over.

Exercises and research problems

1. Obtain the analogue of Theorem 3 for the case where the distribution function of demand varies from stage to stage.

2. Consider the case where there are fixed costs in both initial purchasing and the penalty cost and the distribution of demand has the form $\varphi(x) = 1/k$, $0 \leq x \leq k$, $\varphi(x) = 0$, $x > k$.

3. Consider the process with a fixed cost in the case where there are only two levels of demand, high and low. Can one generalize the result obtained here to the case of an arbitrary finite number of different demands?

4. Obtain the analogues of Theorems 1, 2, and 3 in the case where there is a storage cost at each stage proportional to the quantity of items stored over the previous stage.

5. Obtain the functional equations corresponding to the process in which both the demands and times of demand are random. Consider the cases where the times of demand have a continuous distribution and a discrete distribution.

6. Obtain the analogue of Theorem 5 for processes with arbitrary time lags.

7. Consider the case where there is fixed cost and determine
   a. The "constant stock level" policy which minimizes expected cost
   b. The "sS"-policy which minimizes expected cost.

8. We are interested in producing a single item over a given number of time periods in order to satisfy known future demands. We wish to do this in such a way as to minimize costs, knowing the costs for production, storage, and change in production rate as functions of time.

Let us consider the discrete version first. Let

T = the number of periods,
r_t = demand at time t,
x_t = amount produced in time interval $[t - 1, t]$,
x_0 = given initial production,
y_t = $x_{t+1} - x_t \geq 0$, the increase in production rate at time t,
u_t = excess of supply over demand at time t.

The costs are

c_i = cost of producing an item in the period $[i - 1, i]$,
d_i = cost of storing an item in excess of demand for one period,
e_i = cost of increasing production rate by one unit per unit of time.

Assume that we wish to minimize the total cost of the T-period process under the condition that the supply must always exceed the demand.

9. Consider the above problem under the condition that production cannot be expanded in an arbitrary fashion. In particular, discuss the two cases

a. $x_t \leq x_{t+1} \leq a x_t,\ 1 < a < \infty$

b. $x_t \leq x_{t+1} \leq x_t + b,\ b > 0.$

10. Consider the case where the demand is stochastic, under the following two alternative assumptions

a. Demand must always be satisfied

b. Demand can be postponed one stage

11. Obtain the functional equations corresponding to the process described in § 2 under the assumption that we desire to minimize the probability that the cost exceeds a given quantity c.

12. Consider the functional equations discussed in the chapter under the assumption that the distribution function $\varphi(s)\,ds$ is replaced by the more general Stieltjes distribution $dG(s)$. Obtain the requisite existence and uniqueness theorems and determine in which ways the theorems established above must be modified in order to remain valid.

13. In what ways is the problem of ordering for a military supply depot different from the problem of ordering items for a department store?

14. Assume that there is no penalty for not being able to meet the demand, but that there is a return of b dollars for each item demanded and supplied. Suppose that this return can be used to increase the quantity available at the next stage. Given an initial stock of x, and a supply of money equal to y, how should one order so as to maximize the total expected return? Consider both finite and infinite processes under the assumption of proportional costs.

15. Consider the equation

$$f(x) = \max_{0 \leq y \leq x} \left[g(y) + h(x - y) + \int_0^y f(y - s)\, k(s)\, ds \right]$$

where

(a) $g(0) = h(0) = 0$

(b) $g'(y) > 0,\ h'(y) > 0,\ g'(0) < h'(0).$

(c) $k(s) \geq 0$

(d) $g''(y) > 0,\ h''(y) > 0$

(e) $h(y) - g(y)$ is monotone increasing in y.

Show that the solution is given by

$$f = h(x),\ 0 \leq x \leq \bar{x}$$
$$= g(x) + \int_0^x f(x-s)\,k(s)\,ds,\ x \geq \bar{x}$$

where x is determined as the non-zero root of

$$h(x) = g(x) + \int_0^x h(x-s)\,k(s)\,ds.$$

16. Consider a situation where one must order items to be sold in anticipation of an uncertain demand which can be taken as a known stochastic variable. Let equal ordering periods be indexed 0, 1, 2, . . ., and the demand be described by a distribution

$F_i(x)$ = probability that the demand is less than or equal to x in period i.

Let p be the unit selling price and $C(y)$, assumed differentiable, be the total ordering cost for y units in any period; I be the inventory at the beginning of the present period (period 0); and suppose that all units ordered at the beginning of the period are immediately available—units may be ordered only once during a period and cannot be disposed of except through sales at price p on demand.

Given any ordering policy, u_i, at each stage there will be a cash return of

$$P_i(x_i, y_i, I_i) = p \min(I_i + y_i, x_i) - C(y_i).$$

Let the purpose of the process be to maximize the expected value of

$$\left[\sum_{i=0}^{\infty} a^i P_i(x_i, y_i, I_i)\right],\ 0 < a < 1.$$

Show that the resultant system of recurrence relations is

$$f_k(I) = \max_{y \geq 0} \left[\int_0^{\infty} [p \min(I+y, x) - C(y) + af_{k+1}(\max(0, I+y-x)]\,dF_k(x)\right],$$

and solve in the case where $C(y) = cy$. (Harlan D. Mills)

17. Consider the equation

$$f(x) = \min_{z \geq 0} \left[kz + a\int_x^{\infty} p(s-x)\,\varphi(s)\,ds + af(z)\int_x^{\infty} \varphi(s)\,ds + a\int_0^x f(x-s+z)\,\varphi(s)\,ds\right],$$

corresponding to a one period lag in supply.

Assuming that the optimal policy is to choose z so that $x + z = L$, for $0 \leq x \leq L$, and $z = 0$ for $x > L$, determine L.

18. Prove or disprove that this is the optimal policy.

19. Examine the conjecture that in the general k period lag case, the optimal policy is to order nothing if the sum of the quantities on order and on hand exceed a certain quantity L, and to order a quantity equal to the difference if L exceeds this sum.

Bibliography and Comments for Chapter V

§ 1. The mathematical model of the inventory problem we consider here originated in the pioneer paper of K. D. Arrow, T. E. Harris and J. Marschak, "Optimal Inventory Policy," *Econometrica*, July, 1951. Stimulated by their investigations, two further papers appeared A. Dvoretsky, J. Kiefer and J. Wolfowitz, "The Inventory Problem I, II," *Econometrica*, vol. 20 (1952), pp. 187–222.

The first of these papers is devoted to existence and uniqueness of the solution of the basic functional equation, and to a discussion of some particular processes. The second paper is more statistical in nature and devoted to the question of determining the distribution of functions of demand as the process continues.

The results of this chapter were obtained in conjunction with I. Glicksberg and O. Gross, R. Bellman, I. Glicksberg and O. Gross, "On the Optimal Inventory Equation," *Management Science*, vol. 2 (1955), pp. 83–104.

Since the appearance of these papers, a large number of papers, both published and privately circulated, have appeared on the topic of inventory control. We suggest that the interested reader thumb through the pages of *Econometrica*, *Jour. Soc. Ind. Appl. Math.*, *Jour. Operations Research Society*, *Management Science*, and *Naval Quarterly Jour. of Logistics*, where he will find further results and references.

§ 3. The results discussed here are in accordance with the remark of an earlier chapter that the derivatives of the return functions, or "marginal returns" possess a simpler structure that the return functions in many cases.

Appendix. Further results concerning renewal equations and functions of similar type may be found in W. Feller, "On the Integral Equation of Renewal Theory," *Ann. Math. Stat.*, vol. 12 (1941), R. Bellman, (with the collaboration of J. M. Danskin), "A Survey of the Theory of Time-Lag, Retarded Control and Hereditary Processes," RAND Corporation, 1954, R–271.

CHAPTER VI

Bottleneck Problems in Multi-Stage Production Processes

§ 1. Introduction

In this chapter we shall discuss a particular class of significant and difficult variational problems arising from the study of multi-stage production processes.

We shall first formulate a discrete version of the process, which under certain assumptions of proportionality of output to input leads us to the problem of determining the maximum of a linear form subject to linear constraints, a basic problem to which the theory of linear programming has made notable contributions in recent years. Although the state of analytic research on this fundamental problem is still in its early stages, a large class of problems arising in applications can be successfully resolved numerically, with the aid of modern computing machines and various iterative techniques such as the "simplex" technique.

The study of bottleneck processes, however, which combine a moderate number of activities at each stage with a large number of stages, encounters the usual difficulty of dimensionality if conventional computational methods are used. As in the treatment of the processes of the previous chapters, we can circumvent this obstacle to some degree by using a formulation in terms of functional equations. Since, however, we are interested in explicit analytic solutions, in order to study the character of optimal policies, we shall formulate continuous versions of processes. It is worth emphasizing that the continuous process may actually be closer to reality than the discrete version in many cases. An essential weapon in our mathematical armory is the use of the dual continuous process, thus exploiting the linearity of the process.

To illustrate the method, we shall treat a simple process in detail, in this chapter, while a more complicated process will be discussed in the subsequent chapter. In many cases, these analytic methods, applied with faith and resolution, permit us to obtain explicit analytic solutions of the maximization problem, together with an explicit description of the optimal policies. Many difficulties, however, remain as far as the construction of a general theory is concerned. Examining the following pages, the reader will quickly see that the mathematical theory of these problems is in its rudimentary stages.

The variational problem is that of determining the maximum over the vector function $z(t)$ of the inner product $(x(T), a)$, where x and z are connected by the vector-matrix differential equation

$$dx/dt = Ax + Bz, \; x(0) = c, \tag{1}$$

and z satisfies the constraint

$$Cz \leq Dx, \tag{2}$$

for $0 \leq t \leq T$.

The techniques we employ to discuss this problem will be further developed and applied to classical problems in the calculus of variations in Chapter 9.

§ 2. A General class of multi-stage production problems

A central problem in the theory and application of mathematical economics is that of integrating a complex of industries, of similar or variegated type, so as to produce a given product in a most efficient manner. Here the criterion of efficiency may be minimum time, or maximum profit, or some combination of both.

As an example, which is quite elementary from the economic point of view, but sufficiently advanced from the mathematical viewpoint to generate problems which we cannot resolve as readily as we would desire, let us consider a simple model of a three-industry production process where the individual industries are the "auto" industry, the "steel" industry, and the "tool" industry.[1]

In this highly condensed or "lumped" model of economic interplay [2] we shall assume that the state of each industry is completely specified at any time by its *stockpile* of raw material and by its *capacity* to produce new quantities using these raw materials. Furthermore, we shall begin by assuming that it is sufficient to consider that changes in these basic quantities, stockpile and capacity, occur only at discrete times $t = 0, 1, 2, \ldots, T$.

[1] Needless to say, these names are used merely to guide our intuition. It is not suggested that any deep significance be attached to them.

[2] This type of lumping is precisely analogous to what is done in the study of electric circuits in the low frequency case, where we introduce the concepts of "resistance", "inductance" and "capacitance".

Let us then define the following state variables:

(1)
$$\begin{aligned} x_1(t) &= \text{number of autos produced up to time } t, \\ x_2(t) &= \text{capacity of auto factories at time } t, \\ x_3(t) &= \text{stockpile of steel at time } t, \\ x_4(t) &= \text{capacity of steel mills at time } t, \\ x_5(t) &= \text{stockpile of tools at time } t, \\ x_6(t) &= \text{capacity of tool factories at time } t, \end{aligned}$$

We make the following assumptions concerning the interdependence of these three industries:

(2) a. An increase in auto, steel or tool capacity requires only steel and tools;
b. Production of autos requires only auto capacity and steel;
c. Production of steel requires only steel capacity;
d. Production of tools requires only tool capacity and steel.

The dynamics of the production process may be described as follows: At the beginning of each unit time period, say t to $t + 1$, we allocate various quantities of steel and tools, taken from their respective stockpiles, for the purposes of producing autos, steel, and tools — which is to say increasing the stockpiles of these quantities—and for the purposes of increasing the auto, steel, and tool capacities.

Let, for $i = 1, 2, \ldots,$

(3) a. $z_i(t)$ = amount of steel allocated at time t for the purpose of increasing $x_i(t)$,
b. $w_i(t)$ = amount of tools allocated at time t for the purpose of increasing $x_i(t)$.

Upon referring to the assumptions in (2) we see that

(4)
$$\begin{aligned} &\text{a.} \quad z_3(t) = 0 \\ &\text{b.} \quad w_1(t) = w_3(t) = w_5(t) = 0 \end{aligned}$$

In order to obtain relations connecting $x_i(t + 1)$ with $x_i(t)$, $z_i(t)$ and $w_i(t)$, we must make some further assumptions concerning the relations between output and input. The simplest assumption to make is that we have a linear production process with output of an item always directly proportional to the *minimum input* of required resources.[3] Thus, produc-

[3] As we have observed in the preface, this may not actually be the simplest for mathematical purposes. A more realistic assumption predicated upon a law of diminishing returns, involving nonlinear functions, may actually lead to a simpler mathematical problem. The reason for this is that nonlinear functions take more kindly to a variational approach. On the other hand, linear problems may be more readily treated numerically, in some cases.

tion is directly proportional to capacity whenever there is an abundance of raw materials, i.e., stockpile, and directly proportional to the minimum quantity of raw materials whenever there is an abundance of capacity.

It is because of this dependence upon the minimum resource that we use the name "bottleneck problems."

As an illustration, the increase in the number of autos from t to $t + 1$ will depend upon the capacity of auto factories at t, $x_1(t)$, and the quantity $z_1(t)$ of steel, as defined above in (3a). Since production depends upon the minimum of capacity and supply of raw material, we obtain the equation

$$x_1(t+1) = x_1(t) + \text{Min}\,(\gamma_1 x_2(t), \alpha_1 z_1(t))\,, \tag{5}$$

where γ_1 and α_1 are taken to be known positive constants.

In a similar fashion, combining the assumptions in with those of the previous paragraph, we obtain the following equations which relate $x_i(t+1)$ to $x_i(t)$, $z_i(t)$, and $w_i(t)$:[4]

$$\begin{aligned}
(6)\quad x_1(t+1) &= x_1(t) + \text{Min}\,(\gamma_1 x_2(t), \alpha_1 z_1(t))\\
x_2(t+1) &= x_2(t) + \text{Min}\,(\alpha_2 z_2(t), \beta_2 w_2(t))\\
x_3(t+1) &= x_3(t) - z_1(t) - z_2(t) - z_4(t) - z_5(t) - z_6(t) + \gamma_2 x_4(t)\\
x_4(t+1) &= x_4(t) + \text{Min}\,(\alpha_4 z_4(t), \beta_4 w_4(t))\\
x_5(t+1) &= x_5(t) - w_2(t) - w_4(t) - w_6(t) + \text{Min}\,[\gamma_5 x_6(t), \alpha_5 z_5(t)]\\
x_6(t+1) &= x_6(t) + \text{Min}\,(\alpha_6 z_6(t), \beta_6 w_6(t)),
\end{aligned}$$

where α_i, β_i, and γ_i are constants.

The constraints upon z_i and w_i are obviously

$$\begin{aligned}
(7)\qquad &\text{(a)}\quad z_i, w_i \geq 0\\
&\text{(b)}\quad z_1 + z_2 + z_4 + z_5 + z_6 \leq x_3\\
&\text{(c)}\quad w_2 + w_4 + w_6 \leq x_5
\end{aligned}$$

together with the "common-sense" constraints

$$\begin{aligned}
(8)\qquad &\text{(a)}\quad a_1 z_1 \leq \gamma_1 x_2\\
&\text{(c)}\quad a_2 z_2 = \beta_2 w_2\\
&\text{(b)}\quad a_4 z_4 = \beta_4 w_4\\
&\text{(d)}\quad a_5 z_5 \leq \gamma_5 x_6\\
&\text{(e)}\quad a_6 z_6 = \beta_6 w_6
\end{aligned}$$

[4] All these equations are conservation equations which state that the quantity of an item at time $t + 1$ is the quantity at time t, minus the quantity used over $[t, t+1]$, plus the quantity produced over $[t, t+1]$.

The meaning of these equations is that there is no advantage to any allocation beyond the capacity of production, and again that the minimum resource determines the production level.

By means of these additional constraints we may eliminate the variables w_i completely, obtaining in place of (4.6) the system of equations:

$$
\begin{aligned}
x_1(t+1) &= x_1(t) + \alpha_1 z_1(t),\ x_1(0) = c_1,\\
x_2(t+1) &= x_2(t) + \alpha_2 z_2(t),\ x_2(0) = c_2,\\
x_3(t+1) &= x_3(t) - z_1(t) - z_2(t) - z_4(t) - z_5(t) - z_6(t)\\
&\qquad + \gamma_2 x_4(t),\ x_3(0) = c_3,\\
x_4(t+1) &= x_4(t) + \alpha_4 z_4(t),\ x_4(0) = c_4,\\
x_5(t+1) &= x_5(t) - \varepsilon_2 z_2(t) - \varepsilon_4 z_4(t) - \varepsilon_6 z_6(t) + \alpha_5 z_5(t),\\
&\qquad \varepsilon_i = \alpha_i/\beta_i,\ x_5(0) = c_5,\\
x_6(t+1) &= x_6(t) + \alpha_6 z_6(t),\ x_6(0) = c_6.
\end{aligned} \tag{9}
$$

The constraints, in turn, have the form, for each t:

$$
\begin{aligned}
&\text{(a)} \quad z_i \geq 0\\
&\text{(b)} \quad z_1 + z_2 + z_4 + z_5 + z_6 \leq x_3\\
&\text{(c)} \quad \gamma_2 z_2 + \gamma_4 z_4 + \gamma_6 z_6 \leq x_5\\
&\text{(d)} \quad z_1 \leq f_2 x_2\\
&\text{(e)} \quad z_5 \leq f_6 x_6.
\end{aligned} \tag{10}
$$

We must now choose the $z_i(t)$ for $t = 0, 1, 2, \ldots, T-1$, subject to the above constraints, so as to maximize $x_1(T)$.

§ 3. Discussion of the preceding model

It is easy to see that $x_1(T)$, the total number of autos produced over the time period $[0, T]$, may be written as a linear expression in the quantities $z_i(t)$, $t = 0, 1, 2, \ldots, T-1$, $i = 1, 2, \ldots, 6$. The problem of maximizing $x_1(T)$ subject to the linear constraints of (2.10) is consequently within the domain of linear programming. It may be solved computationally for explicit values of the coefficients and the time T, by iterative processes of various types, provided that T is not too large. In particular, for dynamic processes of the kind considered here, a number of important simplifications are possible.

However, in general, in analyses of the type presented here, we are not so much interested in the numerical solution corresponding to any particular set of constants as we are in the complete set of numerical values obtained from a range of parameter values. In other words, in most cases the whole interest of the investigation lies in a "sensitivity analysis," or equivalently a "stability analysis," of the solution.

This sensitivity analysis is required because of the many assumptions we have made such as linearity of output, the crude description of industries in terms of lumped capacities and stockpiles we have employed, the absence of time lag or "lead time" in production, and so on. Any conclusions concerning the structure of optimal policies that may be drawn from the simplified mathematical model can have validity only if these conclusions are relatively insensitive to the precise values of various parameters which occur.

It is clear from what we have said above that the numerical work involved in performing any reliable sensitivity analysis using purely computational techniques, involving as it does a probing of many-dimensional space, will be tedious, time consuming, and inevitably incomplete.

The question arises then as to whether or not it is possible to determine the intrinsic structure of an optimal policy, regardless of any numerical values we may subsequently assign to the parameters. This knowledge is not only of importance in itself, in allowing us to make a complete sensitivity analysis of the solution, but is also extremely helpful in determining approximate solutions in cases where explicit analysis seems hopeless, and in furnishing analytic clues to the solution of more complicated processes.

As a first step towards obtaining the solution, both analytically and computationally, we shall reformulate the problem in terms of functional equations.

§ 4. Functional equations

It is clear that the total output of cars obtained using an optimal allocation policy is a function only of the initial resources, $c_1, \ldots, c_6$, and the duration of the process, T. Furthermore, c_1 need not be explicitly mentioned.

Let us then define for $T = 1, 2, \ldots$

(1) $f(c_2, c_3, \ldots, c_6; T) =$ The total output obtained over a time interval T starting with initial resources c_i, $i = 2, 3, \ldots, 6$, and employing an optimal policy

Employing the principle of optimality, we obtain the following functional equation for $f(c_2, c_3, \ldots, c_6; T)$:

$$(2) \quad f(c_2, c_3, \ldots, c_6; T+1) = \operatorname*{Max}_{z} [\alpha_1 z_1 + f(c_2', c_3', \ldots, c_6'; T)],$$

where

$$
\begin{aligned}
c_2' &= c_2 + a_2 z_2 \\
c_3' &= c_3 - z_1 - z_2 - z_4 - z_5 - z_6 + \gamma_2 c_4 \\
c_4' &= c_4 + a_4 z_4 \qquad\qquad (3) \\
c_5' &= c_5 - \varepsilon_2 z_2 - \varepsilon_4 z_4 - \varepsilon_6 z_6 + a_5 z_5 \\
c_6' &= c_6 + a_6 z_6,
\end{aligned}
$$

and Z denotes the region in the $(z_1, z_2, z_4, z_5, z_6)$-space defined by the following inequalities

$$
(4) \quad
\begin{aligned}
&\text{(a)} \quad z_i \geq 0 \\
&\text{(b)} \quad z_1 + z_2 + z_4 + z_5 + z_6 \leq c_3 \\
&\text{(c)} \quad \gamma_2 z_2 + \gamma_4 z_4 + \gamma_6 z_6 \leq c_5 \\
&\text{(d)} \quad z_1 \leq f_2 c_2 \\
&\text{(e)} \quad z_5 \leq f_6 c_6
\end{aligned}
$$

The analytic problem of determining f, and, more importantly, the nature of the optimal policy is still one of great difficulty. The computational problem is also formidable involving as it does the tabulation of a function of five variables for each value of T. The homogeneity of the process enables us to reduce this to a problem involving four variables. We shall refer to this fact again in following sections.

The computational problem involved in determining the maximum over the region Z, a polyhedral region bounded by planes, may be greatly simplified by observing that the maximum occurs at vertices.

§ 5. A Continuous version [5]

To simplify the analytic problem, we shall transform the discrete process into a continuous process. In so doing, our purpose is to avail ourselves of the combination of the powerful methods of calculus, together with the resources of linear algebra. It is very often true, in dealing with the physical world, that continuous models are far simpler to discuss than discrete models.

To obtain a continuous version, we assume that decisions are made at times $0, \Delta t, 2\Delta t$, and so on, and that the allocations $z_i(t)$, $w_i(t)$ previously made over the time interval $[t, t+1]$ are replaced by allocations $z_i(t)\Delta t$, $w_i(t)\Delta t$ over the interval $[t, t+\Delta t]$. The quantities $z_i(t)$ and $w_i(t)$ are now *rates of allocation* of resources.

Turning to the equations in (2.9) describing the discrete process and allowing Δt to approach zero, the new equations take the form

[5] Chapter VIII is devoted to a similar continuous version of the discrete process of Chapter II.

$$(1)\quad \begin{aligned}
&\dot{x}_1(t) = a_1 z_1(t),\ x_1(0) = c_1,\\
&\dot{x}_2(t) = a_2 z_2(t),\ x_2(0) = c_2,\\
&\dot{x}_3(t) = -z_1(t) - z_3(t) - z_4(t) - z_5(t) - z_6(t) + \gamma_2 x_4(t),\\
&\qquad\qquad\qquad\qquad\qquad\qquad\qquad\qquad x_3(0) = c_3,\\
&\dot{x}_4(t) = a_4 z_4(t),\ x_4(0) = c_4,\\
&\dot{x}_5(t) = -\varepsilon_2 z_2(t) - \varepsilon_4 z_4(t) - \varepsilon_6 z_6(t) + a_5 z_5(t),\ x_5(0) = c_5,\\
&\dot{x}_6(t) = a_6 z_6(t),\ x_6(0) = c_6.
\end{aligned}$$

($\dot{}$ signifies differentiation with respect to t).

The constraints upon the z_i are now

$$(2)\quad \begin{aligned}
&(a)\quad z_i \geq 0\\
&(b)\quad z_1 + z_2 + z_4 + z_5 + z_6 \leq \infty\\
&(c)\quad \gamma_2 z_2 + \gamma_4 z_4 + \gamma_6 z_6 \leq \infty\\
&(d)\quad z_1 \leq f_2 x_2\\
&(e)\quad z_5 \leq f_6 x_6
\end{aligned}$$

This means that the constraints of (2b) and (2c) disappear. Two conditions which were automatically satisfied before must now be added. These are the conditions that the stockpiles be non-negative at all times,

$$(3)\quad \begin{aligned}
&(b')\ x_3 \geq 0\\
&(c')\ x_5 \geq 0,
\end{aligned}$$

From these constraints we see that whenever $x_3 = 0$ we must have

$$(4)\quad z_1 + z_3 + z_4 + z_5 + z_6 \leq \gamma_2 x_4$$

and similarly when $x_5 = 0$ we must have

$$(5)\quad \varepsilon_2 z_2 + \varepsilon_4 z_4 + \varepsilon_6 z_6 \leq a_5 z_5$$

It follows that z_2, z_3, z_4, and z_6 are unbounded whenever x_3 and x_5 are positive. This means that delta-function type solutions may occur. This point will be discussed in more detail in the subsequent chapter where an example involving this type of solution is discussed. However, a rigorous discussion of this feature of a solution will be postponed until the second volume. We shall proceed essentially in a formal manner in this and the following chapter at various points where a rigorous discussion would take us too far afield.[6]

[6] It is important to point out that the continuous process is *described* by the above equations. A detailed discussion of this point is given in Chapter VIII, where we also discuss the connection between the discrete and continuous processes.

The problem is now to maximize $x_1(T)$ subject to the above constraints.

After some discussion of notation, we shall approach this problem using the functional equation approach of dynamic programming.

§ 6. Notation

Let us introduce vector-matrix notation which will greatly simplify the notation and thus be of considerable help in presenting the general theoretical approach, unclouded by a superabundance of superscripts. Following the discussion of the basic concepts, we shall consider a particular example, to illustrate the analytic minutiae, which are not trivial.

Let $x(t)$, $z(t)$, and c denote respectively the n-dimensional column vectors

$$x(t) = \begin{pmatrix} x_1(t) \\ x_2(t) \\ \cdot \\ \cdot \\ \cdot \\ x_n(t) \end{pmatrix}, \quad z(t) = \begin{pmatrix} z_1(t) \\ z_2(t) \\ \cdot \\ \cdot \\ \cdot \\ z_n(t) \end{pmatrix}, \quad c = \begin{pmatrix} c_1 \\ c_2 \\ \cdot \\ \cdot \\ \cdot \\ c_n \end{pmatrix}$$

and A_i, B_j, for such values of i and j as occur, denote $n \times m$ matrices.

We shall be dealing only with vectors x and z whose components are non-negative. To indicate this fact we use the notation $x \geq 0$ to denote the relations $x_i \geq 0$, $i = 1, 2, \ldots, n$. The inequality $x \geq y$ is equivalent to $x - y \geq 0$.

Turning to the equation in (5.1), we see that it may be written

(1) $$dx/dt = A_1 x + A_2 z, \; x(0) = c$$

where A_1 and A_2 are matrices determined by the coefficients in (5.1). Similarly, the constraints in (5.2)—(5.5) take the form

(2) $$\begin{aligned} &z \geq 0 \\ &B_1 z \leq B_2 x \end{aligned}$$

The problem of maximizing $x_1(T)$ is a particular case of the problem of maximizing a linear combination, $\sum_{i=1}^{n} c_i x_i(T)$. To express this in simple form, we introduce the inner product of two vectors x and y, namely

(3) $$(x, y) = \sum_{i=1}^{n} x_i y_i$$

The general problem is then that of choosing $z(t)$ so as to maximize $(x(T), a)$ where a is a given vector, subject to the relations given above in (1) and (2).

One of the difficulties that arises in the continuous case, and not in the discrete process, is that this maximum may not exist if we restrict $z(t)$ to be a function in the usual sense. We shall proceed on the assumption that the constraints in (6.2) are sufficient to ensure the existence of a maximizing function. This will be the case if (6.2) has the form $z \leq B_3 x$, where B_3 is a positive matrix. A complete treatment will require the use of Stieltjes integrals.

§ 7. Dynamic programming formulation

Since the forms of the equations in (6.2) are time-independent, it follows that $\operatorname*{Max}_z (x(T), a)$ (where we shall assume throughout the remainder of this expository chapter that the maximum actually exists) is a function only of T and the components of c, which is to say, of the initial stockpiles and capacities, the state variables and the duration of the process.

Let us then write

$$(1) \qquad \operatorname*{Max}_z (x(T), a) = f(c, T) \equiv f(c_1, c_2, \ldots, c_n; T).$$

§ 8. The basic functional equation

We shall now derive a functional equation for f using the Principle of Optimality [7], which in this case states that the nature of any optimal allocation policy over the interval $[0, T]$, which is to say, one which yields the maximum of $(x(T), a)$, is such that its continuation over any final sub-interval $[S, T]$ must be an optimal policy for a process of duration $T - S$ starting from the initial state $c(S)$.

Here $c(S)$ is the vector $x(S)$ obtained from (6.1) using an allocation policy over $[0, S]$.

The mathematical transliteration of the verbal principle yields the functional equation

$$(1) \qquad f(c, S + T) = f(c(S), T)$$

for an optimal policy over $[0, S + T]$.

It follows that the policy over $[0, S]$ is determined by the equation

$$(2) \qquad f(c, S + T) = \operatorname*{Max}_{[0, S]} f(c(S), T),$$

where we maximize over all feasible policies over $[0, S]$, that is to say, over all $z(t)$ satisfying the constraints.

Equation (2), together with the initial condition $f(c, 0) = (c, a)$, is the basic functional equation governing the process.

[7] Chapter III, § 3.

§ 9. The resultant nonlinear partial differential equation

Let us now use the basic equation in (8.2) to derive a partial differential equation for f, on the assumption that f and x possess the requisite differentiability properties. As we shall see below, it is quite permissible to proceed formally at this point since we shall derive a technique for verifying the validity of any proposed solution.

Let us take S to be an infinitesimal. Then we have

$$\text{(1)}\quad \begin{aligned} &\text{(a)}\quad f(c, S+T) = f(c, T) + S f_T + o(S), \\ &\text{(b)}\quad c(S) = c + S[A_1 c + A_2 z(0)] + o(S), \\ &\text{(c)}\quad f(c(S), T) = f(c + S[A_1 c + A_2 z(0)], T) + o(S) \\ &\qquad\qquad = f(c, T) + S\left((A_1 c + A_2 z(0), \frac{\partial f}{\partial c}\right) + o(S), \end{aligned}$$

where $\partial f/\partial c$ denotes the vector

$$\text{(2)}\qquad \frac{\partial f}{\partial c} = \begin{pmatrix} \dfrac{\partial f}{\partial c_1} \\ \dfrac{\partial f}{\partial c_2} \\ \cdot \\ \cdot \\ \cdot \\ \dfrac{\partial f}{\partial c_N} \end{pmatrix}$$

As S shrinks to 0, the maximum over the interval $[0, S]$ shrinks to a maximum at $S = 0$, or a maximum over $z(0)$, under our assumptions of continuity. With reference to the expansions in (1) above, we see that the infinitesimal analogue of (1) is the nonlinear partial differential equation

$$\text{(3)}\qquad \partial f/\partial T = \operatorname*{Max}_{z(0)} \left[\left(A_1 c + A_2 z(0), \frac{\partial f}{\partial c}\right)\right]$$

where $z(0)$ is constrained by the equations in (6.2).

§ 10. Application of the partial differential equation

The importance of the equation in (9.3) resides in the fact that it permits us to determine the solution over $[0, T + \Delta T]$ if the solution has already been determined over $[0, T]$ for *all* initial states.

It turns out to be true that in many of these problems the difficulties are readily resolved for small T, since for processes of short duration, the obvious, crude, policies are optimal. It follows then that we have, in theory, a systematic means of continuing the solution up to any desired value of T. Although systematic, the details are by no means trivial, as we shall see in the next chapter.

In the next section, we shall go through the analysis involved in resolving a relatively simple problem. Much of the analysis can be discarded, once we have ascertained the structure of the solution, which in many cases is plausible on economic grounds.

§ 11. A Particular example

As an application of the general approach presented above, let us now consider the problem of maximizing $x_2(T)$, where

$$\text{(1)} \qquad \begin{aligned} dx_1/dt &= a_1 z_1,\ x_1(0) = c_1, \\ dx_2/dt &= a_2 z_2 - z_1,\ x_2(0) = c_2, \end{aligned}$$

and z_1, z_2, the rates of allocation, as functions of t are subject to the following constraints:

$$\text{(2)} \qquad \begin{aligned} &\text{(a)} \quad z_1, z_2 \geq 0, \\ &\text{(b)} \quad z_1 + z_2 \leq x_2, \\ &\text{(c)} \quad z_2 \leq x_1, \\ &\text{(d)} \quad x_2 \geq 0. \end{aligned}$$

for $0 \leq t \leq T$.

In this case, the rates z_1 and z_2 are uniformly bounded, and it is easy to see, using either a direct weak convergence argument, or relying upon classical theorems in the calculus of variations, that the maximum is assumed. Hence we may set in rigorous fashion,

$$\text{(3)} \qquad f(c_1, c_2, T) = \underset{[0, T]}{\text{Max}}\ x_2(T).$$

As in the general case, f satisfies the functional equation

$$\text{(4)} \qquad f(c_1, c_2, S + T) = \underset{[0, S]}{\text{Max}}\ f(x_1(S), x_2(S), T),$$

which, in the limit as $S \to 0$, yields the partial differential equation

$$\text{(5)} \qquad \frac{\partial f}{\partial T} = \underset{z(0)}{\text{Max}} \left[a_1 z_1 \frac{\partial f}{\partial c_1} + (a_2 z_2 - z_1) \frac{\partial f}{\partial c_2} \right].$$

which, at the moment, we recall, is purely formal, since we do not know whether or not f has the requisite continuity properties.

The maximum is taken over the region defined by

(6) $$\begin{aligned}&\text{(a)}\quad 0 \leq z_1, z_2,\\ &\text{(b)}\quad z_1 + z_2 \leq c_2,\\ &\text{(c)}\quad z_2 \qquad \leq c_1,\end{aligned}$$

with the additional constraint

(7) $$a_2 z_2 - z_1 \geq 0,$$

if $x_2 = 0$. The variables are now $z_1 = z_1(0)$, $z_2 = z_2(0)$.

Let us now sketch the analytic procedure that will yield a solution. We begin with the most complicated case, that where $c_2 < c_1$. For a process of short duration, the solution is trivial. We have

(8) $$\begin{aligned}&z_1 = 0,\ z_2 = x_2,\\ &f = c_2 e^{a_2 T}.\end{aligned}$$

This policy is pursued until a "bottleneck" develops, which is to say, c_2 exceeds c_1. Using the optimal policy described in (8) we see that this situation will occur as soon as T exceeds $T_1 = \log(c_1/c_2)/a_2$.

To obtain the solution for $T > T_1$, we rewrite (5) in the form

(9) $$\frac{\partial f}{\partial T} = \operatorname*{Max}_{z(0)} \left[z_1 \left(a_1 \frac{\partial f}{\partial c_1} - \frac{\partial f}{\partial c_2} \right) + a_2 \frac{\partial f}{\partial c_2} z_2 \right].$$

The location of the maximizing point $(z_1(0), z_2(0))$ will depend upon the sign and magnitude of the coefficients of z_1 and z_2. For $T < T_1$ we have

(10) $$a_1 \frac{\partial f}{\partial c_1} - \frac{\partial f}{\partial c_2} = -e^{a_2 T}, \qquad a_2 \frac{\partial f}{\partial c_2} = a_2 e^{a_2 T}.$$

Using our assumption concerning the continuity of $\partial f/\partial c_1$, $\partial f/dc_2$, we suspect that the solution for T slightly greater than T_1 will have the form

(11) $$\begin{aligned}&\text{(a)}\quad z_1 = 0,\ z_2 = x_2 \text{ for } 0 \leq S \leq T_1\\ &\text{(b)}\quad z_1 = 0,\ z_2 = x_1 \text{ for } T_1 < S \leq T.\end{aligned}$$

Applying this policy, f takes the form

(12) $$f = c_1 + (T - T_1) a_2 c_1,$$

where T_1 is as above. In order to determine how long this policy endures when $T > T_1$, we consider the process as starting from $S = T_1$. In terms of $c_1' = c_1(T_1)$, $c_2' = c_2(T_1)$, f has the form

(13) $$f = c_2' + a_2 c_1' (T - T_1)$$

The equation which replaces (9) has precisely the same form with c_1, c_2 replaced by c_1', c_2', namely

$$(14) \qquad \frac{\partial f}{\partial T} = \operatorname*{Max}_{z(T_1)} \left[z_1 \left(a_1 \frac{\partial f}{\partial c_1'} - \frac{\partial f}{\partial c_2'} \right) + a_2 \frac{\partial f}{\partial c_2'} z_2 \right].$$

We have, using (13),

$$(15) \qquad a_1 \frac{\partial f}{\partial c_1'} - \frac{\partial f}{\partial c_2'} = a_1 a_2 (T - T_1) - 1,$$

$$a_2 \frac{\partial f}{\partial c_2'} = a_2 .$$

The coefficient of z_1 is negative for $T < T^* = T_1 + 1/a_1 a_2$, 0 *at* T^*, and positive thereafter.

It follows that the new policy given by (11) remains optimal for $T_1 \leq T \leq T^*$.

Furthermore, since $T^* - T_1$ is independent of c_1 and c_2, we see that we know the form of the optimal policy over a tail interval.

It remains to determine what the policy is in the middle of the interval $[0, T]$ in the general case when T exceeds T^*. We suspect from an examination of the vertices in the figure below that it has the form

$$(16) \qquad z_1 = x_2 - x_1, \; z_2 = x_1 .$$

It is instructive to consider the region determined by the constraints in (6) when $c_2 > c_1$

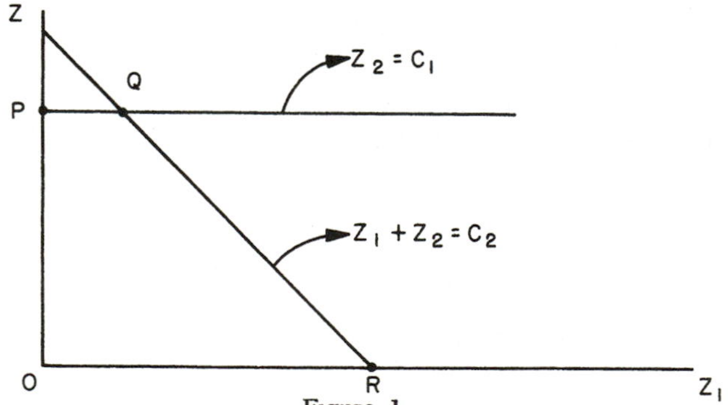

Figure 1

When maximizing over z, the three crucial points are the vertices P, Q and R, where $P = P(0, c_1)$, $Q = (c_2 - c_1, c_1)$, $R = (c_2, 0)$. It is the principle of continuity which leads us to choose Q as the maximizing vertex as soon as c_2 surpasses c_1.

Instead of verifying this directly, which may be done, we shall describe in the next section a more elegant technique which exploits the linearity of the process. This technique serves not only as a means of verifying proposed solutions, but also as a theoretical tool for the determination of the nature of optimal policies.

§ 12. A Dual problem

Let us, to illustrate the principles we shall employ, take our basic equation to have the form

(1) $$dx/dt = Az, \; z(0) = c,$$

with constraints of the form

(2) $$\begin{aligned} &(a) \quad z \geq 0 \\ &(b) \quad Bz \leq x \\ &(c) \quad x \geq 0. \end{aligned}$$

Note that the equation in (6.1) may always be written in the form of (1), if $A_1 \geq 0$, by first writing it in the form

(3) $$dx/dt = A_1 w + A_2 z, \; x(0) = c,$$

with the constraints

(4) $$\begin{aligned} &(a) \quad z \geq 0 \\ &(b) \quad Bz \leq x \\ &(c) \quad w \leq x. \end{aligned}$$

and then combining the vectors w and z into one. However, an equation of the type appearing in (6.1) may also be treated directly by these methods.

Since $x = c + \int_0^t Azdt$, the constraint of (2b) may be written

(5) $$Bz + \int_0^t Czdt \leq c, \quad (C = -A)$$

The problem of maximizing $(x(T), a)$ is equivalent to that of maximizing $\int_0^T (Az, a)\, dt = \int_0^T (z, a')\, dt$, where $a' = A'a$. Here A' denotes the transpose of A.

Beginning all over again, we start with the problem of maximizing $J = \int_0^T (z, a')\, dt$ over all z satisfying the constraints

$$\text{(6)} \qquad \text{(a)} \quad z \geq 0$$

$$\text{(b)} \quad Bz + \int_0^t Czdt_1 \leq c$$

Let $w\ (t)$ be a non-negative vector of the same dimension as c. Then by virtue of (6b) we have

$$\text{(7)} \qquad \int_0^T (w, Bz + \int_0^t Czdt_1)\ dt \leq \int_0^T (w, c)\ dt$$

Let, as above, B' denote the transpose of B. Then, as is easily seen, $(Bz, w) = (z, B'\ w)$. Integration by parts yields, for any constant matrix C,

$$\text{(8)} \qquad \int_0^T (w, \int_0^t Czdt_1)\ dt = \int_0^T (\int_t^T C'wdt_1, z)\ dt$$

Combining these two results, we have

$$\text{(9)} \qquad \int_0^T (w, Bz + \int_0^t Czdt_1)\ dt = \int_0^T (B'w + \int_t^T C'wdt_1, z)\ dt$$

Let us now assume that it is possible to find a vector $w = w\ (t)$ which is non-negative and satisfies the inequality

$$\text{(10)} \qquad B'w + \int_t^T C'wdt_1 \geq \alpha'$$

We then have the chain of equalities and inequalities:

$$\text{(11)} \qquad \int_0^T (w, c)\ dt \geq \int_0^T (w, Bz + \int_0^t Czdt_1)\ dt$$

$$= \int_0^T (B'w + \int_t^T C'wdt, z)\ dt \geq \int_0^T (\alpha', z)\ dt$$

From this it is clear that

$$\text{(12)} \qquad \operatorname*{Inf}_w \int_0^T (w, c)\ dt \geq \operatorname*{Sup}_z \int_0^T (z, \alpha')\ dt$$

where the infimum and supremum are taken over all w and z satisfying the inequalities of (10) and (6b). If the minimum and maximum are assumed, the details are as above. If, however, the minimum and maximum are not assumed, then delta-functions will occur, which is to say, we must reformulate the problems in terms of Stieltjes integrals. A number of interesting and difficult problems arise in this way, which we shall not discuss here.[8]

If the two extremes in (11) are equal, we see that the following relations must hold [9]

$$w_i = 0 \text{ if } c_i > (Bz + \int_0^t Czdt)_i \tag{13}$$

$$z_j = 0 \text{ if } a_j' < (B'w + \int_t^T C'wdt)_j$$

The important fact which we now wish to establish is that, conversely, any pair of non-negative z and w satisfying (13) and the original constraints will furnish solutions to the maximum and minimum problems.

To demonstrate this, let us note that if (13) holds, all the relations in (11) are equalities. Assume now that $\bar{z}$ is another vector satisfying all the constraints and for which

$$\int_0^T (z, a')\,dt < \int_0^T (\bar{z}, a')\,dt \tag{14}$$

Then with the w associated with $\bar{z}$ we have

$$\begin{aligned} \int_0^T (\bar{z}, a')\,dt &\leq \int_0^T (\bar{z}, B'w + \int_t^T C'wdt_1)\,dt \\ &= \int_0^T (B\bar{z} + \int_0^t C\bar{z}dt_1, w)\,dt \leq \int_0^T (c, w)\,dt \\ &= \int_0^T (z, a')\,dt, \end{aligned} \tag{15}$$

a contradiction.

It follows then that we have a procedure for verifying a conjectured solution. Given z, we seek to determine w by means of (13). Having obtained w, we test to see whether or not w satisfies the given constraints. In the next section we shall carry through the details for the problem of § 11. This procedure will encounter difficulties if w is not uniquely determined by (13). In this case, various alternative solutions must be considered.

[8] In particular, we shall not discuss the connection with a min-max result in the theory of games, a result corresponding to known results for the discrete problem.

[9] Apart from sets of measure zero.

§ 13. Verification of the solution given in § 11

Applying the techniques described above, we find that the dual of the problem posed in § 11 is the problem of minimizing $\int_0^T (c_1 w_1 + c_2 w_2)\, dt$ over $w_1(t)$ and $w_2(t)$, where y and w are connected by the equations,

$$\text{(1)} \qquad \begin{aligned} dy_1/dt &= -a_1 w_1 + w_2, & y_1(T) &= -1, \\ dy_2/dt &= -a_2 w_2, & y_2(T) &= a_2\,^{10}, \end{aligned}$$

and the constraints have the form

$$\text{(2)} \qquad \begin{aligned} &\text{(a)} \quad w_1, w_2 \geq 0 \\ &\text{(b)} \quad w_1 + w_2 \geq y_2 \\ &\text{(c)} \quad w_2 \geq y_1 \end{aligned}$$

The equations of (12.13) are now:
If

$$\text{(3)} \qquad \begin{aligned} &\text{(a)} \quad z_2 < x_1, && \text{then } w_1 = 0 \\ &\text{(b)} \quad z_1 + z_2 < x_2, && \text{then } w_2 = 0 \\ &\text{(c)} \quad w_2 > y_1, && \text{then } z_1 = 0 \\ &\text{(d)} \quad w_1 + w_2 > y_2, && \text{then } z_2 = 0 \end{aligned}$$

We have omitted the conditions corresponding to $x_2 \geq 0$ since we suspect that the proposed optimal allocation policy automatically keeps $x_2 \geq 0$. This is actually the case.

We wish to verify that the policy which maximizes $x(T)$ is the following:

(4) (a) For $T - 1/a_1 a_2 < t \leq T$, $z_1(t) = 0$, $z_2 = \text{Min}(x_1, x_2)$

(b) For $0 \leq t \leq T - 1/a_1 a_2$, (1) if $x_2 \leq x_1$, $z_1 = 0$, $z_2 = x_2$
(2) if $x_2 \geq x_1$, $z_1 = x_2 - x_1$, $z_2 = x_1$

It is easily seen that this is a permissible policy in that $z_1 = x_2 - x_1$ is actually non-negative when z_1 and z_2 have the above values.

Having prescribed z, we can determine w using (3) and then test for consistency. There are two cases to consider, depending upon whether x_2 ever exceeds x_1 or not.

Let us assume then that $T \geq T_1$, in which case x_2 can exceed x_1 if appropriate policies are used.

[10] Observe that the dual process proceeds backwards in time.

Case I: $T - 1/a_1 a_2 < T_1 < T$. The solution is given by

(5) $$\text{for } t < T_1: z_1 = 0, z_2 = x_2$$
$$\text{for } t \geq T_1: z_1 = 0, z_2 = x_1$$

For $t < T_1$ these results yield, in conjunction with (2) and (3)

(6) $$\text{for } T_1 < T: w_1(t) = 0, w_2(t) = y_2(t)$$
$$\text{for } T_1 \geq T: w_2(t) = 0, w_1(t) = y_2(t)$$

For $t > T_1$ we obtain, using (1)

(7) $$y_2(t) = a_2, y_1(t) = -1 + a_1 a_2 (T - t) < 0$$

while for $t < T_1$, we have

(8) $$y_2(t) = a_2 e^{a_2(T_1 - t)} > 0$$
$$y_1(t) = a_1 a_2 (T - T_1) - e^{a_2(T_1 - t)} < 0$$

Hence, the inequalities $w_1, w_2 \geq 0$, $w_2 \geq y_1$, $w_1 \geq y_2$ are satisfied in their respective intervals.

Case II: $T_1 < T - 1/a_1 a_2$. This is the most interesting case. The vectors z and w are now determined as follows:

$$\text{for } T - 1/a_1 a_2 \leq t \leq T: \quad z_1 = 0, w_2 = 0$$
$$z_2 = x_1, w_1 = y_2$$

(9) $$\text{for } T_1 \leq t \leq T - 1/a_1 a_2: \quad z_1 = x_2 - x_1, w_2 = y_1$$
$$z_2 = x_1, w_1 = y_2 - y_1$$
$$\text{for } 0 \leq t \leq T_1: \quad z_1 = 0, w_1 = 0$$
$$z_2 = x_2, w_2 = y_2$$

For $T - 1/a_1 a_2 \leq t < T$ we have

(10) $$y_2(t) = a_2, y_1(t) = -1 + a_1 a_2 (T - t)$$

Hence, in this interval $y_1(t) \leq 0 = w_2$. Note that $y_1(T - 1/a_1 a_2) = 0$. In the range $T_1 < t \leq T - 1/a_1 a_2$, we have the equations

(11) $$dy_1/dt = -a_2 y_2 + (a_1 + 1) y_1$$
$$dy_2/dt = -a_2 y_2$$

Let us show that $y_1 \geq 0$ and $y_2 \geq y_1$ in this range. Starting from $t = T - 1/a_1 a_2$ where the inequalities are satisfied, let us reverse the time. The backward equations are

(12) $$dy_1/dt = a_2 y_2 - (1 + a_1) y_1$$
$$dy_2/dt = a_2 y_2$$

From this we obtain

$$d/dt\,(y_2 - y_1) = (1 + a_1)\,y_1 \tag{13}$$

Hence, if y_1 remains non-negative, we will have $y_2 - y_1 \geq 0$. It is clear that dy_1/dt starts out positive and stays positive as long as (y_1, y_2) remains above $a_2 y_2 - (1 + a_1) y_1 = 0$. If it hits the line we have $dy_1/dt = 0$, which means that y_1 has a maximum or a point of inflection. Both are excluded, since

$$\frac{d^2 y_1}{dt^2} = a_2 \frac{dy_2}{dt} - (1 + a_1) \frac{dy_1}{dt} = a_2^2 y_2 > 0 \tag{14}$$

This shows that w_1 and w_2 remain non-negative in this interval.

Finally, for $t < T_1$ we have

$$dy_1/dt = y_2,\ dy_1/dt = -a_2 y_2 \tag{15}$$

As t decreases, y_2 increases and y_1 decreases. Hence, $y_2 \geq y_1$ remains valid.

This completes the verification.

§ 14. Computational solution

The problem of maximizing x_N, where

$$\begin{aligned} x_{k+1} &= a_{11} x_k + a_{12} y_k + b_{11} z_k + b_{12} w_k,\ x_0 = c_1, \\ y_{k+1} &= a_{21} x_k + a_{22} y_k + b_{21} z_k + b_{22} w_k,\ y_0 = c_2, \end{aligned} \tag{1}$$

over sequences $\{z_k\}$ and $\{w_k\}$ subject to constraints of the form

$$d_{i1} z_k + d_{i2} w_k \leq d_{i3} x_k + d_{i4} y_k,\ i = 1, 2, \ldots, M, \tag{2}$$

may be reduced, as we know, to the computation of the sequence $\{f_k(c_1, c_2)\}$, $k = 1, 2, \ldots, N$, where

$$\begin{aligned} f_1(c_1, c_2) &= c_1, \\ f_{N+1}(c_1, c_2) &= \operatorname*{Max}_R \, [f_N(a_{11} c_1 + a_{12} c_2 + b_{11} z + b_{12} w, \\ &\qquad a_{21} c_1 + a_{22} c_2 + b_{21} z + b_{22} w)], \end{aligned} \tag{3}$$

where R is the region defined by

$$d_{i1} z + d_{i2} w \leq d_{i3} c_1 + d_{i4} c_2,\ i = 1, 2, \ldots, M. \tag{4}$$

Although it is not difficult to show that the maximum value is attained at a vertex of the region defined by (4), an exercise we recommend to the reader, which means that the maximization at each stage is trivial computationally, we are still faced with the problem of the tabulation of a sequence of functions of two variables. What seems to make the problem particularly onerous in this case is the fact that we have a possibly increasing grid in the (c_1, c_2) plane. In other words, if we wish to compute $f_N(c_1, c_2)$ in the region $0 \leq c_1 \leq \bar{c}_1$, $0 \leq c_2 \leq \bar{c}_2$, we may have to calculate f_{N-1} in a larger region, f_{N-2} in a still larger region and so on.

It is clear that whenever a situation of this type arises, we have a very costly and time-consuming computation.

Let us now show that we can simultaneously reduce the computation of the sequence $\{f_N(c_1, c_2)\}$ to the computation of two sequences of functions of one variable, and to the case where we have a fixed grid.

Our basic tool is the following homogeneity property of $f_N(c_1, c_2)$,

$$\begin{aligned} f_N(c_1, c_2) &= c_1 f_N(1, c_2/c_1), \\ &= c_2 f_N(c_1/c_2, 1), \end{aligned} \tag{5}$$

for $c_1, c_2 > 0$.

We may thus write (3) in the form

$$\begin{aligned} f_{N+1}(c_1, c_2) &= \operatorname*{Max}_R \left[(a_{11} c_1 + a_{12} c_2 + b_{11} z + b_{12} w) \right. \\ &\qquad \left. f_N\left(1, \frac{a_{21} c_1 + a_{22} c_2 + b_{21} z + b_{22} w}{a_{11} c_1 + a_{12} c_2 + b_{11} z + b_{12} w}\right)\right] \\ &= \operatorname*{Max}_R \left[(a_{21} c_1 + a_{22} c_2 + b_{21} z + b_{22} w) \right. \\ &\qquad \left. f_N\left(\frac{a_{11} c_1 + a_{12} c_2 + b_{11} z + b_{12} w}{a_{21} c_1 + a_{22} c_2 + b_{21} z + b_{22} w}, 1\right)\right] \end{aligned} \tag{6}$$

We see then that the calculation of $f_{N+1}(c_1, c_2)$ can be effected if we know the two functions

$$\begin{aligned} g_N(x) &= f_N(x, 1), \ 0 \leq x \leq 1, \\ h_N(x) &= f_N(1, x), \ 0 \leq x \leq 1. \end{aligned} \tag{7}$$

Hence the computation of the sequence $\{f_N(c_1, c_2)\}$ may be reduced to the computation of the two sequences $\{g_N(x)\}$, $\{h_N(x)\}$.

§ 15. Nonlinear problems

A variety of problems in analysis, and in applications to control problems arising in engineering and mathematical economics, reduce to the maximization or minimization of an integral of the form

$$J(z) = \int_0^T F(x_1, x_2, \ldots, x_n; z_1, z_2, \ldots, z_m)\, dt, \tag{1}$$

over all functions $z_i(t)$

subject to a number of constraints of the form

(2) (a) $dx_i/dt = G_i(x, z), i = 1, \ldots, k$

(b) $R_j(x, z) \leq 0, j = 1, 2, \ldots l.$

In some cases, the nonlinearity leads to a more complete analysis, since it permits us to determine the extremal by classical variational techniques, rather than test vertices as we must do in linear problems. In cases where constraints of the type above enter, we must combine the two approaches. In either situation, the functional equation technique may be utilized for both analytic and numerical purposes.

Problems of this type will be discussed in Chapter 9.

Exercises and Research Problems for Chapter VI

1. Consider the problem of maximizing the linear form $L(x) = \sum_{i=1}^{n} b_i x_i$, subject to the constraint $x_i \geq 0$ and $\sum_{j=1}^{n} a_{ij} x_j \leq c_i, i = 1, 2, \ldots, M$, where we assume that the coefficients a_{ij} and b_i are positive. Let

$$f_n(c_1, c_2 \ldots, c_M) = \operatorname*{Max}_{\{x_i\}} L(x).$$

Show that

$$\begin{aligned} f_1(c_1, c_2, \ldots, c_M) &= b_1 \operatorname*{Min}_i c_i/a_{i1}, \\ f_{n+1}(c_1, c_2, \ldots, c_M) &= \operatorname*{Max}_x [b_{n+1} x \\ &\quad + f_n(c_1 - a_{1n+1} x, \ldots, c_M - a_{Mn+1} x)], \end{aligned}$$

where $0 \leq x \leq \operatorname*{Min}_i [c_i/a_{in+1}]$.

2. Show that $f_n(c_1, c_2, \ldots, c_n)$ is a concave function of the c_i for $c_i = 0$.

3. What conclusion can be drawn from this result concerning the number of the maximizing x_i which are non-zero?

4. Consider the above problem for the case where $M = 1, 2$, or 3, and determine the dependence of the maximizing x_i upon the parameters c_j, and the analytic form of f_n.

5. Show that the tabulation of the function $f_n(c_1, c_2, \ldots, c_M)$ can always be reduced to the tabulation of the function $f_n(c_1, c_2, \ldots, 1)$. Establish the corresponding result for the bottleneck process discussed above.

6. Consider the problem of maximizing $u(T)$ where

$$du/dt = au + v,\ u(0) = c,$$

over all function $v(t)$ satisfying the constraint $0 \leq v \leq u$ for $0 \leq t \leq T$. Here all the quantities involved are scalars.

7. Solve the general problem of maximizing $(x(T), a)$ where

$$dx/dt = Ax + By,\ x(0) = c,$$

over all vectors $y(t)$ satisfying the constraint $0 \leq y \leq x$ for $0 \leq t \leq T$. Here x, y, c and a are vectors, while A and B are matrices.

8. Show that the problem of maximizing $x_2(T)$ under the conditions

(a) $dx_2/dt = a_2 z_2 - z_3,\ x_2(0) = c_2,$

$dx_3/dt = b_3 \gamma_3 z_3 - \gamma_2 z_2,\ x_3(0) = c_3,$

where z_2 and z_3 are functions of t subject to the restraints

(b) 1. $z_2 + z_3 \leq x_2,$

2. $\gamma_2 z_2 + \gamma_3 z_3 \leq x_3,$

3. $z_2, z_3 \geq 0,$

is equivalent to solving the partial differential equation

(c) $$\frac{\partial f}{\partial t} = \operatorname*{Max}_{D(z)} \left[\left(a_2 \frac{\partial f}{\partial c_2} - \gamma_2 \frac{\partial f}{\partial c_3} \right) z_2 + b_3 \gamma_3 \frac{\partial f}{\partial c_3} - \frac{\partial f}{\partial c_2} z_3 \right].$$

where $D(z)$ is the region determined by (b), under appropriate assumptions of continuity.

All parameters appearing are assumed to be non-negative and $f = f(c_2, c_3, t)$.

9. Show that optimal policies depend only upon the ratio $r = c_2/c_3$, or x_2/x_2, and T the time remaining.

10. Determine the form of the solution for small T.

11. Solve the problem in the special case where $b_3 = 0$.

Bibliography and Comments for Chapter VI

§ 1. A discussion of the theory of linear programming may be found in Activity Analysis of Production and Allocation, Edited by T. C. Koopmans, Cowles Commission, U. of Chicago, **1951**, where there is an account of the "simplex" technique of G. Dantzig, and a number of applications. An account of an iterative technique of a different type, the "flooding" technique of A. Boldyreff may be found in A. Boldyreff, Determination of the Maximal Steady Flow of Traffic Through a Railroad Network, RAND Corporation, P–687, 1955. Both of these are "relaxation techniques" of the kind brought into prominence by R. V. Southwell.

§ 5. The methods and results of this and the following section were announced in R. Bellman, "Bottleneck Problems and Dynamic Programming," *Proc. Nat. Acad. Sci.*, vol. 39 (1953), and presented in detail by R. Bellman, "Bottleneck Problems, Functional Equations and Dynamic Programming," *Econometrica*, vol. 23 (1955), pp. 73–87.

§ 9. A rigorous theory of these variational problems will involve at least Lebesgue-Stieltjes integrals and, most likely, the theory of distributions of L. Schwarz. It may well be that this will serve as a motivation for the study of variational problems involving distributions.

§ 12. As in the discrete case, the dual problem is most logically discussed by treating the min–max problem containing both the original and the dual process. A number of results can be established concerning the existence of a value of the corresponding game and the equivalence, min-max = max-min, using existing results in the theory of continuous games, in the case where the policy functions are uniformly bounded as a consequence of the constraints. The general case however, awaits a theory of games over the space of the distribution functions of L. Schwarz.

It is remarkable that so much can be obtained using only the easily derived result of (12.12).

§ 13. R. S. Lehman has found a continuous version of the "simplex" method of Dantzig which can be used to obtain the solutions of variational problems of this type in a systematic fashion. A preliminary account of his results may be found in R. S. Lehman, "On the Continuous Simplex Method" RM–1386, RAND Corporation, 1954.

CHAPTER VII

Bottleneck Problems: Examples

§ 1. Introduction

In the previous chapter, we discussed a multi-stage production process involving three industries, which we called the auto, steel, and tool industries. Taking this problem as our motivation, we were led to a general theoretical formulation of a class of continuous multi-stage production processes in terms of the concepts and techniques of the theory of dynamic programming.

The purpose of the present chapter is to show by grinding through the details of a particular example that this new approach may be utilized to provide explicit analytic solutions of problems of this general type. The analysis is decidedly difficult and it cannot be said that these problems have in any sense been tamed.

We shall consider a lumped two-industry process, involving what we call the auto and steel industries. The high degree of lumping (or more pedantically "conglomeration") is indicated by the fact that at any time t we assume that the state of the industrial system is completely specified by the following quantities:

(1)
$$x_1(t) = \text{auto stockpile at time } t$$
$$x_2(t) = \text{auto capacity at time } t$$
$$x_3(t) = \text{steel stockpile at time } t$$
$$x_4(t) = \text{steel capacity at time } t$$

Taking t to be a continuous variable, at each instant we must determine rates of allocation of the steel stockpile towards three distinct objectives:

(2)

a. Production of autos

b. Building of auto factories,
i.e., increase of auto capacity

c. Building of steel mills,
i.e., increase of steel capacity

The last two of these three objectives are to be sublimated to the primary objective of maximizing the total number of autos produced over a time-period T, which is to say, the quantity $x_1(T)$.

The basic assumptions of our model are the following: The measures of stockpile and capacity are chosen so that one unit of capacity, either auto or steel, is required for the production of one unit of stockpile in unit time. We assume that b_1 units of steel are required to make one unit of autos, b_2 units of steel are required to increase auto capacity by one unit, and b_4 units of steel are required to increase steel capacity by one unit. However, we shall assume that no steel is required to produce additional steel.

A very important assumption is that there is no time-lag between allocation and increase in capacity of production. The problems which arise when time-lag is considered are an order of magnitude more difficult and will not be discussed here.

Let

(3)

$$\begin{aligned}&(a)\quad z_1(t) = \text{rate of production of autos}\\&(b)\quad z_2(t) = \text{rate of increase of auto capacity}\\&(c)\quad z_3(t) = \text{rate of production of steel}\\&(d)\quad z_4(t) = \text{rate of increase of steel capacity}\end{aligned}$$

We derive, following the lines of the argumentation of the previous chapter, the following system of equations

(4)

$$\begin{aligned}dx_1/dt &= z_1(t), & x_1(0) &= c_1\\dx_2/dt &= z_2(t), & x_2(0) &= c_2\\dx_3/dt &= z_3(t) - b_1 z_1(t) - b_2 z_2(t) - b_4 z_4(t), & x_3(0) &= c_3\\dx_4/dt &= z_4(t), & x_4(0) &= c_4\end{aligned}$$

where the z_i and x_i are subject to the following constraints

(5)

$$\begin{aligned}&(a)\quad z_1(t) \leq x_2(t)\\&(b)\quad z_3(t) \leq x_4(t)\\&(c)\quad z_i(t) \geq 0, \quad i = 1, 2, 3, 4,\\&(d)\quad x_3(t) \geq 0\end{aligned}$$

The first two constraints are capacity constraints, i.e., limitations of bottleneck type; the third is a statement that rates of production must be non-negative, i.e., no scrapping or "cannibalization," and the fourth asserts that the steel stockpile must be non-negative, i.e., no borrowing.

The problem is now to determine the $z_i(t)$, satisfying the restrictions of

(5), which maximize $x_1(T)$. Because of the lack of any explicit upper bound on z_2 and z_4, various difficulties arise which must be surmounted by the use of delta functions.

§ 2. Preliminaries

In § 1, we formulated in mathematical terms the problem of utilizing the steel and auto industries so as to maximize auto production. Let us continue from equations (1.4) and (1.5).

The equations can be combined to provide an equivalent system of integral inequalities:

$$(1)\qquad z_1 \leq x_2: \quad z_1(t) - \int_0^t z_2(s)\,ds \leq c_2,$$

$$0 \leq x_3: \quad \int_0^t (-z_3(s) + b_1 z_1(s) + b_2 z_2(s) + b_4 z_4(s))\,ds \leq c_3,$$

$$z_3 \leq x_4: \quad z_3(t) - \int_0^t z_4(s)\,ds \leq c_4$$

Our problem is a special case of the following more general problem. Let Z be the set of all vector functions $z(t)$ which satisfy the conditions

$$(2)\qquad \text{(a)}\quad z(t) \geq 0$$

$$\text{(b)}\quad Bz(t) + \int_0^t Cz(s)\,ds \leq c$$

where B and C are matrices and c is a constant vector. We now wish to find a vector function $z(t)$ in Z which maximizes

$$(3)\qquad \int_0^T (z(t), a)\,dt$$

This is the problem we discussed in the previous chapter. It was shown there that there is a dual problem which furnishes a sufficient condition that a $z(t)$ belonging to Z be a maximizing vector, or in other words, that a feasible solution be optimal.

Let W be the set of vector functions $w(t)$ for which

$$(4)\qquad w(t) \geq 0$$

$$B'w(t) + C' \int_t^T w(s)\,ds \geq a$$

where B' and C' are the transposes of B and C. The dual problem is that of finding the minimum of $\int_0^T (w(t), c)\,dt$, for $w \in W$.

As we showed in § 11 of Chapter 4, we have for all z and w in the respective classes Z and W, the inequality

$$\int_0^T (z(t), a)\, dt \leq \int_0^T (w(t), c)\, dt \tag{5}$$

If we can find two vector functions z and w for which (2.5) holds with equality, they must yield the maximum and minimum, respectively, for the two problems. Two such vector functions for which equality holds will be said to be paired with each other. Thus, a sufficient condition that a z belonging to Z be optimal is that it can be paired with some w in W.

For the auto-steel problem formulated above we have

$$B = \begin{pmatrix} 1 & 0 & 0 & 0 \\ 0 & 0 & 0 & 0 \\ 0 & 0 & 1 & 0 \end{pmatrix} \quad C = \begin{pmatrix} 0 & -1 & 0 & 0 \\ b_1 & b_2 & -1 & b_4 \\ 0 & 0 & 0 & -1 \end{pmatrix} \quad a = \begin{pmatrix} 1 \\ 0 \\ 0 \end{pmatrix} \tag{6}$$

The dual system of inequalities is therefore

$$\begin{aligned} l_1 &\equiv w_2(t) + b_1 \int_t^T w_3(s)\, ds - 1 \geq 0 \\ l_2 &\equiv -\int_t^T w_2(s)\, ds + b_2 \int_t^T w_3(s)\, ds \geq 0 \\ l_3 &\equiv w_4(t) - \int_t^T w_3(s)\, ds \geq 0 \\ l_4 &\equiv b_4 \int_t^T w_3(s)\, ds - \int_t^T w_4(s)\, ds \geq 0. \end{aligned} \tag{7}$$

We have chosen to call the components of w, w_2, w_3 and w_4 in order to keep the connection with the inequalities $z_1 \leq x_2$, $0 \leq x_3$, $z_3 \leq x_4$ clear.

The optimality conditions, i.e., the conditions that (2.5) hold with equality, are:

(8)
If $z_i(t) > 0$, then $l_i(t) = 0$, $(i = 1, 2, 3, 4)$
If $z_1(t) < x_2(t)$, then $w_2(t) = 0$
If $0 < x_3(t)$, then $w_3(t) = 0$
If $z_3(t) < x_4(t)$, then $w_4(t) = 0$

The following are equivalent to the optimality conditions:

(9)
If $l_i(t) > 0$, then $z_i(t) = 0$, $(i = 1, 2, 3, 4)$
If $w_2(t) > 0$, then $z_1(t) = x_2(t)$
If $w_3(t) > 0$, then $0 = x_3(t)$
If $w_4(t) > 0$, then $z_3(t) = x_4(t)$

§ 3. Delta-functions

Before we proceed to determine the solution, let us discuss the use that we will make of delta functions. It can easily happen that the general problems discussed above have *no solutions* if the sets Z and W are composed only of vectors having components which are integrable functions. In fact, as we shall later see, this is the usual case in the auto-steel problem. This difficulty can be evaded by enlarging the sets Z and W so that they contain vector "functions" whose components are sums of integrable functions and "delta functions." In these enlarged classes the problems have solutions. By a delta function concentrated at t_o with weight ω, which we denote by $\omega\delta\,(t - t_o)$, we mean an improper function such that

$$\int_o^t \omega\delta\,(s - t_o)\,\varphi\,(s)\,ds = \begin{matrix} 0 \text{ if } t < t_o \\ \omega\varphi\,(t_o) \text{ if } t > t_o \end{matrix}$$

for every function φ continuous at t_o. (For $t = t_o$ the integral in undefined except when $\varphi\,(t_o) = 0$, in which case it is defined to be 0.)

The use of delta functions can be justified rigorously either by the alternative use of Stieltjes integrals, or by regarding the delta functions as obtained by completing the space of integrable functions by a process similar to that used in obtaining the real numbers from the rationals.

The optimality conditions remain the same even when Z and W are enlarged in the above way. We observe that there is no harm in the violation of the optimality conditions at isolated points or even in sets of measure zero when only measurable functions are allowed as components of z and w. But, when one of the vectors, w, for example, has a component w_i which is a delta function at the point t_o, then for a z to be paired with w, the corresponding optimality conditions must be satisfied at the point t_o.

We shall find that we never have to use delta functions concentrated at any point other than 0 to obtain an optimal z. Intuitively, this means that discontinuous changes are not necessary except at the beginning.

§ 4. The solution

The procedure that we use will be to construct a number of w-solutions which we can pair with z's belonging to Z and hence obtain solutions of our problem. The chief difficulty occurs in constructing w-solutions with suitable properties. In this we are guided by a combination of guesswork and observation of properties that an optimal z should have. Guesswork could be eliminated at the expense of considering a very much larger number of cases.

First of all, it is clear that we should always have $z_3 = x_4$. To produce too much steel is not harmful. This tells us that we should have $l_3\,(t)$

$= 0$ for all t; i.e., $w_4(t) = \int_t^T w_3(s)\,ds$. The remaining inequalities of (2.7) then become

$$l_1 \equiv w_2(t) + b_1 w_4(t) - 1 \geq 0 \tag{1}$$

$$l_2 \equiv b_2 w_4(t) - \int_t^T w_2(s)\,ds \geq 0$$

$$l_4 \equiv b_4 w_4(t) - \int_t^T w_4(s)\,ds \geq 0$$

Shortly before T it is clear that we should be producing autos since x_1 is the quantity we wish to maximize at time T. Hence, we will have $z_1 > 0$, which implies that $l_1 = 0$. This alone will not give us sufficient information to determine w_2 and w_4.

We first construct a w solution, which we shall call the basic w-solution, with the property that $l_2 = 0$ near the end. This means that we must have $w_4(T) = 0$. Then by (4.1) we have

$$w_4(t) = \frac{1}{b_1}(1 - e^{-b_1(T-t)/b_2}) \tag{2}$$

$$w_2(t) = e^{-b_1(T-t)/b_2}$$

We see that w_2, w_3, and w_4 all remain positive as t decreases. We must check to see whether the inequality $l_4 \geq 0$ is satisfied. With the above choice of w we have

$$l_4 = \frac{b_4}{b_1}(1 - e^{-b_1(T-t)/b_2}) - \frac{(T-t)}{b_1} + \frac{b_2}{b_1^2}(1 - e^{-b_1(T-t)/b_2}) \tag{3}$$

The quantity on the right side of this equation is positive for $T - t$ small but is negative when $T - t$ is large. Let t_o be the value of t for which the right side becomes zero. Then $T - t_o$ is the solution of the equation

$$T - t_o = \left(b_4 + \frac{b_2}{b_1}\right)(1 - e^{-b_1(T-t_o)/b_2}) \tag{4}$$

Thus we see that at t_o we must abandon one of the equations $l_1 = 0$ and $l_2 = 0$. Let us try to choose w so that $l_1 = 0$ and $l_4 = 0$ before t_o. We have

$$w_4(t) = w_4(t_o)\,e^{(t_o-t)/b_4} \tag{5}$$

$$w_2(t) = 1 - b_1 w_4(t_o)\,e^{(t_o-t)/b_4}$$

To verify that $l_2 \geq 0$ we compute its derivative. We find

$$\frac{dl_2}{dt} = b_2\frac{dw_4}{dt} + w_2 = 1 - \left(b_1 + \frac{b_2}{b_4}\right) w_4(t_o)\,e^{(t_o-t)/b_4} \tag{6}$$

A condition sufficient to insure that $l_2 \geq 0$ is that $dl_2/dt \leq 0$ for all $t \leq t_o$. This will hold if

$$w_4(t_o) \geq \frac{b_4}{b_2 + b_1 b_4} \tag{7}$$

which by (4.2) and (4.4) is equivalent to $T - t_o \geq b_4$. This last inequality can be checked by putting b_4 in place of $T - t$ in (4.3) and verifying that the quantity thus obtained is positive. We have

$$\begin{aligned} &\frac{1}{b_1}\left[\left(b_4 + \frac{b_2}{b_1}\right)(1 - e^{-b_1 b_4/b_2}) - b_4\right] \\ &= \frac{b_2}{b_1{}^2} e^{-b_1 b_4/b_2}\left[e^{b_1 b_4/b_2} - \left(1 + \frac{b_1 b_4}{b_2}\right)\right] > 0. \end{aligned} \tag{8}$$

Hence $l_2 \geq 0$ for all $t \leq t_o$ with the above choice of w.

We also see from (4.5) that w_4 and w_3 remain positive. Thus (4.5) will give a satisfactory choice of w until w_2 becomes zero. Let t_1 be the value of t when this happens. Then, by (4.2) and (4.4) we have

$$e^{(t_o - t_1)/b_4} = \frac{b_4 + b_2/b_1}{(T - t_o)}. \tag{9}$$

Before t_1 let us see whether we can choose $w_2 = 0$ and have $l_4 = 0$. We see that $w_4 > 0$ and $w_3 > 0$. We have $dl_2/dt = b_2\, dw_4/dt < 0$ so that $l_2 > 0$, and $dl_1/dt = b_1\, dw_4/dt < 0$ so that $l_1 > 0$. Hence this choice of w will be valid for all $t \leq t_1$.

Our basic solution is summarized in the following table. This table also lists the properties that a z paired with this w solution must have. Any z with these properties gives a policy which, if feasible (i.e., satisfies the z constraints), is optimal.

(10)

$t < t_1$		$t_1 < t < t_o$		$t_o < t < T$	
$l_1 > 0$	$z_1 = 0$	$l_1 = 0$		$l_1 = 0$	
$l_2 > 0$	$z_2 = 0$	$l_2 > 0$	$z_2 = 0$	$l_2 = 0$	
$l_3 = 0$		$l_3 = 0$		$l_3 = 0$	
$l_4 = 0$		$l_4 = 0$		$l_4 > 0$	$z_4 = 0$
$w_2 = 0$		$w_2 > 0$	$z_1 = x_2$	$w_2 > 0$	$z_1 = x_2$
$w_3 > 0$	$x_3 = 0$	$w_3 > 0$	$x_3 = 0$	$w_3 > 0$	$x_3 = 0$
$w_4 > 0$	$z_3 = x_4$	$w_4 > 0$	$z_3 = x_4$	$w_4 > 0$	$z_3 = x_4$

Figure 1

Let us see how this table can be used to obtain a partial solution of the auto-steel problem. For the moment let us assume $c_3 = 0$. For $t < t_1$ we must have

(11) $$z_1 = 0,\ z_2 = 0,\ z_3 = x_4,\ z_4 = x_4/b_4.$$

For $t_1 < t < t_o$ we must choose

(12) $$z_1 = x_2,\ z_2 = 0,\ z_3 = x_4,\ z_4 = \frac{x_4 - b_1 x_2}{b_4}.$$

This can be done if and only if $x_4(t_1) - b_1 x_2(t_1) \geq 0$. Let us assume that this inequality is satisfied. Then for $t_o < t < T$ we must have

(13) $$z_1 = x_2,\ z_2 = \frac{x_4 - b_1 x_2}{b_2},\ z_3 = x_4,\ z_4 = 0$$

which is possible provided $x_4(t_1) - b_1 \dot{x}_2(t_1) \geq 0$. Thus we see that for *certain initial conditions* we can obtain the optimal solution.

§ 5. The modified w solution

As already has been noted, we run into trouble if $x_4(t_1) - b_1 x_2(t_1) < 0$. To handle this case we consider a modification of the basic w solution of Fig. 1 above. Let u_o be in the interval $[t_1, T]$. For each such u_o we define a solution as follows:

For $u_o < t < T$ we let $w(t)$ be the same as in the basic solution. For $t < u_o$ we choose $w_2(t) = 0$. For $t < u_o$ but near u_o we choose $w_4(t) = 1/b_1$ so that $l_1 = 0$. This choice will keep $l_4 > 0$ for a while before u_o. We define u_1 to be the point where l_4 becomes 0 with this choice of w. For $t < u_1$ we choose w so that $l_4 = 0$. It is easily seen that this choice makes $l_1 > 0$, $l_2 > 0$, $w_3 > 0$ and $w_4 > 0$ for all $t < u_1$. Hence, in this way we obtain a w solution for each u_o in the interval $[t_1, T]$. We observe that for $u_o = t_1$, $u_1 = t_1$ and this solution is identical with our basic solution of Fig. 1. Note that u_1 depends continuously on u_o. Since for $u_o = T$, $u_1 = T - b_4$, there is a w solution for each u_1 in the interval $[t_1, T - b_4]$.

These w solutions together with the properties of the corresponding z solutions, are summarized in the following table:

	$t<u_1$		$u_1<t<u_o$		$u_o<t<T$	$t_1<t<t_o$	$u_o<t<T$	$t_o<t<T$
	$l_1 > 0$	$z_1 = 0$	$l_1 = 0$		$l_1 = 0$		$l_1 = 0$	
	$l_2 > 0$	$z_2 = 0$	$l_2 > 0$	$z_2 = 0$	$l_2 > 0$	$z_2 = 0$	$l_2 = 0$	
	$l_3 = 0$		$l_3 = 0$		$l_3 = 0$		$l_3 = 0$	
(1)	$l_4 = 0$		$l_4 > 0$	$z_4 = 0$	$l_4 = 0$		$l_4 > 0$	$z_4 = 0$
	$w_2 = 0$		$w_2 = 0$		$w_2 > 0$	$z_1 = x_2$	$w_2 > 0$	$z_1 = x_2$
	$w_3 > 0$	$x_3 = 0$	$w_3 = 0$		$w_3 > 0$	$x_3 = 0$	$w_3 > 0$	$x_3 = 0$
	$w_4 > 0$	$z_3 = x_4$	$w_4 > 0$	$z_3 = x_4$	$w_4 > 0$	$z_3 = x_4$	$w_4 > 0$	$z_3 = x_4$

Since w_3 is a delta function at u_o, we must have $x_3(u_o) = 0$.

Figure 2

Note that if $u_o > t_o$, then there is no t satisfying the conditions of the third column; and if $u_o = T$ then there is no t satisfying the conditions of the last column either.

§ 6. The equilibrium solution

A policy which seems plausible in some instances is the following: Make an initial adjustment to bring x_3 down to zero in such a way that after the adjustment $x_4 = b_1 x_2$. If this is done, after the initial adjustment no increase in capacities is necessary and all available steel can be used for auto production. Such a policy would require for the w paired with it that $l_2(0) = 0$ and $l_4(0) = 0$, because in general both z_2 and z_4 will have to be delta functions. We shall construct a w solution with this property.

First, we note that our basic w solution has this property when T is such that $t_o = 0$. This suggests that we try to choose

(1) $$w_4(t) = \alpha e^{-b_1(T-t)/b_2} + \beta$$

where α and β are constants. If w_2 is chosen so that $l_1 = 0$, the inequalities (4.1) become

(2) $$l_2 \equiv b_2 w_4(t) - (T - t) + b_1 \int_t^T w_4(s)\, ds \geq 0$$

$$l_4 \equiv b_4 w_4(t) - \int_t^T w_4(s)\, ds \geq 0.$$

If $l_2(0) = l_4(0) = 0$, then

(3) $$w_4(0) = \frac{T}{b_2 + b_1 b_4}.$$

We set $E = e^{-b_1 T/b_2}$ and from (6.1) — (6.3) derive the following two equations for α and β:

(4) $$b_2 \alpha + (b_2 + b_1 T)\beta = T$$

$$(b_2 + b_1 b_4) E\alpha + (b_2 + b_1 b_4)\beta = T.$$

A solution of these equations will give a w for which $l_2(0) = l_4(0) = 0$. We have

(5) $$\alpha = \frac{T}{\Delta} \begin{vmatrix} 1 & b_2 + b_1 T \\ 1 & b_2 + b_1 b_4 \end{vmatrix} = \frac{T}{\Delta}[b_1(b_4 - T)]$$

$$\beta = \frac{T}{\Delta} \begin{vmatrix} b_2 & 1 \\ (b_2 + b_1 b_4) E & 1 \end{vmatrix} = \frac{T}{\Delta}[b_2(1 - E) - b_1 b_4 E]$$

where

(6) $$\Delta = \begin{vmatrix} b_2 & b_2 + b_1 T \\ (b_2 + b_1 b_4) E & b_2 + b_1 b_4 \end{vmatrix} = (b_2 + b_1 b_4)(b_2 - b_2 E - b_1 ET)$$

Also

(7) $$\begin{aligned} \Delta &= (b_2 + b_1 b_4) E (b_2 e^{b_1 T/b_2} - b_2 - b_1 T) \\ &> (b_2 + b_1 b_4) E (b_2 + b_1 T - b_2 - b_1 T) = 0 \end{aligned}$$

Now let us assume that $T - t_o \geq T \geq b_4$. Then from (6.5) we see that $\alpha \leq 0$. Let us check to be sure that for the w we have defined $w_2(t)$, $w_3(t)$, and $w_4(t)$ are non-negative for $0 \leq t \leq T$. This is equivalent to verifying that $0 \leq w_4(t) \leq 1/b_1$ and $dw_4/dt \leq 0$. We have $dw_4/dt = \alpha\, b_1/b_2\, e^{b_1 (T-t) b_2} \leq 0$. Hence it will be sufficient to check that $w_4(T) \geq 0$ and $w_4(0) \leq 1/b_1$. Since $T - t_o \geq T$, we conclude from (4.4) and (6.3) that

(8) $$w_4(0) = \frac{T}{b_2 + b_1 b_4} \leq \frac{T - t_o}{b_2 + b_1 b_4} < \frac{b_4 + b_2/b_1}{b_2 + b_1 b_4} = 1/b_1$$

and

(9) $$w_4(T) = \alpha + \beta = \frac{T}{\Delta} b_1 [(b_4 + b_2/b_1)(1 - E) - T] \geq 0$$

We also must check that for $0 \leq t \leq T$, $l_2 \geq 0$ and $l_4 \geq 0$. Since

(10) $$dl_2/dt = b_2\, dw_4/dt + 1 - b_1 w_4(t) = 1 - b_1 \beta$$

and $l_2(T) = b_2 w_4(T) \geq 0$, we have $l_2 \geq 0$ for all t in $[0, T]$. Similarly, we know that $l_4(T) = b_4 w_4(T) \geq 0$. Hence, if we show that $d^2 l_4/dt^2 \leq 0$, we will have proved that $l_4 \geq 0$ for all t in $[0, T]$. We have

(11) $$\frac{d^2 l_4}{dt^2} = \alpha \left(b_4 + \frac{b_2}{b_1}\right) \left(\frac{b_1}{b_2}\right)^2 \geq 0.$$

This completes the proof that the w which we have defined is a solution. Its properties, together with those a z paired with it must have, are summarized in the following table:

(12)

$t = 0$	$0 < t < T$	
	$l_1 = 0$	
$l_2 = 0$	$l_2 > 0$	$z_2 = 0$
	$l_3 = 0$	
$l_4 = 0$	$l_4 > 0$	$z_4 = 0$
	$w_2 > 0$	$z_1 = x_2$
	$w_3 > 0$	$x_3 = 0$
	$w_4 > 0$	$x_3 = x_4$
Note: This solution is valid only for $T \geq b_4$.		

Figure 3

§ 7. A short-time w solution

The w solution which we construct next will be useful in finding the solution of our maximum problem when the total time is short, $T < b_4$. This solution differs from those already constructed in that it allows x_3 to be positive and z_2 to be a delta function concentrated at 0.

For $0 \leq t \leq T$ *let* $w_4(t) = \gamma$, $w_2(t) = 1 - b_1\gamma$ where $0 < \gamma < 1/b_1$. Then $l_1(t) = 0$, $l_4(t) > 0$ for $0 \leq t < T$. Also

(1) $l_2(t) = b_2\gamma - (T - t)(1 - b_1\gamma) = [b_2 + b_1(T - t)]\gamma - (T - t)$

Now, if we choose $\gamma = \dfrac{T}{b_2 + b_1 T}$ then $l_2(0) = 0$ and $l_2(t) > 0$ for $t > 0$. Thus we obtain a solution of the system of inequalities (4.1). It is summarized below together with the properties a z paired with it must have.

(2)

$t = 0$	$0 < t < T$	
	$l_1 = 0$	
$l_2 = 0$	$l_2 > 0$	$z_2 = 0$
	$l_3 = 0$	
	$l_4 > 0$	$z_4 = 0$
	$w_2 > 0$	$z_1 = x_2$
	$w_3 = 0$	
	$w_4 > 0$	$z_3 = x_4$
Since w_3 is a delta function at T, $x_3(T) = 0$.		
Note: This solution is valid for $T < b_4$ only.		

Figure 4

§ 8. Description of solution and proof

We now can give the complete solution to the original problem. There are quite a few cases that we must consider separately. The critical values t_o and t_1, which are defined by (4.4) and (4.9), depend on T, but in such a way that for fixed b_1, b_2 and b_4, $T - t_o$ and $T - t_1$ are constants.

Case I: *T is large enough so that $t_1 \geq 0$.* In this case we choose z_4 to be a delta function concentrated at 0 to bring x_3 down to zero immediately. This means that if the total time is long enough we should not keep any steel in storage but should be using it to build more steel plants. The use of the delta function is permissible because $l_4 = 0$ for t near 0. For $0 < t < t_1$ we let

$$z_1 = 0, z_2 = 0, z_3 = x_4, z_4 = x_4/b_4 \tag{1}$$

thus keeping x_3 at zero level. At t_1 we must distinguish different subcases:

$$\begin{aligned} &\text{IA:} \quad x_4(t_1) - b_1 x_2(t_1) \geq 0 \\ &\text{IB:} \quad x_4(t_1) - b_1 x_2(t_1) < 0 \end{aligned} \tag{2}$$

In case IA we can produce autos at capacity without running out of steel. Hence we let

$$z_1 = x_2, z_2 = 0, z_3 = x_4, z_4 = \frac{x_4 - b_1 x_2}{b_4} \tag{3}$$

for $t_1 \leq t \leq t_o$; and for $t_o < t \leq T$ we let

$$z_1 = x_2, z_2 = \frac{x_4 - b_1 x_2}{b_2}, z_3 = x_4, z_4 = 0. \tag{4}$$

This solution for Case IA is optimal because it can be paired with our basic w solution of Fig. 1.

In Case IB we do not have enough steel to produce autos at capacity. Hence we continue to produce no autos for $t > t_1$, i.e.,

$$z_1 = 0, z_2 = 0, z_3 = x_4, z_4 = \frac{x_4}{b_4}. \tag{5}$$

We do this until $x_4 - b_1 x_2$ becomes zero or $t = T - b_4$, whichever happens first. If $x_4 - b_1 x_2$ becomes zero at t' then we choose $z_1 = x_2$, $z_2 = 0$, $z_3 = x_4$, $z_4 = 0$ thereafter. This solution is seen to be optimal by pairing it with the w solution of Fig. 2 for which $u_1 = t'$. As we have already remarked there is such a solution no matter what t' is, so long as $t_1 < t' \leq T - b_4$. If, on the other hand, $x_4(T - b_4) - b_1 x_2(T - b_4) < 0$, then for $T - b_4 < t \leq T$ we choose

$$z_1 = \frac{x_4}{b_1}, z_2 = 0, z_3 = x_4, z_4 = 0. \tag{6}$$

This solution can be seen to be optimal by pairing it with the w solution of Fig. 2, for which $u_o = T$, $u_1 = T - b_4$.

Case II: *T is such that* $t_1 \leq 0 \leq t_o$. As before we choose z_4 to be a delta function concentrated at 0 to bring x_3 down to zero immediately. Thereafter the solution is as before. There are two subcases:

$$\text{(7)} \qquad \begin{aligned} &\text{IIA:} \quad x_4(0) - b_1 x_2(0) \geq 0 \\ &\text{IIB:} \quad x_4(0) - b_1 x_2(0) < 0 \end{aligned}$$

In Case IIA we let $z_1 = x_2$, i.e., produce autos at capacity. We use the remaining steel to increase steel capacity before t_o and to increase auto capacity after t_o. That is, for $0 < t < t_o$ we let

$$\text{(8)} \qquad z_1 = x_2,\ z_2 = 0,\ z_3 = x_4,\ z_4 = \frac{x_4 - b_1 x_2}{b_4}$$

and for $t > t_o$ we let

$$\text{(9)} \qquad z_1 = x_2,\ z_2 = \frac{x_4 - b_1 x_2}{b_4},\ z_3 = x_4,\ z_4 = 0.$$

This solution is optimal because it can be paired with our basic solution of Fig. 1.

Case IIB is similar to IB. The same prescription holds, and the solution is paired with one from Fig. 2.

Case III: *T is such that* $t_o \leq 0 \leq T - b_4$. There are three subcases:

$$\text{(10)} \qquad \begin{aligned} &\text{IIIA:} \quad c_4 - b_1 c_2 \geq b_1 \frac{c_3}{b_2} \\ &\text{IIIB:} \quad c_4 - b_1 c_2 < -\frac{c_3}{b_4} \\ &\text{IIIC:} \quad \frac{-c_3}{b_4} \leq c_4 - b_1 c_2 < \frac{b_1 c_3}{b_2}. \end{aligned}$$

In Case IIIA we use our initial stockpile of steel to increase auto capacity, i.e., we let z_2 be a delta function concentrated at 0 bringing x_3 down to zero. Thereafter, we let $z_1 = x_2$ and use any remaining steel to increase auto capacity, i.e.,

$$\text{(11)} \qquad z_1 = x_2,\ z_2 = \frac{x_4 - b_1 x_2}{b_2},\ z_3 = x_4,\ z_4 = 0.$$

This solution is optimal because it can be paired with the basic w solution of Fig. 1.

In Case IIIB we find ourselves short on steel capacity. The policy and proof are the same as in Case IB.

In Case IIIC we can make an initial adjustment so that x_3 becomes zero and $x_4 = b_1 x_2$. We do this by choosing z_2 and z_4 to be delta functions concentrated at 0. After that we let $z_1 = x_2$, $z_2 = 0$, $z_3 = x_4$, $z_4 = 0$. This solution is optimal because it can be paired with the equilibrium w solution of Fig. 3.

Case IV: $T \leq b_4$. There are three subcases which depend on the initial values:

$$\text{IVA:} \quad c_4 - b_1 c_2 \geq \frac{b_1 c_3}{b_2}$$

$$(12) \qquad \text{IVB:} \quad c_4 - b_1 c_2 > \frac{-c_3}{T}$$

$$\text{IVC:} \quad \frac{-c_3}{b_4} \leq c_4 - b_1 c_2 < \frac{b_1}{b_2} c_3$$

In Case IVA the solution and proof are the same as in Case IIIA.

In Case IVB we choose $z_2 = 0$ and $z_4 = 0$ for all t. As always we let $z_3 = x_4$. We choose z_1 in any way such that $z_1(t) \leq x_2(t)$ and $x_3(T) = 0$. Thus, in this case the solution is not unique. Any solution of this form can be seen to be optimal by pairing it with the w solution of Fig. 2 for which $u_o = T$.

In case IVC we find ourselves in an intermediate case, unable to follow the policies suggested by IVA and B. In this case we make an initial adjustment of the steel stockpile down to the value c_3', using this steel to increase auto capacity. Thereafter we choose $z_1 = x_2$, $z_2 = 0$, $z_3 = x_4$, and $z_4 = 0$. The value c_3' is determined so that $x_3(T) = 0$. It is found that

$$(13) \qquad c_3' = \frac{b_1 c_3 - b_2 (c_4 - b_1 c_2)}{b_2 + b_1 T}$$

has this property. This solution is optimal because it can be paired with the short-time w solution of Fig. 4.

Summary Initial Adjustments

<table>
<tr><th>Cases</th><th>I: $t_1 \geq 0$.</th><th>II: $t_1 \leqslant 0 \leqslant t_o$</th><th>III: $t_o \leqslant 0 \leqslant T - b_4$</th><th>IV: $T < b_4$</th></tr>
<tr><td>A</td><td colspan="2">Adjust x_3 to 0 by increasing x_4.</td><td colspan="2">Bring x_3 to 0 by increasing x_2 "Build auto capacity".</td></tr>
<tr><td>B</td><td colspan="3">"Build steel capacity".</td><td>No initial adjustments.</td></tr>
<tr><td>C</td><td>No Case</td><td>No Case</td><td>Adjust so that $x_3 = 0$, $x_4 = b_1 x_2$, by increasing x_2 and x_4.</td><td>Adjust x_3 downward, but not to 0, so that $x_3(T) = 0$. Increase x_2.</td></tr>
</table>

After the initial adjustments the optimal policy can be determined by a priority system. Before t_1, building steel capacity, i.e., z_4, has first priority. This continues after t_1 until either $x_4 \geq b_1 x_2$ or $t = b_4$, whichever comes first. When this happens, which may be at t_1, of course, first priority is given to auto production, z_1. This will use up all available steel unless $x_4(t_1) > b_1 x_2(t_1)$. In that case second priority is given to building steel capacity until the time t_o. After t_o second priority is given to building auto capacity.

Bibliography and Comments for Chapter VII

§ 1. The results of this chapter were obtained in collaboration with R. S. Lehman in an unpublished paper; R. Bellman and R. S. Lehman, Studies on Bottleneck Problems in Production Processes, Part I, P–492, RAND Corporation, 1954.

An analysis of similar type, but more intricate, resolving a variational problem in this general class, may be found in R. S. Lehman, Studies in Bottleneck Problems in Production Processes, Part II, P–492, RAND Corporation, 1954.

CHAPTER VIII

A Continuous Stochastic Decision Process

§ 1. Introduction

As we have seen in Chapter II, the formulation of the goldmining problem in its discrete form leads to a number of unsolved problems in connection with the three-choice problem, the non-linear utility problem, and many others we could formulate. We turn, therefore, to a continuous version of the problem in the hopes of overcoming these difficulties by use of the more powerful tools of continuity. As we shall see, we can now resolve the corresponding questions in complete detail and thereby obtain a clear insight into the structure of optimal policies. The information we obtain concerning the structure of policies can now be used to furnish useful approximations to the original discrete process.

One very interesting and significant fact emerges. Whereas the original discrete problem had certain *linear* aspects which made variational analysis difficult, at least in the case where we considered expected return, the continuous version is sufficiently *non-linear* to permit us to employ a variational approach in the classical manner, with certain modifications required by the presence of constraints. However, in carrying through this approach, our knowledge of the form of the solution for the discrete case is of great service in telling us in advance what to expect to find. It is a combination of the two techniques, old and new, which permit a successful attack upon the problem.

Before turning to the method we shall actually employ, we shall discuss two alternative approaches, each possessing certain features of difficulty which render them inappropriate.

It is perhaps equally as important to know which methods fail, and why, as it is to know methods which work. In more general decision processes of this type, a correct formulation of a continuous version is not trivial. Particularly is this true in the case of multi-stage games of continuous type.

There are many different possible formulations, and the correctness of an approach must be judged not only on the grounds of its mathematical rigor, but also on the grounds of analytic difficulty. If we do not have a systematic means of resolving specific problems, we do not have a satisfactory theory.

After this preliminary discussion, we shall turn to the approach we shall actually employ, which is a compromise between the two preliminary methods.

A justification of our approach lies in the fact that we can demonstrate that the limit of the discrete process, in a suitable sense, is the continuous process we discuss. We shall, however, not discuss in this volume these important and interesting questions.

§ 2. Continuous versions—I: A differential approach

Let us now proceed to discuss some possible continuous analogues of the functional equation of (5.1) of Chapter II.

Our basic assumption in this and the following sections will be that each operation is to have a high probability of obtaining a small amount of gold and leaving the machine undamaged. In other words, we renounce any hope of solving our problem for *all* values of the parameters, and consider, instead, a small region of the parameter space, (r_1, r_2, q_1, q_2).

We introduce the quantities

$1 - q_1 \delta =$ the probability of obtaining $r_1 x \delta$ and leaving the machine undamaged if Anaconda is mined,

$1 - q_2 \delta =$ the probability of obtaining $r_2 y \delta$ and leaving the machine undamaged if Bonanza is mined.

where q_1 and q_2 are positive and δ is a small enough positive quantity so that $1 - q_1 \delta$ and $1 - q_2 \delta$ are probabilities, and $r_1 \delta$ and $r_2 \delta$ are less than one.

With $f(x, y)$ as before, we have the functional equation

$$f(x, y) = \operatorname{Max} \begin{bmatrix} A: & (1 - q_1 \delta)(r_1 x \delta + f(x - r_1 x \delta, y)) \\ B: & (1 - q_2 \delta)(r_2 y \delta + f(x, y - r_2 y \delta)) \end{bmatrix} \tag{1}$$

This equation is precisely (5.1) of Chapter 2 for these new parameters. Proceeding formally, on the assumption that f has continuous partial derivatives, we have, for small δ, the approximate equation

$$f(x, y) = \operatorname{Max} \begin{bmatrix} A: & f(x, y) + \delta (r_1 x - q_1 f(x, y) - r_1 x \, \partial f/\partial x) + 0(\delta^2) \\ B: & f(x, y) + \delta (r_2 y - q_2 f(x, y) - r_2 y \, \partial f/\partial y) + 0(\delta^2) \end{bmatrix} \tag{2}$$

The limiting form as $\delta \to 0$ is the equation

$$0 = \operatorname{Max} \begin{bmatrix} A: & r_1 x - q_1 f - r_1 x \, \partial f/\partial x \\ B: & r_2 y - q_2 f - r_2 y \, df/\partial y \end{bmatrix} \tag{3}$$

This approach does not seem to be a fruitful one because of the difficulty of establishing existence and uniqueness theorems for functional equations of this type.

§ 3. Continuous versions—II: An integral approach

Let us now consider a diametrically opposed approach. Let S_N denote some sequence of A (or Anaconda)-choices and B (or Bonanza)-choices totalling N in number. Set

$p_{Nk}(x, y)$ = the probability of surviving N stages and ending in the state represented by (x_{Nk}, y_{Nk}), using S_N, upon starting in state (x, y).

$R_N(x, y)$ = expected return from N stages using S_N, starting in state (x, y).

If S_N actually consists of the first N choices of an optimal policy, we obtain for $f(x, y)$ the functional equation

$$(1) \qquad f(x, y) = R_N(x, y) + \sum_k p_{Nk}(x, y) f(x_{Nk}, y_{Nk})$$

If $N\delta$, where δ is as above, is chosen to remain finite as $\delta \to 0$ and $N \to \infty$, and set equal to t, the analogue of (2.1) is a functional equation of the type

$$(2) \quad f(x, y) = \operatorname*{Max}_S \left[R_S(x, y, t) + \int_{r=0}^{1} \int_{s=0}^{1} f(xr, ys)\, dG_S(r, s, x, y, t)\right]$$

where S denotes a continuous policy over the interval $[0, t]$ and dG_S is a transition probability determined by this policy.

Functional equations of this type occur in the general theory of stochastic processes. We shall not pursue this approach in this volume because of the many difficulties involved in justifying this equation and in defining general continuous policies. Instead, we shall employ an approach intermediate between the differential and the integral approach which yields a functional equation bearing the same relation to (2) as the diffusion or heat equation bears to the Chapman-Kolmogoroff equation in the theory of diffusion processes.

A justification of this approach is the fact that it can be demonstrated that the solution of the discrete process approaches the solution given by the continuous process as $\delta \to 0$. However, as stated above, we shall not discuss this question here.

§ 4. Preliminary discussion

Let us continue to use the simple equation of (2.1) as our model for the following discussion. According to the solution discussed earlier in Chapter II, the A- and B-regions are separated by the boundary curve

$$(1) \qquad L_\delta\colon (1 - q_1\delta)\frac{r_1 x}{q_1} = (1 - q_2\delta)\frac{r_2 y}{q_2}$$

which, as $\delta \to 0$, approaches the line

$$L: \frac{r_1 x}{q_1} = \frac{r_2 y}{q_2} \tag{2}$$

For each $\delta > 0$, the optimal policy has the following form:

"If below L_δ continue using the A-policy until in the B-region, above L_δ. Then use the B-policy until in the A-region, below L_δ, and so on; similarly if above L_δ to start."

Geometrically:

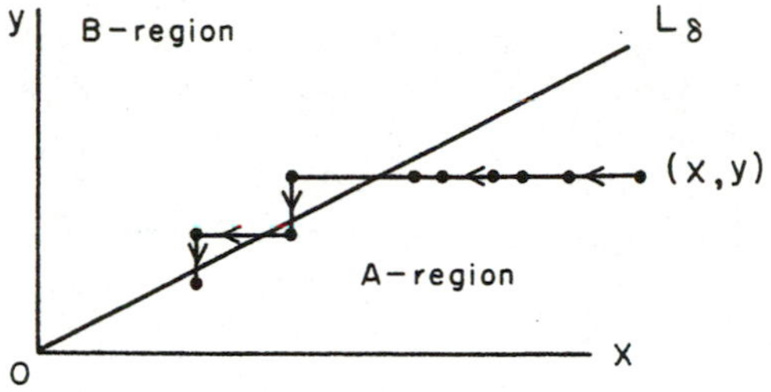

Figure 1

The limiting form of this policy as $\delta \to 0$ is the following:

"If (x, y) is below L, use A until the line L is reached, then continue along L thereafter; if (x, y) is above L, use B until the line L is reached, then continue along L thereafter."

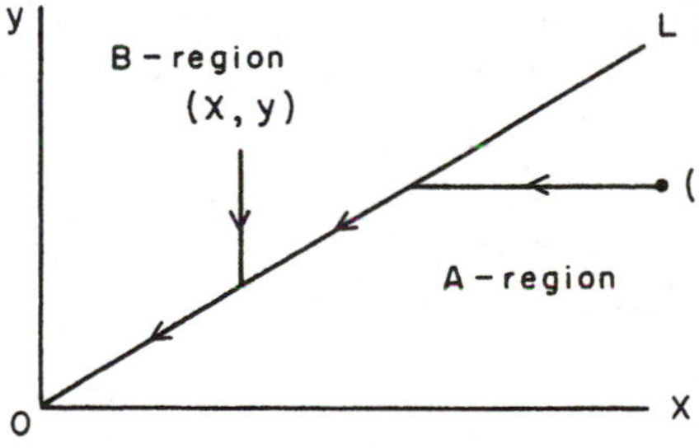

Figure 2

Let us observe that a policy of this type, which requires motion along L, is not included in the set of policies associated with any nonzero δ. These policies, allowing only the use of A or B, yield broken-line paths consisting of horizontal and vertical pieces, as in Fig. 1.

It is clear, however, that a path such as that given in Fig. 2 may be arbitrarily closely approximated by an optimal policy as $\delta \to 0$.

This suggest the important point that a continuous version of the original discrete problem may not possess an optimal policy yielding a maximum return. Instead there may only exist a sequence of policies yielding a supremum-unless we suitably broaden the concept of a policy. *The natural way to accomplish this extension is to allow for the mixing of decisions, in some suitable sense, at each time.*

§ 5. Mixing at a point

The introduction of mixing at a point is, however, with no intention to pun, a mixed blessing, since it carries along with it a number of difficulties of both physical and mathematical nature. Mathematically, we find ourselves confronted by the same difficulties that made us wish to bypass the integral formulation of § 3; physically, we are reluctant to accept a policy which involves mixing decisions as one applicable to a problem where a choice of one or the other decision is required.

To avoid simultaneously the conceptual difficulties of both mathematical and physical origin, let us employ an interpretive device which has been used before in a very similar situation. The essence of this device is the observation that, under certain natural continuity assumptions, mixing decisions at a point is equivalent to mixing decisions over small intervals about the point.

We shall assume then, to construct our mathematical model, that we are considering a process which requires at the times $t = 0, \Delta, 2\Delta$, etc., that we determine the proportion of the following time interval of length Δ which will be devoted to A and B respectively. Thus, over a typical interval $[k\Delta, k\Delta + \Delta]$, we devote the first part, $[k\Delta, k\Delta + \varphi_1 \Delta]$ to the use of A; and over the second part $[k\Delta + \varphi_1\Delta, k\Delta + \Delta]$, B is used:

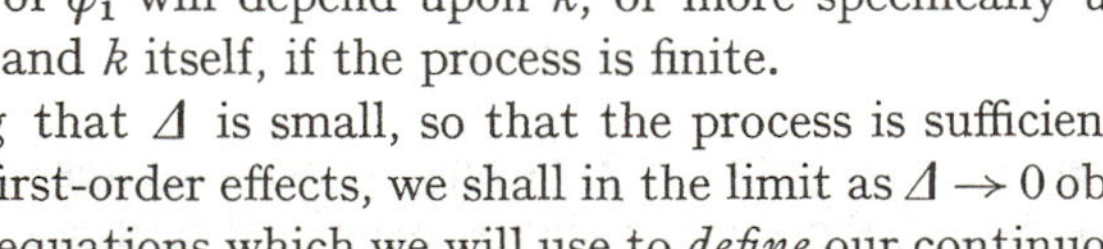

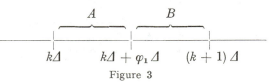

Figure 3

The choice of φ_1 will depend upon k, or more specifically upon $x(k\Delta)$, and $y(k\Delta)$, and k itself, if the process is finite.

Assuming that Δ is small, so that the process is sufficiently well described by first-order effects, we shall in the limit as $\Delta \to 0$ obtain a set of differential equations which we will use to *define* our continuous process.[1] A continuous policy will now be equivalent to a function $\varphi_1(t)$.

In the next chapter, we shall derive the differential equations. To illustrate the power of the method we shall, in turn, solve problems corresponding to the two-choice problem, to the two-choice problem for a

[1] Recall the corresponding comment in Chapter VII.

finite number of stages, to the two-choice problem with a nonlinear utility function, corresponding to the problem discussed in Exercise 1 of Chapter II, and to the three-choice problem of § 13 of that chapter. Although the analysis is quite detailed, the guiding ideas are simple.

To *justify* the use of this formalism, it should be shown that the continuous process obtained in this way is actually the limit of the original discrete process in a natural sense. This will be discussed in the second volume.

§ 6. Reformulation of the gold-mining process

Let us now proceed to carry through the program outlined in the preceding sections. An interesting feature of the mathematics will be the continued interplay between the techniques of the classical calculus of variations and those of dynamic programming.

Let us, to clarify the issue, rephrase the problem we are considering:

"At each of the time instants $t = k\Delta$ we shall have to make a decision concerning the proportion of the following interval of length Δ which will devoted to the use of the machine in mine A and to the use of the machine in mine B. This involves the choice of a fraction φ_1, which depends upon the amounts of gold in the two mines at time t, and upon t itself, if the process is finite.

We arbitrarily assume that once this proportion φ_1 has been chosen, the first part of the interval $[k\Delta, (k + \varphi_1) \Delta]$, is devoted to use of the machine in A, and the second part, $[(k + \varphi_1) \Delta, (k + 1) \Delta]$, to use of the machine in B. If x is the amount of gold in mine A at time $k\Delta$, there is a probability $1 - q_1 \varphi_1 \Delta$ that an amount $r_1 x \varphi_1 \Delta$ is mined, and that the machine is undamaged; and a probability $q_1 \varphi_1 \Delta$ that no gold is mined and that the machine is irretrievably damaged. If mine B contains y at time $k\Delta$ there is a probability $1 - q_2 \varphi_2 \Delta$ that the amount $r_2 y \varphi_2 \Delta$ is obtained, and that the machine is undamaged; and a probability $q_2 \varphi_2 \Delta$ that the operation ceases, where $\varphi_2 = 1 - \varphi_1$.

The problem is to determine the sequence of operations which maximizes the expected amount of gold mined before the machine is damaged."

§ 7. Derivation of the differential equations

It is easily seen that if Δ is small, permuting the order of operations in $[k\Delta, (k + 1) \Delta]$ is a second-order effect. It is this feature which allows mixing over intervals to perform the function of mixing at a point.

A policy now consists of a sequence $\{\varphi_1 (k\Delta)\}$, $k = 0, 1, 2, \ldots$. For any given policy, let

$x(t)$ = amount of gold remaining in A provided the operation has continued to t,

$y(t)$ = amount of gold remaining in B provided the operation has continued to t,

$p(t)$ = probability that the machine survives until t, i.e., that the operation continues until t,

$f(t)$ = expected amount of gold mined up to time t,

where $t = n\Delta$, $n = 0, 1, 2, \ldots$.

Ignoring the second-order terms in Δ, we have

$$(1) \qquad \begin{aligned} x(t+\Delta) &= x(t) - r_1 \varphi_1(t)\, x(t)\, \Delta \\ y(t+\Delta) &= y(t) - r_2 \varphi_2(t)\, y(t)\, \Delta \\ p(t+\Delta) &= p(t)\,(1 - q_1 \varphi_1(t)\, \Delta - q_2 \varphi_2(t)\, \Delta) \\ f(t+\Delta) &= f(t) + p(t)\,[\varphi_1(t)\, r_1\, x(t) + \varphi_2(t)\, r_2\, y(t)]\, \Delta \end{aligned}$$

Letting $\Delta \to 0$, we obtain the system of differential equations

$$(2) \qquad \begin{aligned} dx/dt &= -\varphi_1(t)\, r_1\, x(t), & x(0) &= x_0, \\ dy/dt &= -\varphi_2(t)\, r_2\, y(t), & y(0) &= y_0, \\ dp/dt &= -p(t)\,[\varphi_1(t)\, q_1 + \varphi_2(t)\, q_2], & p(0) &= 1 \\ df/dt &= p(t)\,[\varphi_1(t)\, r_1\, x(t) + \varphi_2(t)\, r_2\, y(t)], & f(0) &= 0 \end{aligned}$$

We now take these equations as the defining equations of our process, and ignore their formal origin. The problem we set ourselves is that of determining $\varphi_1 = \varphi_1(t)$, where

$$(3) \qquad 0 \leq \varphi_1(t) \leq 1, \; \varphi_2(t) = 1 - \varphi_1(t),$$

so as to maximize $f(T)$. A case of particular importance is $T = \infty$.

We shall derive similar equations for the three-choice problem in § 12 below.

§ 8. The variational procedure

Let φ_1 and φ_2 be functions furnishing the maximum,[2] and let

$$(1) \qquad \bar{\varphi}_i = \varphi_i + \varepsilon \beta_i(t),$$

where ε is a small positive quantity, and β_1, β_2 are two functions of t satisfying for all $t \geq 0$ the conditions

$$(2) \qquad 0 \leq \varphi_i + \varepsilon\, \beta_i \leq 1, \; \beta_1 + \beta_2 = 0$$

(which implies $|\beta_i| \leq 1/\varepsilon$), so that the $\bar{\varphi}_i$ are also admissible φ' s.

[2] It is easy to show, as a consequence of the uniform boundedness of the function $\varphi_1(t)$, that the maximum is attained.

It follows that $\beta_i(t) \leq 0$ if $\varphi_i(t) = 1$, $\beta_i(t) \geq 0$ if $\varphi_i(t) = 0$, and β_i can be of either sign if $0 < \varphi_i(t) < 1$, the region where free variation is permitted. Performing the variation, we find readily that

$$(3) \quad \bar{x}(t) = x(t)(1 - \varepsilon r_1 B_1(t)) + o(\varepsilon)^3$$

$$\bar{y}(t) = y(t)(1 - \varepsilon r_2 B_2(t)) + o(\varepsilon)$$

$$\bar{p}(t) = p(t)(1 - \varepsilon q_1 B_1(t) - \varepsilon q_2 B_2(t)) + o(\varepsilon)$$

$$\bar{f}(T) - f(T) = \varepsilon \int_0^T \{-f'(t)(q_1 B_1(t) + q_2 B_2(t)) + r_1 B_1(t) p(t) x'(t) + r_2 B_2(t) p(t) y'(t) + r_1 \beta_1(t) p(t) x(t) + r_2 \beta_2(t) p(t) y(t))\} dt + o(\varepsilon)$$

where we have set

$$(4) \qquad B_i(t) = \int_0^t \beta_i(s)\, ds$$

and the bars refer to the perturbed variables.

Integrating by parts to eliminate the $B_i(t)$, we find

$$(5) \qquad \bar{f}(T) - f(T) = \varepsilon \int_0^T [K_1(t)\beta_1(t) + K_2(t)\beta_2(t)]\, dt + o(\varepsilon)$$

where

$$(6) \qquad K_1(t) = -q_1 \int_t^T f'(s)\, ds + r_1 p(T) x(T) - r_1 \int_t^T p'(s) x(s)\, ds$$

$$K_2(t) = -q_2 \int_t^T f'(s)\, ds + r_2 p(T) y(T) - r_2 \int_t^T p'(s) y(s)\, ds$$

Since $\bar{f}(T) - f(T) \leq 0$, we see that whenever $K_i(t) > K_j(t)$ we must have $\varphi_i(t) = 1$, $\varphi_j(t) = 0$. These relations yield implicit equations for φ_i and φ_j. In the next section we shall discuss the behavior of the K-functions in more detail, in order to determine $\varphi_1(t)$ explicitly.

§ 9. The behavior of K_i.

The fundamental relation is

$$(1) \qquad d/dt\,(K_1 - K_2) = (q_1 - q_2) f'(t) - p'(t)(r_2 y - r_1 x) = p[q_1 r_2 y - q_2 r_1 x].$$

[3] The term $o(\varepsilon)$ denotes a function of t which approaches 0 as $\varepsilon \to 0$ for all t in $[0, T]$.

Thus a "mixed policy" (one for which more than one of the φ_i is positive for a given t, which implies $K_1(t) = K_2(t)$) can be optimal only on the line $q_1 r_2 y = q_2 r_1 x$. This line is precisely the boundary line that one obtains by passage to the limit from the solution in the discrete case as $\Delta \to 0$, as in § 4.[4]

If a mixed policy is pursued along the line, φ_1 and φ_2 must be chosen to stay on this line, which means that the slope, $s = y/x$, must be kept constant. Since

$$d/dt\,(y/x) = y'/x - (x'/x)\, s(t) = [r_1 \varphi_1 - r_2 \varphi_2]\, s \tag{2}$$

we see that we must have

$$\varphi_1 = \frac{r_2}{r_1 + r_2}, \varphi_2 = \frac{r_1}{r_1 + r_2} \tag{3}$$

§ 10. The solution for $T = \infty$

With these preliminaries out of the way, let us determine the optimal policy for the infinite process, $T = \infty$. The infinite problem is, as usual, simpler than the finite case because of the homogeneity introduced by infinite time; after any initial actions, we are confronted by a problem of the same type, with different initial values. Let us note that a consequence of this, and the homogeneity of the equations with respect to x and y, is that the decision at any point is a function only of the slope $s = y/x$.

Let us begin by observing that if policy A is ever used above the line $q_1 r_2 y = q_2 r_1 x$ in the (x, y)-plane, it is used thereafter. This follows immediately from (9.1) which shows that $K_1 - K_2$ is increasing when $q_1 r_2 y - q_2 r_1 x > 0$. Since use of A decreases x and leaves y unchanged, once $K_1 > K_2$ the use of A maintains the inequality.

Near the y-axis, however, the use of A continually is not as rewarding as continual use of B. For with $\varphi_1 = 1, \varphi_2 = 0$, for $t \geq 0$, we have

$$\begin{aligned} x(t) &= x_0 e^{-r_1 t} \\ y(t) &= y_0 \\ p(t) &= e^{-q_1 t} \\ f(t) &= \int_0^t r_1 x_0 e^{-r_1 s} e^{-q_1 s}\, ds \end{aligned} \tag{1}$$

and thus

$$f_A(\infty) = r_1 x_0/(q_1 + r_1)\,.$$

[4] Having been led to expect the appearance of this line as a consequence of the analysis of the discrete case, it is relatively easy to spot it.

However, $\varphi_1 = 0$, $\varphi_2 = 1$ for all t yields similarly $f_B(\infty) = r_2 y_0/(q_2 + r_2)$. For y_0/x_0 sufficiently large $f_B(\infty) > f_A(\infty)$. Thus, there is a region near the y-axis where B is used.

This region where B is used extends down to the line $q_1 r_2 y = q_2 r_1 x$. To prove this we observe that a mixed policy cannot be pursued above the line, and that if A is ever used above the line it is always used thereafter. Using A indefinitely, however, would eventually take (x, y) into the region near the y-axis where B is known to be optimal, a contradiction. Hence B is always used above the line. Similarly, below the line A is always used.

When the line $q_1 r_2 y = q_2 r_1 x$ is reached, the point (x, y) must remain on the line thereafter. For if not, then an A policy must be used in a B region or vice versa, which is impossible. Hence, on the line itself the mixed policy of (9.3) must be employed.

We have thus demonstrated

THEOREM 1. *With reference to the equations* (7.2) *and the constraints* (7.3), *the maximum value of* $f(\infty)$ *is attained by use of the policy*

(2)
$$\varphi_1 = 1 \text{ for } q_1 r_2 y < q_2 r_1 x\,,$$
$$\varphi_2 = 1 \text{ for } q_1 r_2 y > q_2 r_1 x\,,$$
$$\varphi_1 = \frac{r_2}{r_1 + r_2},\ \varphi_2 = \frac{r_1}{r_1 + r_2} \text{ for } q_1 r_2 y = q_2 r_1 x\,.$$

Note that φ_1 and φ_2 are determined almost everywhere by the above arguments, and hence are essentially unique. The above constructive derivation of the solution furnishes an alternative existence proof.

§ 11. Solution for finite total time

In finding the solution for finite T, we shall begin by determining what policy is used last. Since an optimal policy has the property that its continuation after any initial part is also optimal, we shall consider first the case where T is small. We have

(1)
$$f(T) = \int_0^T p(s)\,[\varphi_1(s)\, r_1 x(s) + \varphi_2(s)\, r_2 y(s)]\, ds$$
$$= r_1 x_0 \int_0^T \varphi_1(s)\, ds + r_2 y_0 \int_0^T \varphi_2(s)\, ds + o(T)$$

for T close to 0.

It follows then that for small T the maximum is obtained by taking $\varphi_1(s) = 1$, $\varphi_2(s) = 0$ for $r_1 x_0 > r_2 y_0$ and $\varphi_1(s) = 0$, $\varphi_2(s) = 1$ for $r_2 y_0 > r_1 x_0$. As is to be expected, for processes of small duration expected gain, without worry about termination, is the determining factor.

If $q_1 = q_2$ the lines $r_2 y = r_1 x$ and $q_1 r_2 y = q_2 r_1 x$ coincide, and the optimal policy is easily found to be the same as that for $T = \infty$.

Let us consider the general case where $q_1 \neq q_2$. Assume, without loss of generality, that the line $r_2 y = r_1 x$ lies above the line $q_1 r_2 y = q_2 r_1 x$. The positive quadrant then is divided into three regions, which we label I, II, III. (Fig. 4).

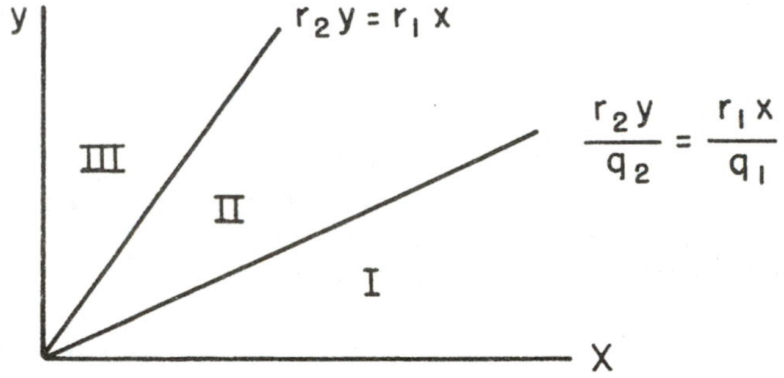

Figure 4

As before, it follows that in region I a B-policy once used must be continued thereafter, while in regions II and III the same holds for an A-policy. Also, in regions I and II an A-policy is used if the time remaining is sufficiently small, and in III a B-policy under the same conditions. From this we conclude that an A-policy is always used in I, and a B-policy always while in III.

Let us now establish that an optimal policy never switches from A to B. Let us suppose otherwise and let t_0 be the time at which the change occurs. Since at t_0, A is terminated, the point $(x(t_0), y(t_0))$ must be in region I, or on the boundary between I and II. Using B will keep the point $(x(t), y(t))$ in I for all $t > t_0$ since we know that B once used in I must be continued. However, this contradicts the fact that A is used in I whenever the time remaining is sufficiently small. Similarly, the combination of using the mixed policy and then B cannot occur, since the change-over must occur on the boundary between I and II, and then B is used thereafter in region I, a contradiction.

This reduces the number of types of solutions to six: A always; B always; the mixed policy followed by A; A then the mixed policy and finally B; B then the mixed policy and then A; B followed by A.

Let t_0 be the value of t at which the last change of policy is made in an optimal strategy, if such a change occurs. For $t_0 < t \leq T$, we must have $\varphi_1(t) = 1$, $\varphi_2(t) = 0$. We now compute the value of $K_1(t_0) - K_2(t_0)$. We have for $t_0 < t \leq T$,

(2)
$$x(t) = x(t_0)\, e^{-r_1 (t-t_0)},\ y(t) = y(t_0)$$
$$p(t) = p(t_0)\, e^{-q_1 (t-t_0)},$$
$$f'(t) = p(t_0)\, e^{-(q_1+r_1)(t-t_0)}\, r_1\, x(t_0)$$

and, after some simplification,

(3)
$$K_1(t_0) - K_2(t_0) = p(t_0)\, r_1\, x(t_0) \left[\left(1 - \frac{q_2}{q_1+r_1}\right) e^{-(q_1+r_1)(T-t_0)} + \frac{q_2}{q_1+r_1} - \frac{r_2\, y(t_0)}{r_1\, x(t_0)}\right]$$

For any fixed point $(x(t_0), y(t_0))$ in II, the right side is positive for $T - t_0$ small, and negative for $T - t_0$ large. It is equal to zero for precisely one value of $T - t_0$. This zero determines when the changeover occurs. When it occurs, A is used for the remaining time, with any of the six beginnings above, depending upon the location of the initial point.

§ 12. The three-choice problem

The continuous version of the three-choice problem mentioned above in § 13 of Chapter II leads via the same formal process as given in § 7 to the following. Given

(1)
$$dx/dt = -[\varphi_1(t)\, r_1 + \varphi_3(t)\, r_3]\, x(t), \qquad x(0) = x_0$$
$$dy/dt = -[\varphi_2(t)\, r_2 + \varphi_3(t)\, r_4]\, y(t), \qquad y(0) = y_0$$
$$dp/dt = -p(t)\, [\varphi_1(t)\, q_1 + \varphi_2(t)\, q_2 + \varphi_3(t)\, q_3], \quad p(0) = 1,$$
$$df/dt = p(t)\, [(\varphi_1(t)\, r_1 + \varphi_3(t)\, r_3) x(t) + (\varphi_2(t)\, r_2 + \varphi_3(t)\, r_4)\, y(t)]$$
$$f(0) = 0,$$

where, for all t,

(2)
$$\varphi_1 + \varphi_2 + \varphi_3 = 1, \quad \varphi_i \geq 0,$$

It is required to determine the $\varphi_i(t)$ so as to maximize $f(T)$.

We shall consider only the case where $T = \infty$.

As before, let us set $\bar{\varphi}_i = \varphi_i + \varepsilon\beta_i$, and $B_i(t) = \int_0^t \beta_i(s)\, ds$

We obtain

(3)
$$\bar{x}(t) = x(t)\,(1 - \varepsilon\, r_1 B_1(t) - \varepsilon\, r_3 B_3(t)) + o(\varepsilon)$$
$$\bar{y}(t) = y(t)\,(1 - \varepsilon\, r_2 B_2(t) - \varepsilon\, r_3 B_3(t)) + o(\varepsilon)$$
$$\bar{p}(t) = p(t)\,(1 - \varepsilon \sum_{i=1}^{3} q_i B_i(t)) + o(\varepsilon)$$
$$d\bar{f}/dt = \bar{p}\, [(\bar{\varphi}_1\, r_1 + \bar{\varphi}_3\, r_3)\, \bar{x} + (\bar{\varphi}_2\, r_2 + \bar{\varphi}_3\, r_4)\, \bar{y}]$$

Consequently, following the same technique as before, we obtain

$$\bar{f}(T) - f(T) = \varepsilon \int_0^T [K_1 \beta_1 + K_2 \beta_2 + K_3 \beta_3]\, dt + o(\varepsilon) \tag{4}$$

where

$$K_1(t) = -q_1 \int_t^T f'(s)\, ds + r_1 p(T)\, x(T) - r_1 \int_t^T p'(s)\, x(s)\, ds \tag{5}$$

$$K_2(t) = -q_2 \int_t^T f'(s)\, ds + r_2 p(T)\, y(T) - r_2 \int_t^T p'(s)\, y(s)\, ds$$

$$K_3(t) = -q_3 \int_t^T f'(s)\, ds + p(T)[r_3 x(T) + r_4 y(T)] - \int_t^T p'(s)[r_3 x(s) + r_4 y(s)]\, ds$$

§ 13. Some lemmas and preliminary results

The statements in the lemmas below concerning the dependence of the φ_i upon the K_i are, of course, taken to hold almost everywhere.

LEMMA 1. *If* $K_i(t) > K_j(t)$, *then* $\varphi_i(t) = 1$ *or* $\varphi_j(t) = 0$.

PROOF: Let E be the set of t for which the assertion does not hold. Let $\beta_i = 1$, $\beta_j = -1$ for t in E, and let the β's be zero otherwise. The variation is admissible for ε sufficiently small and makes $\bar{f}(T) - f(T)$ positive if $m(E) > 0$.

LEMMA 2. *If* $K_i(t) > K_j(t)$ *for* $j \neq i$, *then* $\varphi_i = 1$.

The proof follows immediately from the above.

LEMMA 3. *If there is a* j *such that* $K_i(t) < K_j(t)$, *then* $\varphi_i = 0$.

Again a simple consequence of Lemma 1.

Let us now compute the derivatives of the K_i. A straight-forward calculation yields the symmetric results

$$\begin{aligned} K_1'(t) &= p[C_1 \varphi_2 + C_2 \varphi_3] \\ K_2'(t) &= p[-C_1 \varphi_1 - C_3 \varphi_3] \\ K_3'(t) &= p[-C_2 \varphi_1 + C_3 \varphi_2] \end{aligned} \tag{1}$$

where we have set

$$\begin{aligned} C_1 &= q_1 r_2 y - q_2 r_1 x \\ C_2 &= q_1 r_4 y - (q_3 r_1 - q_1 r_3) x \\ C_3 &= (q_3 r_2 - q_2 r_4) y - q_2 r_3 x \end{aligned} \tag{2}$$

The relative positions of the three lines $C_i = 0$ are determined by the quantity

$$D = q_1 r_2 r_3 + q_2 r_1 r_4 - q_3 r_1 r_2 \tag{3}$$

If we assume that all three lines lie in the positive quadrant, a straightforward calculation shows that if $D > 0$ the lines have the position shown in Fig. 5, while if $D < 0$ they lie as shown in Fig. 6.

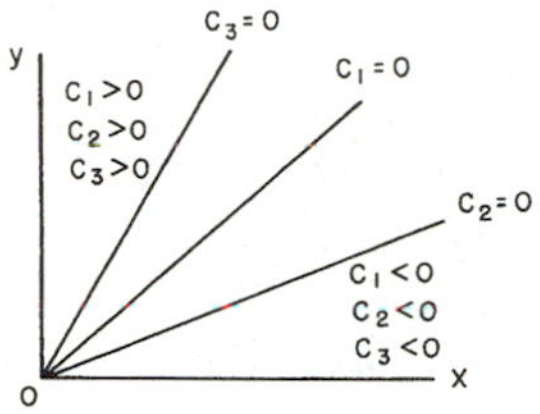

Figure 5

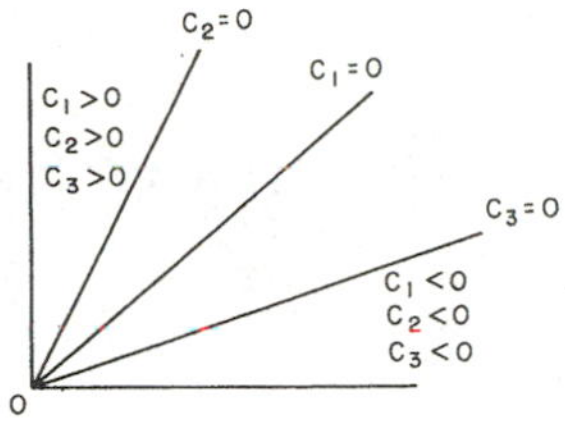

Figure 6

It is possible for both cases $D > 0$, $D < 0$ to occur. The case where one of the lines $C_2 = 0$, $C_3 = 0$ lies outside the positive quadrant yields an immediate simplification of the following arguments without changing the over-all structure. Consequently, we shall discuss in detail only the above cases.

§ 14. Mixed policies

As above, we denote by the term "mixed policy" a situation in which some of the φ_i have values different from 0 and 1. By an A-policy we shall mean $\varphi_1 = 1$, a B-policy $\varphi_2 = 1$, and a C-policy $\varphi_3 = 1$. Let us prove

LEMMA 4. *No optimal policy contains a mixture of A, B, and C policies.*

PROOF: Let us assume that in some interval we have simultaneously $\varphi_1, \varphi_2, \varphi_3 > 0$. In this interval we must have $K_1 = K_2 = K_3$.
This yields

$$\begin{aligned} &\varphi_1 + \varphi_2 + \varphi_3 = 1 \\ &K_1' - K_2' = p\,[C_1 \varphi_1 + C_1 \varphi_2 + (C_2 + C_3)\, \varphi_3] = 0 \\ &K_1' - K_3' = p\,[C_2 \varphi_1 + (C_1 - C_3)\, \varphi_2 + C_2 \varphi_3] = 0 \end{aligned} \tag{1}$$

The solution for $\varphi_1, \varphi_2, \varphi_3$ is, if $C_1 - C_2 - C_3 \neq 0$,

$$\varphi_1 = \frac{-C_3}{C_1 - C_2 - C_3}, \quad \varphi_2 = \frac{-C_2}{C_1 - C_2 - C_3}, \quad \varphi_3 = \frac{C_1}{C_1 - C_2 - C_3} \tag{2}$$

Since the φ_i must be positive in this interval, we must have C_1, $-C_2$, and $-C_3$ all of the same sign. It is easily verified upon referring to Figs. 2 and 3 that in both cases $D > 0$, $D < 0$, this can never occur.

Furthermore, $C_1 - C_2 - C_3 = 0$ only if the lines $C_1 = 0$, $C_2 = 0$, $C_3 = 0$ coincide. When this occurs the problem is equivalent to the two-choice problem.

Let us now investigate the possibility of using mixed policies involving only two of the three policies, A, B, or C.

LEMMA 5. *Concerning the mixing of two and only two policies, we have the following results:*

(3) (a) *A mixture of A and B is permissible only along*

$$C_1 = 0, \text{ where } \varphi_1 = r_2/(r_1 + r_2),\ \varphi_2 = r_1/(r_1 + r_2).$$

(b) *A mixture of A and C is permissible only along $C_2 = 0$, where*

$$\varphi_1 = \frac{r_4 - r_3}{r_1 + r_4 - r_3},\quad \varphi_3 = \frac{r_1}{r_1 + r_4 - r_3}$$

(c) *A mixture of B and C is permissible only along $C_3 = 0$, where*

$$\varphi_2 = \frac{r_3 - r_4}{r_2 + r_3 - r_4},\quad \varphi_3 = \frac{r_2}{r_2 + r_3 - r_4}$$

PROOF: If $\varphi_1, \varphi_2 > 0, \varphi_3 = 0$, we must have $K_1 = K_2 > K_3$. In an interval where this occurs,

$$0 = K_1' - K_2' = p\,[C_1\,(\varphi_1 + \varphi_2)] \tag{4}$$

Hence $C_1 = 0$. The values of φ_1 and φ_2 which keep (x, y) on this line are determined as in the two-choice case. The other assertions in Lemma 5 are obtained similarly.

§ 15. The solution for infinite time, D > 0

Having obtained these auxiliary results, we now proceed to find the solution to the problem of maximizing $f(\infty)$. We shall assume that $r_3 > r_4$, since the case $r_4 > r_3$ can be handled by interchanging the roles of x and y and A and B. The degenerate case, $r_3 = r_4$, will be discussed separately.

Let us make an initial observation that when $r_3 > r_4$ the mixed AC policy is never used, for by (14.3) φ_1 and φ_3 cannot both be positive. The solution takes two distinct forms depending upon whether $D > 0$ or

$D < 0$. Let us begin by considering $D > 0$. We shall establish the principal results in a series of lemmas.

LEMMA 6. *In an optimal policy, B is used near the y-axis.*

PROOF: There is a region near the y-axis where A is not used. For if $C_1 > 0, C_2 > 0$ and A is used, i.e., $\varphi_1(t) = 1$, we have $K_1' = 0, K_2' < 0, K_3' < 0$. This means that K_1 remains the largest for $t_1 \geq t$. Hence, if A is used in this region, it must be pursued thereafter. Let us now compute the results of a continued A-policy, a continued B-policy, and a continued C-policy. We have

$$(1) \qquad \begin{aligned} f_A(\infty) &= r_1 x_0/(q_1 + r_1) \\ f_B(\infty) &= r_2 y_0/(q_2 + r_2) \\ f_C(\infty) &= \frac{r_3 x_0}{q_2 + r_3} + \frac{r_4 y_0}{q_3 + r_4} \end{aligned}$$

A comparison of $f_A(\infty)$ and $f_B(\infty)$ shows that $f_B(\infty) > f_A(\infty)$ for y_0/x_0 sufficiently large.

Let us now show that in the region above the line $C_3 = 0$, if C is used it is used continually thereafter. Using C increases the slope $s(t) = y(t)/x(t)$, for with $\varphi_3 = 1$ we have

$$(2) \qquad s'(t) = s(t)(r_3 - r_4) > 0$$

On the other hand, using B decreases the slope. Hence, we cannot use B after C, for to do so would return us to a region where C was to be used. We have already shown that A cannot be used after C when close to the y-axis. A comparison of $f_B(\infty)$ and $f_C(\infty)$ shows that it is better to use B rather than C near the y-axis if $r_2 y/(q_2 + r_2) > r_4 y/(q_3 + r_4)$, or $q_3 r_2 - q_2 r_4 > 0$. This, however, is precisely equivalent to the condition that $C_3 = 0$ lie within the positive quadrant, which we have assumed.

It follows that there is a region near the y-axis where neither A nor C is used. Since by Lemma 5 no mixed policy is used above the line $C_3 = 0$, we conclude that there is a region adjoining the y-axis where B must be used.

LEMMA 7. *The lower boundary of the B-region adjoining the y-axis is the line $C_3 = 0$. On that line a mixed BC-policy is employed. Below $C_3 = 0$, B is never used.*

PROOF: Let us begin with initial values (x_0, y_0) near the y-axis in the region where B is used and consider what form an optimal strategy can have. B cannot be used indefinitely since this would eventually take (x, y) near the x-axis where comparison of $f_A(\infty)$ and $f_B(\infty)$ shows that A is superior. However, since both A and C increase the slope y/x, B cannot be followed by A or C since both of these would immediately put the point (x, y) back into a region where B is to be used. Consequently, B must be followed by one of the mixed policies.

As we have already seen, for $r_3 > r_4$ the mixed policy AC is never used in an optimal strategy. We assert that if a mixed policy is used in an optimal strategy, then continuing the mixed policy forever is optimal. For let (t_o, t_1) be an interval on which the mixed policy is pursued. Since the point $(x(t_1), y(t_1))$ lies on the same ray as $(x(t_o), y(t_o))$, because of the homogeneity the same policy, continued for an equal length of time, is optimal. Hence the mixed policy may be continued forever. Taking this remark into account, we can show that for $D > 0$ the mixture AB never occurs in an optimal strategy. By Lemma 5a, AB could only be used on the line $C_1 = 0$. If AB were used there, we would have

$$K_3' = p[C_3\varphi_2 - C_2\varphi_1] < 0$$

since $C_2 > 0$ and $C_3 < 0$ there (cf. Fig. 2). Since $K_1(\infty) = K_2(\infty) = K_3(\infty) = 0$ and $K_1 = K_2 = 0$ while AB is being used, it follows that $K_3 > K_1 = K_2$ while the AB-mixture is being used. This, however, implies that $\varphi_3 = 1, \varphi_1 = \varphi_2 = 0$, which is a contradiction.

The remaining possibility then is that BC is used after B on the line $C_3 = 0$. B cannot be used below this line as a consequence of the above arguments.

LEMMA 8. *There is a line $L = 0$ between $C_2 = 0$ and the x-axis such that C is used in the region between $C_3 = 0$ and $L = 0$, and the policy A is used in the region below $L = 0$.*

PROOF: By the results already established we know that the only policies which can be used in the region below the line $C_3 = 0$ are A and C. Since both of these policies increase the slope exponentially, eventually the point (x, y) will reach the line $C_3 = 0$ where the mixed policy BC is employed.

Let us investigate the possibilities of changes from A to C and from C to A. By (13.1) we have

$$K_1'(t) - K_3'(t) = p[C_1\varphi_2 + C_2\varphi_3 + C_2\varphi_1 - C_3\varphi_2]$$

and hence when only C or A is used,

$$K_1'(t) - K_3'(t) = p\,C_2[\varphi_1 + \varphi_3] \tag{3}$$

which is positive above $C_2 = 0$ and negative below. Now in a changeover from C to A we must have $K_1' - K_3' \geq 0$. Consequently, a change from

C to A cannot occur below $C_2 = 0$. Similarly, we observe that a change from A to C cannot occur above $C_2 = 0$. Also there cannot be a change from A to BC because when A is used above $C_2 = 0$, $K_1 - K_3$ is positive and increases; hence BC, which requires $K_3 > K_1$, cannot be used. Thus the assumption that A can be used above $C_2 = 0$ leads to a contradiction, since, as we know, BC must be used eventually.

We also can prove that a change from A to C cannot occur on the line $C_2 = 0$. For suppose that such a change occurred. At this time of change we would have $K_1 = K_3$. The C-policy will then take the point (x, y) above the line $C_2 = 0$ where $K_1' - K_3' > 0$, hence $K_1 > K_3$, which means that A must be used, a contradiction.

There are now two possible cases:

(1) C is used in the entire region below $C_3 = 0$.

(2) There is a line $L = 0$ lying between the x-axis and $C_2 = 0$ such that A is used below $L = 0$ and C is used above.

The following proof by contradiction shows that the first case does not occur. Let (x_0, y_0) be a point below $C_3 = 0$. By assumption C and BC are the only policies used so that we must have $K_3'(t) = 0$ for all $t \geq 0$. Since $K_3(\infty) = 0$, we have $K_3(0) = 0$. Because C is preferable at (x_0, y_0), we must have $0 = K_3(0) \geq K_1(0)$. Hence, since $K_1(\infty) = 0$, we have by (13.1)

$$(4) \qquad 0 \leq K_1(\infty) - K_1(0) = \int_0^{t'} p(t)\, C_2\, dt + \int_{t'}^{\infty} p(t)\,[C_1 \varphi_2 + C_2 \varphi_3]\, dt$$

where t' is the time of changeover from C to BC. Keeping x_0 fixed, let $y_0 \to 0$. This entails $t' \to \infty$. Since $C_1 \varphi_2 + C_2 \varphi_3$ is uniformly bounded, the second integral tends to zero. We have then, using the expressions for x, y, p, obtained from a C-policy

$$(5) \qquad \lim_{y_o \to 0} \int_o^{t'} e^{-q_3 t}\,[q_1 r_4 y_0 e^{-r_4 t} - (q_3 r_1 - q_1 r_3)\, x_0 e^{-r_3 t}]\, dt \geq 0$$

or

$$(6) \qquad -\int_0^{\infty} (q_3 r_1 - q_1 r_3)\, x_0 e^{-(q_3 + r_3)\, t}\, dt = -\frac{(q_3 r_1 - q_1 r_3)}{q_3 + r_3}\, x_0 \geq 0,$$

which contradicts the assumption that the line $C_2 = 0$ passes through the positive quadrant.

This completes the consideration of the case $D > 0$ when both $C_2 = 0$ and $C_3 = 0$ are contained in the positive quadrant. The complete result is

THEOREM 2. *If* $D = q_1 r_2 r_3 + q_2 r_1 r_4 - q_3 r_1 r_2 > 0$, *the solution to the problem of maximizing* $f(\infty)$ *subject to* (12.1) *is given schematically by Fig.* 7. It does not seem possible to specify L in any simple way.

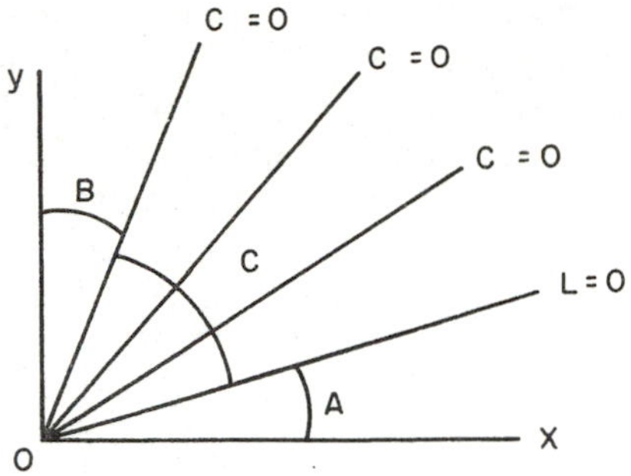

Figure 7

Finally, let us discuss the degenerate cases in which $C_3 = 0$ or $C_2 = 0$ do not lie in the positive quadrant. If $C_3 = 0$ lies outside, the C-region extends all the way to the y-axis.

§ 16. D < 0

Let us now consider the case in which $D < 0$. In this case it turns out that C is never used, which means that the solution is as given in the two-choice problem

Lemma 11. *B is used near the y-axis.*

Proof: Precisely as before.

Lemma 12. *The lower boundary of the B-region adjoining the y-axis is $C_1 = 0$. On that line AB is used. Below the line B is not used.*

Proof: As in the case $D > 0$ we conclude that a B-policy must be followed by one of the mixed policies AB or BC. However, in the present case where $D < 0$, the mixed policy BC cannot be used in an optimal strategy. For when BC is used, we have

$$K_1'(t) = p\,[C_1\varphi_2 + C_2\varphi_3] < 0 \tag{1}$$

because $C_3 = 0$ is below $C_2 = 0$ and $C_1 = 0$. Also $K_1(\infty) = K_2(\infty) = K_3(\infty) = 0$, and $K_2'(t) = K_3'(t) = 0$ when the mixed policy BC is used. Hence $K_1(t) > K_2(t) = K_3(t)$ when the BC-mix is used. This, however, is a contradiction since it implies that $\varphi_1 = 1$, $\varphi_2 = \varphi_3 = 0$. Hence, a B-policy must be followed by use of AB on $C_1 = 0$.

Again the same argument as above shows that B is not used below $C_1 = 0$.

Lemma 12. *A is used in the entire region between $C_1 = 0$ and the x-axis.*

Proof: First, C is not used just before the AB-mixture. While AB is employed, $K_1'(t) = K_2'(t) = 0$, and $K_3'(t) = p\,[-C_2\varphi_1 + C_3\varphi_2] > 0$, as can be seen from Fig. 7. It follows that $K_3 < K_2$ and $K_3 < K_1$ immediately before the changeover to AB occurs. Hence C is not used immediately before AB.

It follows then that there is a region below $C_1 = 0$ and adjoining this line, where A is used. However, it is impossible to use another choice before A is an optimal policy. When A is used below C_1, we have

$$(2) \qquad K_1'(t) = 0, K_2'(t) = -pC_1 > 0, K_3'(t) = -pC_2 > 0$$

Hence, K_1 is the largest for all smaller t, and the A-region extends to the x-axis.

Collecting the above results, we have

THEOREM 8. *If* $D = q_1 r_2 r_3 + q_2 r_1 r_4 - q_3 r_1 r_2 < 0$, *the solution to the problem of maximizing* $f(\infty)$ *never uses a C-policy and has the two-choice form:*

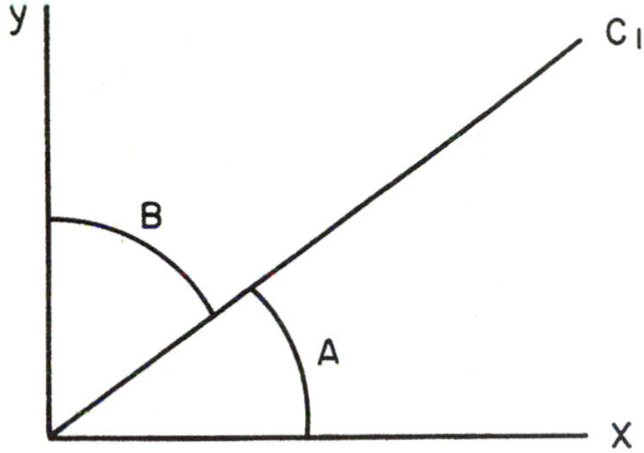

Figure 8

§ 17. The case $r_3 = r_4$

Some of the preceding arguments fail in this case because the C-policy keeps the slope y/x constant. It follows from (14.3b) and (14.3c) that neither of the mixed policies AC or BC is ever used.

Let us first of all show that if $D < 0$, C is never used. To do this we compare the result of using AB repeatedly with that obtained from using C.

When AB is used continually, an easy calculation yields

$$(1) \qquad f_{AB}(\infty) = \frac{r}{r+s}(x_0 + y_0)$$

where

$$(2) \qquad r = \frac{r_1 r_2}{r_1 + r_2}, \quad s = \frac{q_1 r_2 + q_2 r_1}{r_1 r_2}$$

Similarly the result of using C continually is

$$(3) \qquad f_C(\infty) = \frac{r_3}{q_3 + r_3}(x_o + y_o)$$

The inequality $f_{AB}(\infty) > f_C(\infty)$ is equivalent to $D < 0$.

If $D > 0$, the above argument proves that no mixed policies are pursued. Different cases arise depending upon which of the lines $C_2 = 0$, $C_3 = 0$ pass through the positive quadrant. As before, it can be established that if $C_3 = 0$ is the positive quadrant, it is better to use B rather than C near the y-axis. Let us now determine where the changeover from B to C can be made. Let t_o be the time of changeover. For $t_o < t < \infty$, we have

$$(4) \qquad K_1'(t) = -C_2, \; K_2'(t) = -pC_3, \; K_3'(t) = 0$$

Also, we must have $K_1(t_o) \leq K_2(t_o) = K_3(t_o)$. Using again the remark that $K_1(\infty) = K_2(\infty) = K_3(\infty)$, we see that for $t \geq t_o$, we must have $C_3 = 0$. Thus, B is followed until the line $C_3 = 0$ is encountered and then C is followed. In this degenerate case C plays the role of BC. Similarly, changeover from A to C occurs when $C_2 = 0$ is reached. If C_3 does not lie within the positive quadrant, C is used up to the y-axis. If $C_2 = 0$ does not lie within, C is used up to the x-axis.

§ 18. Nonlinear utility—two-choice problem

Let us now consider briefly the two-choice problem discussed in § 6—10 under the condition that we wish to maximize the expected value of some function u of the total return R.

In view of the results obtained for the discrete problem, or rather of the lack of results, it is somewhat surprising to find that for every utility function u, which is strictly increasing and has a continuous derivative, the optimal policy is precisely the same as that for the linear utility problem solved above. This alone should be sufficient to warn the unwary that continuous versions should not be used without close attention to the kind of approximation they afford.

Since any monotone-increasing utility function can be approximated arbitrarily closely by a function of the above type, it follows that this policy is optimal for any monotone-increasing utility function, although not necessarily unique. A function of this class of great theoretical and practical importance is

$$(1) \qquad \begin{aligned} u(R) &= 0 \text{ for } 0 \leq R < R_o \\ &= 1 \text{ for } R \geq R_o \end{aligned}$$

The expected value of $u(R)$ is the probability that R is greater than or equal to R_o.

Let the variables have their previous connotations; we obtain as before

$$
\begin{aligned}
dx/dt &= -\varphi_2(t)\, r_1\, x(t), & x(0) &= x_o \\
dy/dt &= -\varphi_2(t)\, r_2\, y(t), & y(0) &= y_o \\
dp/dt &= -p(t)\,[\varphi_1(t)\, q_1 + \varphi_2(t)\, q_2], & p(0) &= 1
\end{aligned} \tag{2}
$$

Let $z(t) = x_o + y_o - x(t) - y(t)$, the quantity which represents the total amount of gold mined up to t if the machine has survived up to this time. The expected value of $u(R)$ is given by the integral

$$G = -\int_o^\infty u(z(t))\, dp(t) \tag{3}$$

This is easiest seen by considering that we are paid for the total amount of gold that the machine has mined up to the time that the machine is damaged.

Our aim is to find the functions $\varphi_1(t), \varphi_2(t)$ subject to the constraints

$$0 \leq \varphi_i \leq 1, \varphi_1 + \varphi_2 = 1 \tag{4}$$

which maximize G.

Pursuing the same perturbation techniques as above, we obtain after some straightforward calculation

$$\overline{G} - G = \varepsilon \int_o^\infty [K_1(t)\,\beta_1(t) + K_2(t)\,\beta_2(t)]\, dt + o(\varepsilon) \tag{5}$$

where

$$
\begin{aligned}
K_1 &= q_1\, p(t)\, u(z(t)) - \int_o^\infty [p'(s)\, u'(z(s))\, r_1\, x(s) \\
&\qquad - q_1\, p'(s)\, u(z(s))]\, ds \\
K_2 &= q_2\, p(t)\, u(z(t)) - \int_t^\infty [p'(s)\, u'(z(s))\, r_2\, y(s) \\
&\qquad - q_2\, p'(s)\, u(z(s))]\, ds
\end{aligned} \tag{6}
$$

Furthermore,

$$K_1'(t) - K_2'(t) = p(t)\, u'(z(t))\, [q_1\, r_2\, y(t) - q_2\, r_1\, x(t)] \tag{7}$$

It follows that if we assume that $u'(z) > 0$ when $z > 0$, the arguments and results of the linear case carry over with very slight modifications.

Bibliography and Comments for Chapter VIII

§ 1. The results of this chapter were obtained in collaboration with R. S. Lehman: R. Bellman and R. S. Lehman, "On the Continuous Gold-Mining Equation," *Proc. Nat. Acad. Sci.*, vol. **40** (**1954**), pp. **115–119**, R. Bellman and R. S. Lehman, "On a Functional Equation in the Theory of Dynamic Programming and its Generalizations," *(unpublished)*.

§ 2. Equation (2.3) can be used in a formal manner to obtain the character of the solution, but nothing as yet has been done in the study of nonlinear partial differential equations of this type.

§ 3. The modern theory of stochastic processes, as expounded in J. L. Doob, Stochastic Processes, J. Wiley and Sons, (1953), furnishes a basis for a rigorous theory of such equations, but the mathematics required is definitely on an exalted plane.

§ 5. The concept of mixing at a point being replaced by mixing over a small interval was used in R. Bellman and D. Blackwell, "Some two-person games involving bluffing," *Proc. Nat. Acad. Sci.*, vol. **35** (**1949**), pp. **600–5**, in the study of some simple two-person poker games; cf. also R. Bellman, "On Games Involving Bluffing," *Rend. Circ. Mat. Palermo* (2), vol. **1** (**1952**), pp. 1–18.

Extensive results concerning the convergence of the discrete process to the continuous version will be found in H. Osborn, "On the Convergence of Discrete Stochastic Processes to their Continuous Analogues," RM–1368, RAND Corporation, (1955), H. Osborn, "The Problem of Continuous Programs," P–718, Rand Corporation, (1955) ; see also, *Pac. Jour. Math.*, Vol. **6** (1956), pp. **721–731.**

CHAPTER IX

A New Formalism in the Calculus of Variations

§ 1. Introduction

In two previous chapters, in our treatment of multi-stage production processes, we encountered the problem of maximizing the functional $(x(T), a)$ over all functions $z(t)$ subject to the relations

$$(1) \qquad \begin{aligned} &\text{a.} \quad dx/dt = Az, \quad x(0) = c, \\ &\text{b.} \quad Bz \leq Cx, \\ &\text{c.} \quad z \geq 0. \end{aligned}$$

Utilizing the fact that the maximum, which we assume is attained, is a function only of the initial vector c and the duration of the process T, we obtained a functional equation for $f(c, T) = \underset{z}{\text{Max}}\,(x(T), a)$, which we converted into a partial differential equation. As we mentioned at the end of Chapter 7, this same approach is equally available for the study of other classes of problems in the calculus of variations.

We shall pursue the investigation in this chapter, devoting our attention to two particular classes of problems. The first is that of determining the maximum or minimum of functionals of the form

$$(2) \qquad J(z) = \int_0^T F(x_1, x_2, \ldots, x_n, z_1, z_2, \ldots, z_m)\, dt,$$

subject to relations and constraints of the form

$$(3) \qquad \begin{aligned} &\text{a.} \quad dx_i/dt = G_i(x, z), \; x_i(0) = c_i, \; i = 1, 2, \ldots, n, \\ &\text{b.} \quad R_k(x, z) \leq 0, \; k = 1, 2, \ldots, l. \end{aligned}$$

The second is the eigenvalue problem associated with the equation

$$(4) \qquad u'' + \lambda^2 \varphi(t)\, u = 0, \qquad u(0) = u(1) = 0.$$

Since this problem is, under reasonable assumptions concerning $\varphi(t)$, equivalent to the problem of determining the relative minima of

$$(5) \qquad J(u) = \int_0^1 u'^2\, dt,$$

subject to the constraints

(6) a. $\int_0^1 \varphi(t)\, u^2\, dt = 1\,,$

b. $u(0) = u(1) = 0\,,$

we have a problem closely related to that described in equations (2) and (3). The two-point boundary condition, however, introduces features of novelty and difficulty.

Following our usual approach, we shall introduce suitable state variables and derive a functional equation for the minimum of $J(u)$ as a function of these variables. The limiting form of this functional equation will be a partial differential equation.

We shall then turn to a discussion of the numerical solution of these equations. After indicating the conventional solution by means of partial difference equations, we shall show how difference equations can enter along another route. The importance of this alternate approach lies in the fact that it enables us to bypass a number of thorny, analytic difficulties native to the domain of the calculus of variations. It also enables us to avoid a number of difficulties associated with the stability of computational techniques.

Using this approach, we shall consider also some problems involving a Cebycev functional

$$J(z) = \operatorname*{Max}_{0 \le t \le T} F(x_1, x_2, \ldots, x_n; z_1, z_2, \ldots, z_m)$$

In any case, we shall throughout the chapter consistently adopt a purely formal viewpoint. In this introductory, expository account we are primarily interested in presenting the basic principles of the functional equation method. A rigorous account, necessarily of a higher level of difficulty, will be reserved for the second volume.

§ 2. A new approach

Before embarking upon the high seas of analysis, let us discuss the basic idea of this new approach to continuous variational problems.

The classic technique in the calculus of variations, patterned directly upon the finite dimensional techniques of calculus, depends upon the concept of a function yielding an extremum as a point in function space, and the characterization of this point by means of variational properties.

We shall instead consider the calculus of variations as consisting of a particular class of multi-stage decision processes of continuous type. A function yielding an extremum may then be considered to be a continuous policy.

Let us give some simple examples which may serve to illustrate this idea more clearly than any abstract discussion.

EXAMPLE 1. *Determine the curve connecting two points, P and Q, having the property that a particle travelling along the curve under the influence of gravity will go from P to Q in minimum time.*

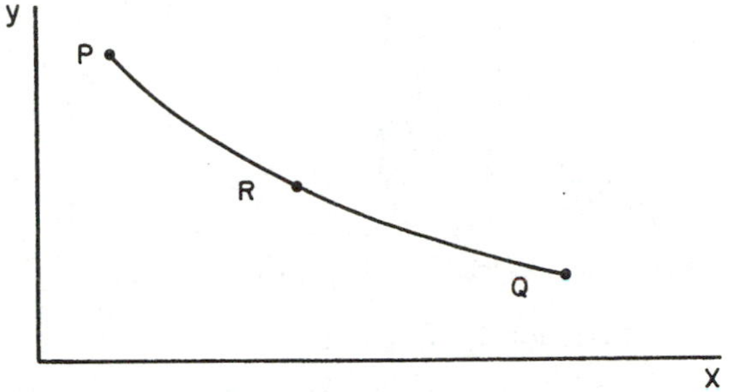

Figure 1

(*the classical brachistochrone problem*)

It is clear that along an extremal, whatever the path between P and some intermediate point R, the path between R and Q must be such as to minimize the time required to traverse RQ, given the left-hand velocity at R.

At each point on the curve, we determine a direction of motion, which is to say a tangent to the curve. The optimal policy or extremal may be expressed not only by means of an equation for y in terms of x, the usual approach, but also by means of an equation for dy/dx in terms of y and the given left-hand velocity at (x, y).

EXAMPLE 2. *Suppose that we are presented with the problem of drawing a curve passing through P and Q, as in the figure below, of fixed length L, which will include a maximum area in the curvilinear quadrilateral bounded by the curve, the perpendiculars PP', QQ', and the segment $P'Q'$ of the x-axis.*

It is clear that along an extremal, whatever the path between P and R, and whatever the shaded area obtained in this way, the continuation from R to Q must maximize the area $RR'\,Q'Q$ subject to the restriction that the curve RQ have length $L - L'$.

The optimal policy may be expressed by means of an equation for dy/dx in terms of y and $L - L'$, rather than by an equation for y in terms of x.

Both of the conclusions in these two examples are applications of the "principle of optimality" discussed in Chapter 3, and applied in all of the preceding chapters. The mathematical expression of this principle will yield our new approach to the calculus of variations.

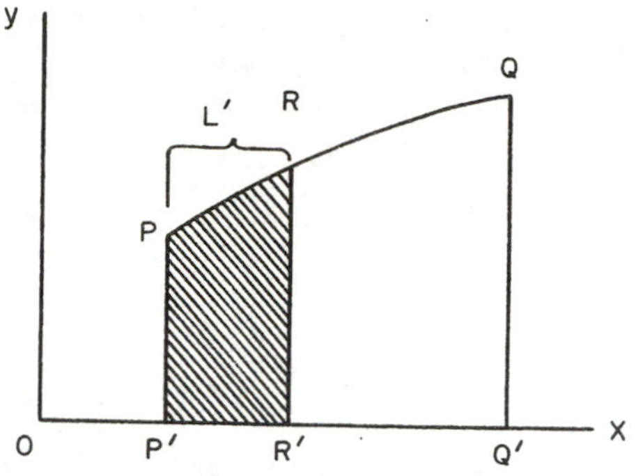

Figure 2 (the classical isoperimetric problem)

An advantage of this new approach lies in the fact that very often in the determination of optimal policies for multistage processes, the determination of the next move in terms of the current state of the process is in many ways a simpler, more natural and even more important piece of information than the determination of the entire sequence of moves in an optimal policy to be followed from some fixed initial position.

Speaking in geometrical terms, we seek to determine the intrinsic equations of extremal curves. In place of considering the curve as the locus of points, we regard it as the envelope of tangents, a dual approach to the classical treatment.[1]

In general, as is always to be expected, the combination of the two approaches, local and global, will be most powerful, since some aspects of an extremal are most simply described in point coordinates, and others in tangential coordinates.

We shall in the following sections apply these ideas to a number of representative problems, and discuss the application of this approach to the computation of solutions.

§ 3. $\operatorname*{Max}_{y} \int_0^\infty F(x, y)\, dt$

In Chapter 1 we considered the discrete process which gave rise to the functional equation

$$(1) \quad f(x) = \operatorname*{Max}_{0 \le y \le x} [g(y) + h(x-y) + f(ay + b(x-y))], f(0) = 0 .$$

[1] In the terminology of game theory, there may be a considerable advantage to viewing a process in its extensive rather than normal form. Essentially, only then do we take full advantage of the intrinsic structure of a process and thus differentiate it from other multi-stage processes and other multi-dimensional maximization problems.

A continuous version of this process gives rise to the problem of maximizing the functional

$$(2) \qquad J(y) = \int_0^\infty [g(y) + h(x - y)]\, dt\,,$$

with respect to $y(t)$, where

$$(3) \qquad \begin{aligned} &\text{a.} \quad dx/dt = -ay - b(x - y),\ a, b > 0,\ x(0) = c\,, \\ &\text{b.} \quad 0 \leq y(t) \leq x(t),\ t \geq 0\,. \end{aligned}$$

Let us then, to introduce our method, consider as our first example, the problem of maximizing an integral of the form

$$(4) \qquad J(y) = \int_0^\infty F(x, y)\, dt\,,$$

subject to the relation between x and y,

$$(5) \qquad dx/dt = G(x, y),\ x(0) = c\,.$$

To begin with, let us omit any constraint such as (3b).

Let us once again repeat that we shall proceed formally since we are interested here only in presenting the mechanics of our approach. This is to say, we shall consistently assume that maxima and minima exist, and that the extremals possess the requisite differentiability properties we shall need. The problem of establishing these properties rigorously is quite distinct from that of deriving the formalism and will not be considered here. Furthermore, as we shall indicate below, in a number of cases, we can pursue a path which eliminates any necessity for obtaining a priori results concerning the nature of the maximizing y.

Returning to the maximization problem posed above, we observe that the maximum value of $J(y)$ will be a function only of the initial value of x, namely c. Let us therefore write

$$(6) \qquad \operatorname*{Max}_{y} J(y) = f(c)\,,$$

and proceed to derive a functional equation for $f(c)$.

Let $y = y(t)$ be a function yielding the maximum of $J(y)$. We have then

$$(7) \qquad f(c) = \int_0^S F(x, y)\, dt + \int_S^\infty F(x, y)\, dt\,,$$

for any $S > 0$.

Consider the second integral. The effect of any initial choice of $y(t)$, for t in the interval $[0, S]$, will be, by way of the differential equation of (5), to convert c into the value of x at S, which we call $c(S)$. It follows then, that whatever the initial choice of y over $[0, S]$, we will have over the remaining interval, $[S, \infty]$, a problem of precisely the same form as the original, with the difference that c is now $c(S) \equiv x(S)$. Since the integrand is independent of t, and also the differential equation, the new interval may be considered to be $[0, \infty]$, with $x(0) = c(S)$.

It follows then, invoking the principle of optimality, that equation (7) may be rewritten

$$f(c) = \int_0^S F(x, y)\, dt + f(c(S)). \tag{8}$$

Since the choice of the function y must be made so as to yield the maximum value $f(c)$, we obtain the basic functional equation

$$f(c) = \operatorname*{Max}_{y\,[0,\,S]} \left[\int_0^S F(x, y)\, dt + f(c(S)) \right], \tag{9}$$

for any $S > 0$.

From this equation we shall derive a differential equation for $f(c)$ by letting S approach 0. For small S we have, under appropriate assumptions of continuity,

$$f(c) = \operatorname*{Max}_{y\,[0,\,S]} [F(c, y(0))\, S + f(c + SG(c, y(0)) + o(S)]. \tag{10}$$

As the interval $[0, S]$ shrinks to zero, a choice of y over $[0, S]$ becomes ultimately a choice of $y(0)$. Let us, for notational simplicity, set $v = y(0)$. Then (10) leads to

$$f(c) = \operatorname*{Max}_{v} [F(c, v)\, S + f(c) + SG(c, v) f'(c)] + o(S), \tag{11}$$

which in the limit as $S \to 0$ yields

$$0 = \operatorname*{Max}_{v} [F(c, v) + G(c, v) f'(c)]. \tag{12}$$

Applying calculus to determine this maximum, we obtain the *two* equations

$$\begin{aligned} 0 &= F(c, v) + G(c, v) f'(c), \\ 0 &= F_v(c, v) + G_v(c, v) f'(c). \end{aligned} \tag{13}$$

Elimination of $f'(c)$ between these two equations yields the determinantal equation

$$\begin{vmatrix} F(c, v) & G(c, v) \\ F_v(c, v) & G_v(c, v) \end{vmatrix} = 0, \tag{14}$$

which determines v as a function of c.

Having determined v as a function of c, which is to say, y as a function of x, we return to the differential equation of (5) and find x, and subsequently y, as functions of t by solving the differential equation

$$dx/dt = G[x, y(x)], \; x(0) = c. \tag{15}$$

From this we see that relatively simple policy, $y = \varphi(x)$, may yield a relatively complicated extremal function, $x = x(t)$.

§ 4. Discussion

Let us take $G(x, y)$ to be uniformly negative and equal to $-A(x, y)$ so that we may consider the above to represent a continuous allocation process where the rate of return is $F(x, y)$ and the rate of expenditure of resources is $A(x, y)$. Starting from the basic equation

$$0 = \operatorname*{Max}_v [F(c, v) - A(c, v) f'(c)], \tag{1}$$

we have, for all v,

$$0 \geq F(c, v) - A(c, v) f'(c), \tag{2}$$

and thus

$$f'(c) \geq F(c, v)/A(c, v). \tag{3}$$

Since there is equality for at least one value of v, we obtain the equation

$$f'(c) = \operatorname*{Max}_v \frac{F(c, v)}{A(c, v)}. \tag{4}$$

This equation tells us that the policy which maximizes the overall return proceeds locally to maximize the ratio of the rate of return to the rate of expenditure of resources, a policy we have encountered before, cf. Exer. 18 of Chapter 1, § 8 of Chapter 2.

This is a very interesting interpretation of the Euler equation for variational problems of the above simple form. We leave it to the reader to verify that (14) of § 3 is a first integral of the Euler equation obtained in the classical manner.

§ 5. The two dimensional case

We leave as an exercise the proof of the result that the same technique applied to the problem of determining the maximum of

$$\int_0^\infty F(x_1, x_2, y_1, y_2)\, dt, \tag{1}$$

over all functions $y_1(t)$ and $y_2(t)$ subject to

$$\begin{aligned} dx_1/dt &= G(x_1, x_2, y_1, y_2), & x_1(0) &= c_1, \\ dx_2/dt &= H(x_1, x_2, y_1, y_2), & x_2(0) &= c_2, \end{aligned} \tag{2}$$

yields the determinantal equation

$$(3)\qquad \begin{vmatrix} F(c_1, c_2, u, v) & G(c_1, c_2, u, v) & H'(c_1, c_2, u, v) \\ F_u & G_u & H_u \\ F_v & G_v & H_v \end{vmatrix} = 0,$$

connecting the values $y_1(0) = u(c_1, c_2)$, $y_2(0) = v(c_1, c_2)$.

It is an open problem as to whether or not a solution to the above variational problem can be obtained in the same form as in the one-dimensional case, i.e., in the form $y_1 = \varphi_1(x_1, x_2)$, $y_2 = \varphi_2(x_1, x_2)$.

§ 6. $\max_y \int_0^T F(x, y)\, dt$.

Let us now consider the more general problem of determining the maximum of

$$(1)\qquad J(y) = \int_0^T F(x, y)\, dt$$

subject to the relation connecting x and y,

$$(2)\qquad dx/dt = G(x, y), \qquad x(0) = c.$$

As we shall point out again below, there are certain advantages to considering the finite problem, despite the complication caused by an additional parameter.

The two state variables are now c and T. In many applications, c represents the initial quantity of resources and T the duration of the process. We now write

$$(3)\qquad \max_y J(y) = f(c, T).$$

Employing precisely the same reasoning as in the previous section, we obtain the functional equation

$$(4)\qquad f(c, T) = \max_{y[0, S]} \left[\int_0^S F(x, y)\, dt + f(c(S), T - S) \right],$$

which leads, in the limit as $S \to 0$, to the nonlinear partial differential equation

$$(5)\qquad 0 = \max_v [F(c, v) + G(c, v) f_c - f_T].$$

This, in turn, leads to the simultaneous equations

$$(6)\qquad \begin{aligned} f_T &= F(c, v) + G(c, v) f_c \\ 0 &= F_v(c, v) + G_v(c, v) f_c. \end{aligned}$$

Solving for f_c and f_T we have

$$f_c = -F_v(c, v)/G_v(c, v) \equiv P(c, v) \tag{7}$$
$$f_T = F - GF_v/G_v \equiv Q(c, v)$$

To obtain an equation for v, the fundamental variable, we equate f_{Tc} with f_{cT} and obtain the equation

$$P_v v_T = Q_v v_c + Q_c. \tag{8}$$

This is a first order linear partial differential equation for $v = v(c, T)$, which may be solved by means of the method of characteristics, a point we shall mention again below in § 14, or by numerical means, given $v(0, T)$ or $v(c, 0)$.

It is here the advantage of a T-dependent formulation becomes clear. We can determine v as a function of c for $T = 0$ quite readily, since for small T, we have

$$f(c, T) = \operatorname*{Max}_v [F(c, v) T + o(T)]. \tag{9}$$

Consequently, for $T = 0$, $v = v(c, 0)$ is determined by the condition that it maximizes $F(c, v)$.

§ 7. $\operatorname*{Max}_y \int_0^T F(x, y)\, dt$ under the Constraint $0 \leq y \leq x$

Let us now consider the problem of determining the maximum of $J(y) = \int_0^T F(x, y)\, dt$ subject to the relations

$$\text{(a)} \quad dx/dt = G(x, y), \; x(0) = c, \tag{1}$$
$$\text{(b)} \quad 0 \leq y \leq x.$$

As far as the classical approach is concerned, the difficulty of the problem resides in the fact that y cannot be determined, in general, by means of an unrestricted variation. When $0 < y < x$, we may vary freely, and in intervals where this inequality holds, y must satisfy the Euler equation. However, when $y = 0$ or x, we merely have an Euler inequality. The heart of the problem lies in determining how to fit together the three types of solution, $y = 0$, $y = x$, and y a solution of the Euler equation. This is equivalent to determining the transition points where two types of solution join.

At the present time, there exists no *uniform* technique for solving these problems in explicit analytic form. Certain classes of problems of this type do have a simple structure to their solution, as we shall briefly discuss below.

Let us now see how the functional equation technique applies to this problem. Define

(2) $$f(c, T) = \underset{y}{\text{Max}}\, J(y)$$

As above, we derive the partial differential equation

(3) $$f_T = \underset{0 \leq v \leq c}{\text{Max}} \left[F(c, v) + G(c, v) f_c\right].$$

The original constraint $0 \leq y \leq x$ has been translated into the constraint $0 \leq v \leq c$. The initial condition is

(4) $$f(c, 0) = 0 \,,$$

for all c.

We see that the constraint $0 \leq v \leq c$ prevents us from differentiating freely with respect to v. In § 10, we shall show how (3) can be used to derive the structure of the solution, under certain assumptions concerning F and G.

§ 8. Computational solution

Let us examine the nonlinear partial differential equation

(1) $$f_T = \underset{v}{\text{Max}} \left[F(c, v) + G(c, v) f_c\right],$$

with $f(c, 0) = 0$ and sketch a procedure that may be used to compute the solution.

In place of allowing the variables T and c continuous variation, we restrict their range to the set of values

(2) $$\begin{aligned} T &= 0, \Delta, 2\Delta, \ldots, k\Delta, \ldots \\ c &= 0, \pm\delta, \pm 2\delta, \ldots, \pm k\delta, \ldots \end{aligned}$$

where Δ and δ are both positive quantities.

The partial derivatives f_T and f_c are now approximated to by the difference quotients

(3) $$\begin{aligned} f_T &\cong \frac{f(c, T+\Delta) - f(c, T)}{\Delta} \\ f_c &\cong \frac{f(c+\delta, T) - f(c-\delta, T)}{2\delta} \end{aligned}$$

with the result that the nonlinear differential equation in (1) assumes the approximate form

$$(4)\quad f(c, T+\Delta) = f(c, T) + \Delta\left[\operatorname*{Max}_{v}\left\{F(c, v) + G(c, v)\,\frac{f(c+\delta, T) - f(c-\delta, T)}{2\delta}\right\}\right]$$

$$f(c, 0) = 0$$

Starting with the known values for $f(c, 0)$ we can compute successively the values of $f(c, \Delta), f(c, 2\Delta), \ldots$, and so on.

Although this method is conceptually very simple, there are great difficulties encountered in actual computing practice. Essentially the main question is how to choose the quantities Δ and δ. The convergence of the process and the stability of the numerical solution depend upon the proper ch oice ofthese parameters. For the linear equations that appear when the maximum is removed, there is a fairly complete and satisfying theory of these matters. For nonlinear equations, however, practically no theory exists and the matter rests in the realm of art and experience.[2]

It is interesting to observe that the numerical solution of (7.3), an equation with a constraint, is easier to obtain than the numerical solution of (1) above, due to the fact that the existence of the constraint narrows the range that must be examined to determine the maximum. Consequently, in many cases, the more realistic process will possess a simpler computational solution.

In § 11 we shall discuss an alternate computational scheme, also based upon difference equations, which in practice seems to be more efficient and which enables us to proceed in a rigorous fashion, without having to enter difficult domains of the calculus of variations.

§ 9. Discussion

We have mentioned above the difficulties that may arise in solving a variational problem subject to restraints, and also the fact that certain cases may be completely resolved.

Let us show how the functional equation in (8.1) may be used to yield information concerning the structure of the solution. We shall consider only the case where $F(c, v)$ is strictly concave in v for all c, and $G(c, v)$ is linear in v. The nonlinear partial differential equation then has the form

$$(1)\qquad f_T = \operatorname*{Max}_{0 \le v \le c}\left[F(c, v) + (g(c) + h(c)\,v)\,f_c\right]$$

[2] There is also the problem of choosing a suitable difference-quotient approximation. In (3), we choose a symmetric approximation for f_c and an asymmetric one for f_T. For the case of linear equations, stability considerations may often be helpful. For nonlinear equations, practically nothing is known.

The function $F(c, v) + (g(c) + h(c)v) f_c$ is strictly concave in v for all values of c and T, and the maximum over v is uniquely assumed. It may, however, occur at $v = 0$, $v = c$ or at an interior point.

Assuming that all the functions involved vary continuously with c and T, we can make the following important observation. Since the function $F(c, v) + (g(c) + h(c)v) f_c$ varies continuously and is strictly concave, the maximum cannot shift from $v = 0$ to $v = c$ without passing through interior points of the interval $[0, c]$ first.

This is a particular case of the fact that the maximum value for v depends continuously upon c and T. This remark can be used to shorten greatly the time involved in the computational solution of these processes, and furthermore, it makes feasible the numerical solution of multidimensional processes.[3]

It follows that any extremal must have the following structure. An interval where $y = 0$ must be followed and preceded by an interval in which $0 < y < x$, and similarly for an interval where $y = x$.

The question arises as to how often the solution can switch from one type to another. In order to answer this, we must make further assumptions concerning the functions which appear. It is not difficult to construct examples showing that there may be an arbitrarily large number of such transitions if F is chosen suitably. In the example considered in the next section we will carry through the discussion in greater detail.

§ 10. An example

Let us consider the problem of determining the maximum of

(1) $$J(y) = \int_0^T (x - y)\, dt$$

under the conditions

(2) $$\text{a.}\quad dx/dt = b(y),\; x(0) = c$$

$$\text{b.}\quad 0 \le y \le x$$

The basic equation is

(3) $$f_T = \operatorname*{Max}_{0 \le v \le c} [c - v + b(v) f_c].$$

Let us now assume that $b(y)$ satisfies the conditions

(4) $$\text{a.}\quad b(0) = 0, \qquad b'(0) = \infty$$

$$\text{b.}\quad b'(y) > 0, \qquad b'(y) \to 0 \text{ as } y \to \infty$$

$$\text{c.}\quad b''(y) < 0,$$

[3] Cf. the remarks in § 22 and § 23 of Chapter I.

A simple function satisfying these conditions is $y^{1/2}$.

Let us assume, as is quite plausible, that $f_c > 0$[4]. Then, turning to the determination of the maximum of $K(v) = c - v + b(v) f_c$, we see that the derivative with respect to v, $K'(v) = -1 + b'(v) f_c$, is positive for small v, negative for large v and zero for just one value of v. Let us furthermore assume, as is also plausible in this case, that $f_c = 0$ at $T = 0$ and monotone increasing thereafter as a function of T.

If we allow v to traverse the interval $0 \leq v < \infty$, we see that there will always be a solution of $K'(v) = 0$. However, if v is constrained by the condition that $0 \leq v \leq c$, then if f_c is large, which is to say, if T is large, $K'(v)$ will remain positive throughout the interval $0 \leq v \leq c$. This means that the maximum will be at $v = c$, or $y = x$, for T large compared to c.

It remains to determine the transition curve $T = T(c)$ at which this cross-over in policy occurs. We know that the solution will have the form

$$\text{(5)} \quad \begin{aligned} &\text{a.} \quad y = x\,, && 0 \leq t \leq t_1 \\ &\text{b.} \quad 0 < y < x\,, && t_1 < t < T\,, \end{aligned}$$

The first part of the curve, where $y = x$, will appear only if T is sufficiently large. If T is small, the solution will consist only of the second part, where $0 < y < x$.

Consider then the case where T is small. There are two courses we may pursue. We may first use the fact that the maximum in (3) occurs inside the interval, which means that (3) is equivalent to the two equations

$$\text{(6)} \quad \begin{aligned} f_T &= c - v + b(v) f_c \\ 0 &= -1 + b'(v) f_c \end{aligned}$$

These equations, combined with the boundary values

$$\text{(7)} \quad f(c, 0) = 0,\ v(c, 0) = 0$$

suffice to determine $f(c, T)$, for T small.

Alternatively, we may use the classical variational technique, armed with the knowledge that we can ignore the constraint $0 < y < x$. Setting

$$\text{(8)} \quad J(y) = \int_0^T \left[c + \int_0^t b(y)\, ds - y\right] dt\,,$$

we readily obtain as the variational equation, the Euler equation

$$\text{(9)} \quad (T - t)\, b'(y) - 1 = 0\,.$$

With y determined uniquely by this equation, we can compute $J(y)$ for the extremal and thus $f(c, T)$.

[4] In the following section, we shall show how these results may be derived by a consideration of the discrete process.

As T increases, the critical value of T, as a function of c, is furnished by the value for which the equation

$$-1 + b'(v) f_c = 0 \tag{10}$$

has the solution $v = c$, which is to say the value of T furnished by the equation

$$f_c(c, T) = \frac{1}{b'(c)} \tag{11}$$

If $f_c(c, T)$ is monotone increasing as a function of T, as surmised, this equation has one root $T(c)$. Once we have determined this critical value the solution is completely determined.

§ 11. A discrete version

One of the methods we can employ to make the above arguments rigorous is based upon the discrete approximation to the continuous problem.[5] Considering the problem above in § 10, a discrete version is the problem of determining the maximum of

$$J(y) = J(y_0, y_1, y_2, \ldots, y_N) = \sum_{k=0}^{N} (x_k - y_k) \tag{1}$$

over all y_i subject to the relations

(2) a. $x_{k+1} = x_k + b(y_k)$,

b. $0 \leq y_k \leq x_k, \quad k = 0, 1, 2, \ldots, N$.

If we set

$$u_N(c) = \operatorname*{Max}_{y} J(y), \tag{3}$$

we obtain the recurrence relations

(4) a. $u_0(c) = c$,

b. $u_{N+1}(c) = \operatorname*{Max}_{0 \leq v \leq c} [c - v + u_N(c + b(v))], \quad N = 0, 1, \ldots.$

Using the same methods we have employed in § 12 of Chapter 1, and in our discussion of the optimal inventory equation, it is easy to establish the following result:

THEOREM 1. *For each $N \geq 1$, there exists a function $v_N(c)$ with the following properties:*

[5] They may also be rigorously established using classical techniques. A reference will be found in the bibliography at the end of the chapter.

(5) a. $v_N(c)$ *is monotone decreasing as* c *increases,*

b. $v_{N+1}(c) > v_N(c)$, $N = 1, 2, \ldots$

c. *There is a unique solution to* $v_N(c) = c$ *which we call* c_N; $c_{N+1} > c_N$.

d. *For* $0 \leq c \leq c_N$, $u_N(c) = u_{N-1}(c + b(c))$, i.e. $v = c$.

e. *For* $c_N \leq c$, $u_N(c) = c - v_N(c) + u_{N-1}[c + b(v_N(c))]$,

f. $u_N'(c) \geq u_{N-1}'(c)$, $N = 1, 2, \ldots$, *for* $c \geq 0$.

The proof, which is inductive along the usual lines, we leave to the reader.

A similar result can be obtained for the more general case, corresponding to the problem in § 7, if we impose suitable conditions on $F(x, y)$ and $G(x, y)$. The proof is much more detailed.

As we saw in § 7—8, the problem of determining the maximum of $J(y) = \int_0^T F(x, y)\, dt$ subject to the relations

$$\text{(6)} \qquad \begin{aligned} &\text{(a)} \quad dx/dt = G(x, y),\ x(0) = c \\ &\text{(b)} \quad 0 \leq y \leq x \end{aligned}$$

can be reduced to the problem of solving the nonlinear partial differential equation

$$\text{(7)} \qquad f_T = \operatorname*{Max}_{0 \leq v \leq c} [F(c, v) + G(c, v) f_c],\ f(c, 0) = 0$$

This equation may be approached numerically by converting it into a partial difference equation.

In order to use this method with confidence, we must first establish the fact that the variational problem is equivalent to solving this nonlinear equation, a matter of some difficulty when constraints are imposed, and then that the finite difference method yields an approximate solution to the nonlinear equation, again a complicated question. Both of these problems may be avoided in the following way. We replace the original problem by the problem of determining the maximum over y_k of the function

$$\text{(8)} \qquad F(\{y_k\}) = \Delta \sum_{k=0}^{N} F(x_k, y_k) ,$$

subject to the relations

$$\text{(9)} \qquad \begin{aligned} &\text{(a)} \quad x_{k+1} = x_k + \Delta G(x_k, y_k),\ x_0 = c \\ &\text{(b)} \quad 0 \leq y_k \leq x_k,\ k = 0, 1, 2, \ldots, N , \end{aligned}$$

where $x_k = x(k\Delta)$, $y_k = y(k\Delta)$, $N\Delta = T$.

Setting

$$f_N(c) = \operatorname*{Max}_{y} J(\{y_k\}), \tag{10}$$

we replace the above maximization problem by the recurrence relations

$$\begin{aligned} f_0(c) &= 0, \\ f_{N+1}(c) &= \operatorname*{Max}_{0 \le v \le c} [\Delta F(c, v) + f_N(c + \Delta G(c, v))]. \end{aligned} \tag{11}$$

In cases treated to date this has turned out to be a more reliable computational procedure, and it possesses a number of other attractive features from the numerical point of view as well.

It turns out to be not too difficult to show that

$$\lim_{\Delta \to 0} f_N(c) = f(c, T), \tag{12}$$

under conditions upon F and G that are normally assumed in the calculus of variations. Actually, these conditions can be greatly lightened. However, any discussion of this would take us too far afield.

§ 12. A convergence proof

Since a discussion of the convergence question, even under strong assumptions, becomes quite long-winded in its full generality, without adding much in principle, we shall content ourselves with the proof of a typical result.

Let us set

$$f(c, T) = \operatorname*{Max}_{y} \int_0^T F(x, y)\, dt, \tag{1}$$

subject to the constraints

$$\begin{aligned} &\text{a.} \quad dx/dt = G(x, y), \qquad x(0) = c, \\ &\text{b.} \quad 0 \le y \le x. \end{aligned} \tag{2}$$

It is convenient to set $y = \varphi x$,[6] so that we have, introducing a new F and G,

$$f(c, T) = \operatorname*{Max}_{y} \int_0^T F(x, \varphi)\, dt, \tag{3}$$

[6] This is particularly so in the numerical calculation since this change of independent variable permits the maximization to be over a fixed region, $0 \le \varphi_k \le 1$, rather than over a variable region. On the other hand, there are cases where the variable region is desirable, particularly in connection with shrinking processes.

where

(4) $$\text{a. } dx/dt = G(x, \varphi), \qquad x(0) = c,$$
$$\text{b. } 0 \leq \varphi \leq 1 \text{ for all } t \geq 0.$$

Let us now define, for $n = 1, 2, \ldots$, a sequence of approximating problems: Maximize

(5) $$J_N(\{\varphi_k\}, n) = \sum_{k=0}^{N} F(x_k, \varphi_k)/n, \quad N = [T/n].$$

where x_k and φ_k are related by the equations

(6) $$x_{k+1} - x_k = G(x_k, \varphi_k)/n, \qquad k = 0, 1, \ldots, n-1,$$
$$x_0 = c,$$

and the variables φ_k are constrained by the relations

(7) $$0 \leq \varphi_k \leq 1, \quad k = 0, 1, \ldots, N.$$

Here, as above, $x_k = x(k\Delta)$, $\varphi_k = \varphi(k\Delta)$.
For each c and T, let

(8) $$f(c, T, n) = \text{Max } J_N(\{\varphi_k\}, n).$$

We wish to show that

(9) $$\lim_{n \to \infty} f(c, T, n) = f(c, T).$$

We first require the following

LEMMA: *Let* $G(x, \varphi)$ *satisfy a Lipschitz condition for* $m \leq x \leq M$, $0 \leq \varphi \leq 1$. *Let*

(10) a. $\varphi(t)$ *be a step-function with constant value* φ_k, $0 \leq \varphi_k \leq 1$, *in the interval* $k/n \leq t < (k+1)/n$, $k = 0, 1, \ldots, N$;

b. $\{x_k\}$ *be defined recursively by* (6), *and let the uniform bounds on the sequence be* m *and* M; $m \leq x_k \leq M$.

c. $\bar{x}(t)$ *be step function with constant value* x_k *for* $k/n \leq t < (k+1)/n$,

d. $x(t)$ *be defined as the solution of the differential equation in* (4).

Then there exists a constant k *depending only upon* G *and* T *such that* $x(t) - \bar{x}(t) | \leq k/n$, *for* $0 \leq t \leq N$.

This may be proved by the Cauchy-Lipschitz method, applied in the same way as in the proof of the existence theorem for systems of ordinary differential equations.

Let us now state the limit relation of (9) as

THEOREM 2. *Under the assumptions*

(11) a. *F and G have continuous second partial derivatives.*

b. *There exist constants p, q, r such that $px \leq G(x, y) \leq qx + r$, for $x > 0$ and $0 \leq y \leq x$*

c. *G_y is of one sign: either $G_y > 0$ or $G_y < 0$ for all $x > 0$, and $0 \leq y \leq x$*

we have

(12) $$\lim_{n \to \infty} f(c, T, n) = f(c, T),$$

for all $c \geq 0$, $T > 0$.

PROOF: Given $c > 0$ and $T > 0$, let $N = [T/n]$, as above. The condition of (11b) enables us to assert the uniform boundedness of $x(t)$ for $0 \leq t \leq T$. Let $m \leq x(t) \leq M$, and thus $m \leq x_k \leq M$. Since, by assumption, $F(x, \varphi)$, and $G(x, \varphi)$ satisfy Lipschitz conditions in the region $m \leq x \leq M$, $0 \leq \varphi \leq 1$, by virtue of the Lemma above, there exists a constant B_1 dependent only on c, T, F and G, such that

(13) $$| J(\varphi) - J_N(\{\varphi_k\}, n) | \leq B'/n,$$

whenever $\varphi(t)$ and φ_k are as in Lemma 1 above. It follows that

(14) $$f(c, T, n) \leq f(c, T) + B'/n,$$

for all $n = 1, 2, \ldots$.

Let $\{n_i\}$ be a subsequence of $\{n\}$ for which $\lim_{i \to \infty} f(c, T, n_i) = \liminf_{n \to \infty} f(c, T, n)$. Given $\varepsilon > 0$, let $\varphi(t)$ be chosen so that

(15) $$f(c, T) < J(\varphi) + \varepsilon.$$

Now $\varphi(t)$ is the limit almost everywhere of a sequence $\{\varphi_m(t)\}$ of step-functions for which $0 \leq \varphi_m(t) \leq 1$, and we have $\lim_{m \to \infty} J(\varphi_m) = J(\varphi)$. Hence we may take the φ appearing in (15) to be a step-function, and actually a step-function constant in each interval of the form $k/n \leq t < (k + 1)/n$, for some arbitrarily large $n = n_i$. From (13) we have

(16) $$f(c, T) < J_N(\{\varphi_k\}, n) + B'/n + \varepsilon \leq f(c, T, n) + B'/n + \varepsilon.$$

Hence

(17) $$f(c, T) \leq \lim_{n_i \to \infty} f(c, T, n_i) + \varepsilon = \liminf_{n \to \infty} f(c, T, n) + \varepsilon.$$

On the other hand, using (14) we see that

$$\limsup_{n \to \infty} f(c, T, n) \leq f(c, T) . \tag{18}$$

Since ε is arbitrary, we see that (12) holds.

If $\{\varphi_{kn}\}$ maximizes $J_N(\{\varphi_k\}, n)$, then $\{\varphi_{kn}\}$ determines for each $n = 1, 2, \ldots$, a step function $\varphi_n(t)$ with the property that

$$\lim_{n \to \infty} J(\varphi_n) = f(c, T) . \tag{19}$$

If there is a convergent subsequence which converges almost everywhere to a limit $\varphi(t)$, then $\lim_{n \to \infty} J(\varphi_n) = J(\varphi)$, and $\varphi(t)$ is a maximizing function.

If this function possesses suitable monotonicity properties we can employ Helly's theorem to obtain a convergent sub-sequence. Otherwise, we must use weak convergence arguments or analogous techniques.

§ 13. Max$_y \int_0^T F(x, y, t)\, dt$

So far we have considered time independent processes—those where F and G are independent of t. Let us now treat the more general case, that of maximizing

$$J(y) = \int_0^T F(x, y, t)\, dt , \tag{1}$$

subject to the relation

$$dx/dt = G(x, y, t) , \qquad x(0) = c . \tag{2}$$

In order to apply the functional equation technique, we imbed this problem within the wider problem of determining the maximum of

$$J(y) = \int_a^T F(x, y, t)\, dt , \tag{3}$$

subject to the constraint

$$dx/dt = G(x, y, t) , \qquad x(a) = c . \tag{4}$$

Here a ranges over the interval $[0, T]$.

Keeping T fixed, the two state variables are now a and c, and we may write

$$\operatorname*{Max}_y J(y) = f(a, c) . \tag{5}$$

The functional equation for f is

$$(6)\qquad f(a, c) = \underset{y\,[a,\, a+S]}{\text{Max}} \left[\int_a^{a+S} F(x, y, t)\, dt + f(a+S, c(S)) \right],$$

for $0 < S < T - a$.

Letting $S \to 0$, we obtain the equation

$$(7)\qquad 0 = \underset{v}{\text{Max}} \left[F(c, v, a) + f_a + G(c, v, a) f_c \right],$$

where $v = v(a, c)$ is the value of $y(a)$.

From (7) we obtain the two equations

$$(8)\qquad \begin{aligned} 0 &= F(c, v, a) + f_a + G(c, v, a) f_c \\ 0 &= F_v(c, v, a) \qquad + G_v(c, v, a) f_c \end{aligned}$$

Solving for f_a and f_c, we obtain

$$(9)\qquad \begin{aligned} f_c &= -F_v/G_v = P(c, v, a) \\ f_a &= (FG_v - F_v G)/G_v = Q(c, v, a). \end{aligned}$$

As above, equating the values of f_{ca} and f_{ac}, we obtain the first order partial differential equation for v,

$$(10)\qquad P_v v_a + P_a = Q_v v_c + Q_c.$$

Those who are familiar with quasi-linear partial differential equations of this type will readily verify that the characteristics of this equation are equivalent to the Euler equations obtained by classical variational techniques.

§ 14. Generalization and discussion

If we now consider the problem of determining the maximum of the functional

$$(1)\qquad J(y) = \int_0^T F(x(t), y(t))\, dt,$$

subject to relations

$$(2)\qquad dx/dt = g(x, y), \quad x(0) = c,$$

where x, y, c and g are n-dimensional column vectors, and F is a scalar function of x and y,[7] we can proceed in a similar fashion. Setting

$$(3)\qquad f(c, T) = \underset{y}{\text{Max}}\, J(y),$$

[7] Any explicit dependence upon t can always be removed by consideration of t as a dependent variable x_{n+1}, defined by $dx_{n+1}/dt = 1$, $x_{n+1}(0) = 0$.

the principle of optimality yields the functional equation

$$(4) \qquad f(c, S+T) = \operatorname*{Max}_{y\,[0,\,S]} \left[\int_0^S F(x, y)\, dt + f(c(S), T) \right]$$

The classical transversality conditions fall out as a special case in this equation, as might be expected on the basis of the duality between point and tangential coordinates which we have indicated above.

Carrying through the calculations similar to those in (8) and (9) in § 11 above, we obtain a system of quasi-linear partial differential equations for the vector $v = v(c, T) = y(0)$.

This equation has a characteristic theory, and, as is to be expected, the characteristics are equivalent to the Euler equations of the variational problem. The rigorous proof is quite complicated and will not be presented here.

§ 15. Integral constraints

We considered above in § 7 a variational problem where y was constrained by the condition $0 \leq y \leq x$. Let us now discuss the problem for the case where we impose the additional constraint

$$(1) \qquad \int_0^T y\, dt \leq m.$$

The minimum of $\int_0^T F(x, y)\, dt$ will now be a function of the three state variables c, T and m. Denote it by $f(c, T, m)$. Using the above methods, we see that f satisfies the equation

$$(2) \qquad f_T = \operatorname*{Max}_{0 \leq v \leq c} \left[F(c, v) + G(c, v) f_c - v f_m \right].$$

Problems involving constraints of the type encountered in the preceding sections arise in the consideration of many physical problems if we impose realistic bounds on such quantities as velocity, acceleration, radius of curvature, rate of allocation of resources, and so forth. Integral constraints, such as that appearing above, or a constraint of the form $\int_0^T y'^2\, dt \leq m$, appear if we assume that resources are bounded, that the kinetic energy is bounded, and so on.

Generally speaking, integral constraints are more readily handled than point constraints. Although, theoretically, the Lagrange multiplier method is capable of treating both types of constraints, as well as more general classes, in practice we encounter the difficulty discussed above of determining when the variable is within the domain of variability, and when it is on the boundary.

§ 16. Further remarks concerning numerical solution

Let us consider the problem of determining the maximum of the integral

$$J(x) = \int_0^T F(x, x', t)\, dt, \tag{1}$$

where $x(0) = c$, but x is otherwise unrestrained. Assuming that F satisfies appropriate conditions, the solution is determined by the Euler equation

$$\partial F/\partial x - d/dt\, \partial F/\partial x' = 0. \tag{2}$$

This is a second order equation, of the form

$$x'' = G(x, x', t), \tag{3}$$

which means that two boundary conditions are necessary to determine the solution. One condition is furnished by the original constraint $x(0) = c$, while the other, arising from the variational procedure is

$$\left.\frac{\partial F}{\partial x'}\right|_{t=T} = 0. \tag{4}$$

As we see, one condition is at $t = 0$, and the other at $t = T$. On the other hand, in order to integrate (3) in a convenient fashion, either with a digital computer or an analog computer, we require the values of x and x' at $t = 0$ or at $t = T$. Unfortunately, we do not obtain either of these sets of conditions from the above analysis.

We are thus confronted by the classical difficulty of a *two-point* boundary condition. If G is linear in x and x', we face no particular difficulty; If, however, as is generally true, G is nonlinear, we must face the fact that there is no systematic technique for determining the solution of (3), satisfying (4) and the initial condition.

The usual procedure is to start the integration at $t = 0$, beginning with a range of values of $x'(0)$, and narrowing the range until (4) is sufficiently well approximated. This is a time-consuming procedure, sometimes complicated by stability problems, which becomes rapidly more inefficient as the dimension of the variational problem increases.

We have assumed that F is a function possessing a sufficiently smooth behavior to justify the use of (2). If we allow F to possess terms such as $|x - a|$ or Max $(x - a, x' - b, g(t))$, functions which arise very naturally in economic and engineering processes, the application of the usual variational approach becomes increasingly difficult.

Combine the above complications with those furnished by the existence of constraints, and we see that conventional methods must be supplemented if we wish to resolve a variety of problems arising in a very natural way from the physical world.

Let us note, finally, that the remarks we made concerning the need for a sensitivity or stability analysis, in Chapter 7 and in connection with discrete decision processes, are, of course, equally valid in the context of continuous decision processes.

§ 17. Eigenvalue problem

Let us now devote our attention to the problems of determining the values of λ which permit a non-trival solution of the equation

$$(1) \qquad \begin{aligned} u'' + \lambda^2 \varphi(t)\, u &= 0\,, \\ u(0) = u(1) &= 0\,, \end{aligned}$$

to exist.

The connection between our previous work and this problem, which at first sight seems far removed, arises from the fact that under light conditions on $\varphi(t)$, the eigenvalue problem is equivalent to the problem of determining the relative minima of $\int_0^1 u'^2\, dt$ subject to the constraints

$$(2) \qquad \int_0^1 \varphi(t)\, u^2\, dt = 1, \qquad u(0) = u(1) = 0\,,$$

or, conversely, to that of determining the relative maxima of $\int_0^1 \varphi(t)\, u^2\, dt$ subject to the constraints

$$(3) \qquad \int_0^1 u'^2\, dt = 1\,, \qquad u(0) = u(1) = 0\,.$$

What makes this problem different in quality from those we have considered above is the fact that as we traverse an extremal the condition $u(0) = 0$ is violated. Consequently, we must imbed this problem in a more general class of problems possessing the requisite invariance properties if we wish to employ the functional equation approach. Happily, there are several ways of doing this.

In the first approach, we consider the minimization of

$$(4) \qquad J(u) = \int_a^1 u'^2\, dt\,,$$

over all u satisfying the conditions

$$(5) \qquad \text{(a)} \quad u(a) = k,\ u(1) = 0$$

$$\text{(b)} \quad \int_a^1 \varphi(t)\, u^2\, dt = 1$$

Here the new state variable a satisfies the condition $0 \leq a < 1$. We assume that the function $\varphi(t)$ satisfies the constraint $0 < b_1 \leq \varphi(t) \leq b_2$ for $0 \leq t \leq 1$, and is continuous over $[0, 1]$.

An equivalent problem is that of maximizing

(6) $$K(u) = \int_a^1 \varphi(t)\, u^2\, dt,$$

subject to the constraints

(7) $$\text{(a)} \quad u(a) = k,\ u(1) = 0$$

$$\text{(b)} \quad \int_a^1 u'^2\, dt = 1.$$

A second less obvious formulation that serves the purpose is the following: Minimize

(8) $$J(u) = \int_a^1 u'^2\, dt$$

subject to the constraints

(9) $$\text{(a)} \quad u(a) = 0,\ u(1) = 0,$$

$$\text{(b)} \quad \int_a^1 [\varphi(t)\, u^2 + k(1 - t)\, \varphi(t)\, u]\, dt = 1.$$

§ 18. The first formulation

Let us set

(1) $$f(a, k) = \text{Min} \int_a^1 u'^2\, dt,$$

subject to the constraints

(2) $$\text{(a)} \quad u(a) = k,\ u(1) = 0,$$

$$\text{(b)} \quad \int_a^1 \varphi(t)\, u^2\, dt = 1$$

We write, along an extremal,

(3) $$\text{(a)} \quad \int_{a+s}^1 \varphi(t)\, u^2\, dt = 1 - s\, \varphi(a)\, k^2,$$

$$\text{(b)} \quad u(a + s) = k + sv,$$

$$\text{(c)} \quad f(a, k) = v^2 s + \int_{a+s}^1 u'^2\, dt,$$

to terms in $o(s)$.[8]

[8] In order to simplify the analysis, we shall proced directly to the derivation of the limiting partial differential equation.

Now make the change of variable

$$(4) \qquad u(t) = [1 - s\,\varphi(a)\,k^2/2]\,w(t),\ a + s \leq t \leq 1\,,$$

in order to maintain the condition (2*b*). We then have

$$(5) \qquad \text{(a)} \quad w(a+s) = k + sv + s\,\varphi(a)\,k^3/2\,,$$

$$\text{(b)} \quad f(a,k) = v^2 s + (1 - s\,\varphi(a)\,k^2) \int_{a+s}^{1} w'^2\,dt$$

to terms in $o(s)$.

Combining the above results, we obtain the approximate functional equation

$$(6) \quad f(a,k) = \operatorname*{Min}_{v} [v^2 s + (1 - s\,\varphi(a)\,k^2)\,f(a+s, k+sv + s\,\varphi(a)\,k^3/2)] + o(s)\,.$$

Letting $s \to 0$, the result is the equation

$$(7) \qquad 0 = \operatorname*{Min}_{v} [v^2 + v f_k] + f_a + \frac{\varphi(a)\,k^3}{2} f_k - \varphi(a)\,k^2 f\,,$$

or

$$(8) \qquad f_a = f_k^2/4 - \varphi(a)\,k^3 f_k/2 + \varphi(a)\,k^2 f$$

The initial condition is at $a = 1$, and not trivial, since $f(a,k) \to \infty$ as $a \to 1$. There are two ways to determine this initial condition, as we shall discuss in the next section.

§ 19. An approximate solution

If a is close to 1, and $\varphi(t)$ continuous, as assumed, we may replace the variational problem in (17.1) and (17.2) by the approximate problem: Minimize $\int_a^1 u'^2\,dt$, subject to the constraints

$$(1) \qquad \text{(a)} \quad u(a) = k,\ u(1) = 0$$

$$\text{(b)} \quad \int_a^1 u^2\,dt = 1$$

upon absorbing the factor $\varphi(1)$ into the function $u(t)$.

This problem may be approached in two ways. Using the classical approach, we obtain the Euler equation

$$(2) \qquad u'' + \lambda u = 0\,,$$

which may be resolved explicitly. The unknown parameter is determined by the constraints in (1a) and (1b).

The second method uses (17.8) with $\varphi(a) \equiv 1$. Since the solution of the problem in (1) above is known for $k = 0$, namely

(3) $$f(a, 0) = \frac{\pi^2}{(a-1)^2},$$

we can obtain a solution to (17.8) as a power series in k, for $k \geq 0$. Since we are primarily interested in the solution for small k, this is a useful form of the solution for numerical purposes.

§ 20. Second formulation

We leave the derivation of the corresponding partial differential equation for the variational problem defined by (16.8) and (16.9) as an exercise for the reader, with the hint that the essential point is to renormalize $u(t)$ constantly so as to maintain the initial condition $u(a) = 0$.

§ 21. Discrete approximations

Since the partial differential equation for the minimum $f(a, k)$ possesses certain unpleasant features as far as initial values are concerned, the following discrete formulation may be of value.

Let us consider the problem of minimizing the function

(1) $$F(u_1, u_2, \ldots, u_{N-1}) = \sum_{k=1}^{N} (u_k - u_{k-1})^2,$$

subject to the constraints

(2) $$\text{(a)} \quad \sum_{k=1}^{N-1} \varphi_k u_k^2 = 1,$$

$$\text{(b)} \quad u_0 = a, \; u_N = 0.$$

Corresponding to the use of the state variable R, we consider the sequence $\{f_R(a)\}$ defined as follows

(3) $$f_R(a) = \underset{\{u_k\}}{\text{Min}} \sum_{k=R}^{N} (u_k - u_{k-1})^2,$$

where the u_k are subjected to

(4) $$\text{(a)} \quad \sum_{k=R}^{N} \varphi_k u_k^2 = 1,$$

$$\text{(b)} \quad u_{R-1} = a, \; u_N = 0,$$

for $k = 1, 2, \ldots, N-1$.

Since this involves a variable range for each quantity u_k, let us make a change of variable

(5) $$\varphi_k u_k = v_k,$$

under the assumption that $0 < b_1 \leq \varphi_k \leq b_2 < \infty$ for $k = 0, 1, 2, \ldots, N$. Then

$$(6) \qquad f_R(a) = \underset{\{v_k\}}{\text{Min}} \sum_{k=R}^{N} \left(\frac{v_k}{\varphi_k} - \frac{v_{k-1}}{\varphi_{k-1}} \right)^2,$$

where

$$(7) \qquad \text{(a)} \quad \sum_{k=R}^{N} v_k^2 = 1$$

$$\text{(b)} \quad v_{R-1} = \varphi_{R-1}\, a, \; v_N = 0 .$$

We leave as an exercise the task of determining the recurrence relation for the sequence $\{f_R(a)\}$.

§ 22. Successive approximation

Returning to the equation

$$(1) \qquad f(c) = \underset{D[0,S]}{\text{Max}} \left[\int_0^S F(x, y)\, dt + f(c(S)) \right],$$

obtained in § 3, it is tempting to envisage the use of successive approximations for the solution of the equation. If, however, we choose an initial function $f_0(c)$, and define a second approximation by means of the equation

$$(2) \qquad f_1(c) = \underset{D[0,S]}{\text{Max}} \left[\int_0^S F(x, y)\, dt + f_0(c(S)) \right],$$

we see that in the limit, as $S \to 0$, we must have $f_1(c) = f_0(c)$, provided that $f_0(c)$ is continuous.

At first sight, this would seem to render the use of successive approximations impossible. Actually this is not so. What is true is that we must approximate in *policy space* rather than in *function space*. We must concentrate our attention primarily upon $v = v(c, T)$ rather than upon $f(c, T)$. Nonetheless, $f(c, T)$ still plays an important auxiliary role.

To illustrate this point, let us discuss the problem of maximizing

$$(3) \qquad J(y) = \int_0^T F(x, y)\, dt,$$

subject to the relations

$$(4) \qquad dx/dt = G(x, y), \quad x(0) = c .$$

Then, as in § 6, we obtain the equation

$$(5) \qquad f_T = \underset{v}{\text{Max}} \left[F(c, v) + G(c, v) f_c \right].$$

Let us now choose an initial approximation $v_0 = v_0(c, T)$, which is equivalent to $y_0 = y_0(x, T - t)$, keeping in mind the connection between physical time t, and T, the remaining time for the process. Using this value of y_0, we compute x_0 by means of the differential equation,

$$dx_0/dt = G(x_0, y_0(x_0, T - t)), \qquad x_0(0) = c, \tag{6}$$

and then $f_0(c, T)$ by means of

$$f_0(c, T) = \int_0^T F(x_0, y_0)\, dt. \tag{7}$$

This function, f_0, satisfies the linear partial differential equation

$$f_{0T} = F(c, v_0) + G(c, v_0) f_{oc}. \tag{8}$$

To obtain the next approximation to an extremal y, or an optimal v, we determine $v_1(c, T)$ as a function which maximizes the function

$$F(c, v) + G(c, v) f_{oc}. \tag{9}$$

Using $v_1(c, T)$, we obtain $y_1(x, T - t)$ and then x_1 and f_1 as above. Having obtained f_1 we compute v_2 as a function which maximizes

$$F(c, v) + G(c, v) f_{1c}, \tag{10}$$

and continue in this way, deriving a sequence of approximations to f, $\{f_n\}$, and a sequence of approximations to v, $\{v_n\}$.

§ 23. Monotone approximation

Let us now show that this sequence of approximations to f is *monotone* increasing, a fact which is important theoretically and computationally. We have

$$\begin{aligned} f_{1T} &= F(c, v_1) + G(c, v_1) f_{1c}, \\ f_{0T} &= F(c, v_0) + G(c, v_0) f_{0c} \leq F(c, v_1) + G(c, v_1) f_{oc}. \end{aligned} \tag{1}$$

Hence

$$(f_1 - f_0)_T \geq G(c, v_1)(f_1 - f_0)_c. \tag{2}$$

Since $f_1(c, 0) = f_0(c, 0) = 0$, we see that $f_1 - f_0 \geq 0$ for all $T \geq 0$.

Continuing in this way, we readily establish the monotone property of the sequence $\{f_n\}$. If the sequence is uniformly bounded, we have convergence. However, it is essential to know when the sequence of partial derivatives, $\{f_{nc}\}$, $\{f_{nT}\}$, and the sequence of policies $\{v_n\}$ also, converge. The general question is a difficult one and we shall not enter into it here.

It is interesting to note, however, that we do possess a systematic technique for improving any particular policy.

§ 24. Uniqueness of solution

As we have noted above, we are bypassing any of the rigorous aspects of the derivation of the partial differential equations we have encountered and any study of the *existence* of the solution of these equations. It is, however, worth noting that the *uniqueness* of solution may be established quite readily by means of the same device we have formalized as Lemma 1 in Chapter 3.

Consider, for example, the equation

$$f_T = \operatorname*{Max}_{v} \left[F(c, v) + G(c, v) f_c \right], \tag{1}$$

and assume that there exists another solution of this equation, $g = g(c, T)$, which possesses the same initial value, namely

$$f(c, 0) = g(c, 0) = 0, \tag{2}$$

for all c. Then, we have also

$$g_T = \operatorname*{Max}_{w} \left[F(c, w) + G(c, w) g_c \right]. \tag{3}$$

Let $v = v(c, T)$ be a function which furnishes the maximum in (1) and $w = w(c, T)$ a function which furnishes the maximum in (3). Then we have the inequalities

$$\begin{aligned} f_T &= F(c, v) + G(c, v) f_c \geq F(c, w) + G(c, w) f_c \\ g_T &= F(c, w) + G(c, w) g_c \geq F(c, v) + G(c, v) g_c . \end{aligned} \tag{4}$$

These inequalities yield

$$\begin{aligned} f_T - g_T &\geq G(c, w)(f_c - g_c) \\ &\leq G(c, v)(f_c - g_c) . \end{aligned} \tag{5}$$

Thus, if we set $u = f - g$, we see that u satisfies the inequalities

$$G(c, w)\, u_c \leq u_T \leq G(c, v)\, u_c . \tag{6}$$

Since the solutions of

$$\begin{aligned} x_T - G(c, w)\, x_c &= 0, \quad & x(c, 0) &= 0, \\ y_T - G(c, v)\, y_c &= 0, \quad & y(c, 0) &= 0, \end{aligned} \tag{7}$$

are identically zero, it follows from a comparison theorem that u is identically zero.

§ 25. Minimum maximum deviation

Let us now discuss the numerical solution of a variational problem of the following type:

Minimize

(1) $$\max_{0 \le t \le T} |u - a|,$$

over all functions $v(t)$ satisfying the constraint $-1 \le v \le 1$, where

(2) $$du/dt = g(u, v), \ u(0) = c_1.$$

Consider the corresponding discrete process where

(3) $$u_{k+1} = u_k + g(u_k, v_k)\Delta, \ u_0 = c_1,$$

and $u_k = u(k\Delta)$, $\Delta = T/N$, $v_k = v(k\Delta)$.

Define

(4) $$f_N(c_1) = \min_{\{v_k\}} \max_{0 \le k \le N} |u_k - a|.$$

Then

(5) $$f_0(c_1) = |c_1 - a|,$$

and

(6) $$f_{N+1}(c_1) = \max\left[|c_1 - a|, \min_{|v| \le 1} f_N(c_1 + g(c_1, v)\Delta)\right],$$

for $N = 0, 1, 2, \ldots.$

We have thus reduced the solution of the original variational problem to a computation of a sequence of functions of one variable determined by the foregoing recurrence relation.

Exercises and Research Problems for Chapter IX

1. Obtain functional equations for the following quantities

a. $\max_f \int_o^T f^2\, dt, f(0) = c, \quad \int_0^T f'^2\, dt = 1$

b. $\max_f \int_o^T f^{2m}\, dt, f(0) = c, \quad \int_0^T (f')^{2n}\, dt = 1$

c. $\max_f \int_o^T f dt, f(0) = c, \quad \int_0^T f^2\, dt \le a, \quad \int_0^T t^2 f^2\, dt \le b$

2. Obtain functional equations for the following quantities

a. $\max_f \int_o^T fg dt, f(0) = c$, f monotone increasing, $\int_0^T f^2\, dt \le 1$

b. $\max_f \int_o^T fg dt, f(0) = c$, f monotone increasing and convex (concave), $\int_0^T f^2\, dt = 1.$

3. Carry through the analysis suggested in § 18 and obtain the first few terms of the expansion of $f(a, k)$ as a power series in k, for a close to 1.

4. Follow the same procedure for the second formulation of the eigenvalue problem.

5. Obtain a functional equation for the following quantity

$$\operatorname*{Min}_f \int_0^T [(x - c)^2 + kf^2] \, dt ,$$

$$dx/dt = - ax + f, \; x(0) = c .$$

6. Obtain a corresponding result for the general case

$$\operatorname*{Min}_f \int_0^T [\sum_{k=0}^{N-1} (x^{(k)} - c_k)^2 + f^2] \, dt ,$$

$$x^{(N)} = a_1 x^{(N-1)} + \ldots + a_N x + f ,$$

$$x^{(k)}(0) = c_k, \; k = 0, 1, \ldots, N - 1 .$$

7. Use the functional equation approach to determine the minimum of $\int_0^T (1 - x)^2 \, dt$ over f where $0 \leq f \leq M$, $M > 1$, $dx/dt = -x + f$,

$$x(0) = 1, \text{ and } \int_0^T f dt \leq a < T,$$

8. Under the same conditions determine the minimum of

$$\int_0^T (dx/dt)^2 \, dt.$$

9. Determine the minimum of $\int_0^T f^2 \, dt$ over all f satisfying

$$dx/dt = - x + f, \; x(0) = 1, \; 1 - a \leq x \leq 1 + a \text{ for } 0 \leq t \leq T .$$

10. Determine the minimum of $\int_0^T (x - y) \, dt$ over y, given that

a. $dx/dt = b(y)$, $x(0) = c$, $0 \leq y \leq x$,

b. $b''(y)$ is continuous and $b''(y) < 0$, $b'(y) > 0$

c. $b'(0+) = +\infty$

11. Consider the same problem under the assumption that $b'(0+)$ is finite.

12. Determine the minimum of $\int_0^T [K(y) + L(x - y)]\, dt$ over all y where

a. $K(0) = L(0) = 0$

b. $K''(x), L''(x) \geq 0$ for all x,

c. $dx/dt = - ay - b(x - y),\ x(0) = e,\ b > a > 0$.

13. Consider the problem of minimizing the functional

$$J(x) = \int_s^T [a_1(t)(x - b_1(t))^2 + a_2(t)(x' - b_2(t))^2]\, dt,$$

$0 \leq s \leq T$, over all functions x such that $x(s) = c$, and $\int_s^T x'^2\, dt < \infty$.

Assume that all functions appearing are continuous and that $a_i(t) > 0$ in the interval $[0, T]$.

Define

$$f(c, s) = \text{Min}\, J(x, s).$$

Show that

$$f_s = - a_1(s)(c - b_1(s))^2 + b_2 f_c - f_c^2/4a_2(s),$$

$$f(c, T) = 0 \text{ for all } c.$$

14. Show that $f(c, s) = u(s) + cv(s) + c^2 w(s)$, where u, v and w depend only upon s.

15. Show that u, v and w satisfy the equations

(a) $u'(s) = - a_1(s) b_1^2(s) + b_2(s) v(s) - v^2(s)/4a_2(s)$,

(b) $v'(s) = 2a_1(s) b_1(s) + 2b_2(s) w(s) - v(s) w(s)/a_2(s)$,

(c) $w'(s) = - a_1(s) - w^2(s)/a_2(s)$,

with $u(T) = v(T) = w(T) = 0$.

16. Obtain the corresponding results for the functional

$$J(x) = \int_s^T [a_1(t)(x - b_1(t))^2 + a_2(t)(x' - b_2(t))^2 + a_3(t)(x'' - b_3(t))^2]\, dt$$

17. Consider the following discrete analogue of the problem in 13. We wish to minimize the function

$$J(x) = \sum_{K=1}^{N} [\varphi_K(x_K) + \psi_K(x_K - x_{K-1})],$$

over all possible values of the x_K, $K = 1, 2, \ldots, N$, with $x_0 = x$. As usual, assume that $\varphi_K(x)$ and $\psi_K(x)$ are continuous functions, with appropriate properties at $x = \infty$.

Show that this problem leads to the sequence $\{f_K(x)\}$ defined as follows

$$f_N(x) = \operatorname*{Min}_{x_N} [\varphi_N(x_N) + \psi_N(x_N - x)],$$

$$f_R(x) = \operatorname*{Min}_{x_R} [\varphi_R(x_R) + \psi_R(x_R - x) + f_{R+1}(x_R)].$$

18. Consider, in particular, the case where φ_K and ψ_K are quadratic in x, with $\varphi_K = b_K(x - d_K)^2$, $\psi_K = c_K x^2$. Show that, in this case, $f_N(x) = u_N + v_N x + w_N x^2$, where u_N, v_N and w_N are independent of x.

19. Show that

$$u_{N-1} = [b_{N-1} d^2_{N-1} + u_N - (d_{N-1} b_{N-1} - v_N/2)^2]/[b_{N-1} + e_{N-1} + w_N]$$

$$v_{N-1} = \frac{-2e_{N-1}(d_{N-1} b_{N-1} - v_N/2)}{b_{N-1} + e_{N-1} + w_N}$$

$$w_{N-1} = \frac{e_{N-1}(b_N + w_N)}{b_{N-1} + e_{N-1} + w_N}$$

20. Let $\{x_K\}$ denote the sequence of minimizing x_K's. Show that

$$x_1 = \frac{xe_1 + d_1 b_1 - v_2/2}{b_2 + e_1 + w_2},$$

$$x_K = \frac{x_{K-1} + d_{K-1} b_{K-1} - v_{K+1}/2}{b_K + e_K + w_K}$$

21. Treat in a corresponding manner the problem of minimizing the expression

$$J(x) = \sum_{K=1}^{N} [a_K(x_K - b_K)^2 + e_K(x_K - x_{K-1})^2 + g_K(s_K - d_K)^2],$$

where $s_K = x_1 + x_2 + \ldots + x_K$.

22. Consider the stochastic case where the parameters appearing are stochastic variables, and it is desired to minimize the expected value of $J(x)$.

23. Consider the scalar equation

$$du/dt = g(u, v),\ u(0) = c,$$

where v is to be chosen so as to minimize the functional $J(v) = \int_0^T h(u - c)\, dt$.

Let $f(c, T) = \operatorname*{Min}_{v} J(v)$, and derive a functional equation for f.

24. Consider the process where we wish to minimize

$$\int_0^T h(u - w(t))\, dt .$$

25. Consider the problem where we wish to minimize

$$\operatorname*{Max}_{0 \leq t \leq T} |u - a(t)| ,$$

where $a(t)$ is a known function of t.

26. Consider a corresponding two-dimensional problem, using the equation

$$u'' + b(u^2 - 1)u' + u = f(t) ,$$

with $-1 \leq f(t) \leq 1$.

27. Treat in the same fashion the problem of maximizing

$$J(f, g) = \int_0^T \operatorname{Min}(x, y)\, dt ,$$

where

$$dx/dt = ax + f,\ x(0) = c_1 ,$$
$$dy/dt = by + g,\ y(0) = c_2 ,$$

and the functions $f(t)$ and $g(t)$ satisfied the constraints.

$$f + g = 1, f, g \geq 0 .$$

28. Consider the equation

$$d^2u/dt^2 + a^2 u = f(t),\ u(0) = c_1,\ u'(0) = c_2,\ a^2 < 1.$$

We wish to choose $f(t)$ subject to $-1 \leq f(t) \leq 1$ so as to reduce u to zero in minimum time. What is the corresponding functional equation?

29. Obtain the solutions of the brachistochrone and isoperimetric problems using the functional equation approach.

30. Determine the path of a ray of light through an inhomogeneous medium, assuming that the path minimizes time.

31. Consider the problem of determining the minimum of

$$J_N(w) = \operatorname*{Max}_{0 \leq k \leq N} \operatorname{Max} [|u_k|, |1 - v_k|],$$

over all sequences $\{w_k\}$ satisfying the conditions $|w_k| \leq 1$, where

$$u_{k+1} = g(u_k, v_k, w_k),\ u_0 = c_1,$$
$$v_{k+1} = h(u_k, v_k, w_k),\ v_0 = c_2 .$$

Let

$$f_N(c_1, c_2) = \underset{w}{\text{Min}}\ J_N(w).$$

Show that

$$f_{N+1}(c_1, c_2) = \text{Max}\,[\,|c_1|, |1 - c_2|, \underset{|w| \leq 1}{\text{Min}}\, f_N(g(c_1, c_2, w), h(c_1, c_2, w))\,].$$

32. Obtain the corresponding recurrence relations for the case where

$$J_N(w) = \underset{0 \leq k \leq N}{\text{Max}}\ D_k(u_k, v_k).$$

33. Consider a rocket-powered aircraft moving in level flight with a mass equal to the fixed mass of the aircraft, w, plus the mass of the fuel, m. The force of propulsion is taken to be a known function of the rate of burning of the fuel, the velocity of the aircraft and the mass of the fuel. Equivalently the force of propulsion is a known function of the thrust and drag, which are, in turn, known functions of the burning rate, the velocity, and the mass.

Let

(1) $x(t)$ = the distance along the x-axis from the origin at time t.

$v(t)$ = velocity of the aircraft.

$m(t)$ = mass of fuel.

w = fixed mass of aircraft.

$y(t)$ = burning rate of fuel.

$F(y, v, m)$ = force of propulsion.

Then

$$(w + m)\frac{d^2x}{dt^2} = F(y, v, m),$$

or

$$(w + m)\frac{dv}{dt} = F(y, v, m),\ v(0) = v_0,$$

and

$$\frac{dm}{dt} = -y,\ m(0) = m_0.$$

34. Consider a discrete version of the process described above, and impose a restriction on the burning rate, $0 \leq y(t) \leq R$.

Let

$f(v, m)$ = the range covered starting with initial velocity v and a quantity of fuel m, ending with terminal velocity v_T, using an optimal burning policy.

Show that

$$f(v, m) = v\Delta + \underset{0 \leq y \leq R}{\text{Max}} \left[f\left(v + \frac{F(y, v, m)\,\Delta}{w + m}, m - y\Delta\right)\right]$$

and show how the quantity v_T enters.

(R. Bellman — S. Dreyfus, H. Cartaino — S. Dreyfus)*

35. Similarly, let us define

$f(v, m, d) =$ the time required to traverse a distance d, starting from initial velocity v_0 and a given quantity m of fuel, with a required terminal velocity v_T, using an optimal burning policy.

Show that

$$f(v, m, d) = \Delta + \underset{0 \leq y \leq R}{\text{Min}} \left[f\left(v + \frac{F(y, v, m)\,\Delta}{w + m}, m - y\Delta, d - v\Delta\right)\right].$$

36. Consider the equation

$$\frac{d^2x}{dt^2} + (x^2 - 1)\frac{dx}{dt} + x = r(t) + v\left(x, \frac{dx}{dt}, t\right), \; x(0) = c_1, \; x'(0) = c_2,$$

where the function $v = v\left(x, \frac{dx}{dt}, t\right)$ is to be determined, subject to the constraint $|v| \leq 1$, so as to minimize the expected value of

$$J(x) = \int_0^T x^2\,dt + |x(T)|,$$

over a suitable class of random functions $r(t)$.

Going over to the discrete version, show that we obtain the recurrence relation

$$f_0(c_1, c_2) = \Delta c_1^2 + |c_1 + c_2\Delta|,$$

$$f_N(c_1, c_2) = \underset{|v_0| \leq 1}{\text{Min}} \left[\Delta c_1^2 + \int_{-\infty}^{\infty} f_{N-1}(c_1 + c_2\Delta, c_2 + [-(c_1^2 - 1)c_2 - c_1 + r_0 + v_0]\Delta)\,dG(r_0)\right],$$

where $dG(r_0)$ is the distribution function for the independent random variables $\{r_i\}$.

37. Consider the linear equation

$$\frac{d^2x}{dt^2} + x = r(t) + v\left(x, \frac{dx}{dt}, t\right), \; x(0) = c_1, \; x'(0) = c_2,$$

* H. Cartaino - S. Dreyfus, Application of Dynamic Programming to the Minimum Time-to-Climb Problem, *Aeronautical Engineering Review*, 1957.

where v is to be determined so as to minimize the expected value of

$$J(x) = \int_0^T [x^2 + (dx/dt)^2]\, dt.$$

Find the corresponding recurrence relations, determine the structure of the sequence $\{f_N(c_1, c_2)\}$, and the optimal policy.*

38. Returning to the problem discussed in 7, consider the problem of determining v so as to minimize

$$J = \text{Prob}\,\{ \underset{0 \le t \le T}{\text{Max}} |x| \ge a \}.$$

Show that the discrete version leads to recurrence relation

$$f_{n+1}(c_1, c_2) = \underset{|v_0| \le 1}{\text{Min}} \int_{-\infty}^{\infty} f_n(c_1 + c_2 \Delta, c_2 + [-(c_1^2 - 1) c_2 - c_1 + r_0 + v_0]\Delta)\, dG(r_0)$$

39. Consider the case where the r_i are not independent. Assume, to begin with, that the distribution of r_{n+1} depends on the value of r_n. Define

$f_N(c_1, c_2, r)$ = minimum expected value of J_N, given the initial state (c_1, c_2), and the information that the value of the random variable at the preceding stage was r.

Show that the recurrence relation for the sequence $\{f_N\}$ is

$$f_N(c_1, c_2, r) = \text{Min}\, [\Delta c_1^2 + \int_{-\infty}^{\infty} f_{N-1}(c_1 + c_2 \Delta, c_2 + [-(c_1^2 - 1) c_2 - c_1 + r_0 + v_0]\, \Delta)\, dG(r_0, r)]$$

40. Consider the problem of determining a monotone decreasing sequence of approximations to the first characteristic value of $u'' + \lambda\varphi(t)\, u = 0$, $u(0) = u(1) = 0$. Let $\varphi(t)$ be a continuous positive function of t in $[0, 1]$, so that the first characteristic value is defined by

$$\lambda_1 = \underset{u}{\text{Min}} \frac{\int_0^1 u'^2\, dt}{\int_0^1 \varphi(t)\, u^2\, dt}$$

Let us approximate in policy space by considering functions $u'(t)$ which are constant on the intervals $[k\Delta, (k+1)\Delta]$, $k = 0, 1, 2, \ldots, N-1$, $N\Delta = 1$, i.e.

$$u'(t) = u_k,\ k\Delta \le t < (k+1)\,\Delta.$$

Let $\lambda_1(N)$ denote the minimum over this space. Show that

$$\lambda_1(N) \ge \lambda_1(2N),$$

* R. Bellman, Dynamic Programming and Stochastic Control Processes, *Trans. I.R.E.*, 1957.

and derive a recurrence method for computing $\lambda_1 (N)$.

41. Consider the corresponding problem for the equation

$$u^{(4)} + \lambda\varphi (t) u = 0, \ u (0) = u' (0) = u (1) = u' (1) = 0,$$

corresponding to the variational problem defined by

$$\lambda_1 = \operatorname*{Min}_u \frac{\int_0^1 u''^2 \, dt}{\int_0^1 \varphi (t) u^2 \, dt}$$

Bibliography and Comments for Chapter IX

§ 1. The method presented in this chapter was announced in R. Bellman, "Dynamic Programming and a new formalism in the calculus of variations," *Proc. Nat. Acad. Sci.*, vol. 39 (1953), pp. 1077–1082.

§ 3. A number of applications of the functional equation approach to mathematical economics may be found in R. Bellman, "On the Application of Dynamic Programming to Variational Problems arising in Mathematical Economics," Proc. Symposium on Calculus of Variations and Applications, Chicago, 1956.

§ 7. A treatment of the problem in this section using classical variational techniques may be found in R. Bellman, W. Fleming and D. V. Widder, "Variational Problems Involving Constraints," *Annali di Matematica*, 1956.

§ 11. A discussion of this problem may be found in R. Bellman, "Notes on the Theory of Dynamic Programming," *Jour. Soc. Ind. Appl. Math.*, 1955.

§ 12. A convergence proof of a different type, based upon the functional equations describing the process, may be found in R. Bellman, "Functional Equations in the Theory of Dynamic Programming—VI, A Direct Convergence Proof," *Annals of Math.*, 1957.

The proof given here is due to W. Fleming.

§ 14. A proof of the equivalence of the characteristics of the system of quasilinear partial differential equations for the v_i and the Euler equations obtained in the usual manner has been obtained by H. Osborn. It may be found in Chapter 7 of R. Bellman, "Dynamic Programming of Continuous Processes". The RAND Corporation, R–271, 1954.

§ 21. This last formulation seems the most promising for computational purposes.

§ 22. The results of this section were sketched in R. Bellman, "Monotone Approximation in Dynamic Programming and the Calculus of Variations," *Proc. Nat. Acad. Sci.*, vol. 40 (1954), pp. 1073–5.

§ 25. A discussion of the application of the functional equation technique to some analytic problems arising from control processes may be found in R. Bellman, "On the Application of Dynamic Programming to the Study of Control Processes," Proc. Symposium on Nonlinear Control Processes, Bklyn Polytechnic Inst., 1956.

CHAPTER X

Multi-Stage Games

§ 1. Introduction

In the previous chapters we have discussed a number of decision processes which, although of different origins and varying analytic structure, possess an important feature in common—all decisions are directed towards the single goal of maximizing the value of the criterion function. In this chapter we shall consider a class of multi-stage decision processes where this unanimity of purpose no longer holds true. Some decisions will be made to maximize and some to minimize.

Perhaps the most interesting fashion in which we encounter those cross-purpose processes is in the study of activities in which two animate adversaries counter optimal moves at each stage of the process.

A number of situations in the economic world may be profitably considered in these terms, and the theory of games of chance and skill affords a number of fascinating applications of the general techniques.

Furthermore, in the physical world, in connection with testing and experimentation, it is often useful to conceive of nature, in some vague anthropomorphic fashion, as an opponent attempting to conceal the truth from us. The design of experiments may be conceived of as a game in which we attempt to extract information from a stubborn, but fair, opponent.

The mathematical theory developed in recent years to treat problems characterized by this interplay between divergent aims is the theory of games. Although a good deal of effort had been directed in this direction by E. Borel, the theory rests upon a fundamental result of von Neumann, the celebrated min-max theorem. We shall very briefly discuss the foundations of the theory prior to a discussion of multi-stage games.

These multi-stage games may be considered not only to constitute an extension of the single-stage theory, but in many ways they may be considered to be more fundamental. The single-stage game may be conceived of as a steady-state version of an original dynamic process, namely the multi-stage process.

After these preliminaries, we shall discuss some particular multi-stage games arising from multi-stage allocation processes, and then consider "games of survival" and pursuit games. Following these examples, we shall present a general formulation, along the lines of Chapter 3, and then, as in Chapter 4, prove a number of existence and uniqueness theorems for certain important special classes of equations.

In the main, the techniques used correspond to those employed in the treatment of the one-person processes. Games of survival, however, present special difficulties, requiring more advanced tools for a general treatment. The method we employ is applicable only to a restricted class of equations.

One of the interesting aspects of games of survival is the application of this concept to the study of non-zero-sum games, where the players are no longer in direct opposition. A formulation of these games in terms of survival enables us to remetrize these games so as to make them zero-sum. Furthermore, as we shall show below, a quite reasonable approximation enables us to derive a new metric for non-zero-sum games, one with an associated min-max theorem.

§ 2. A Single-stage discrete game

We shall now consider a class of decision processes involving two persons which we shall call *games*. The two protagonists, whom we shall call *players*, will be named rather prosaically A and B.[1] Let us consider a particular game.

The rules of the game are as follows. The first player, A, has a choice of M different plays, which we shall designate by the numbers 1, 2, . . ., M, and the second player, B has a choice of N different plays, denoted by 1, 2, . . ., N. If A chooses the i—th of his alternatives and B the j—th of his alternatives, A receives a quantity a_{ij} and B a quantity b_{ij}. If these quantities are positive, we may think of them as gains, and if negative as losses.

A convenient way to indicate these returns or *payoffs*, is by means of the two payoff or game matrices

$$M_A = (a_{ij}),\ M_B = (b_{ij}),\ 1 \leq i \leq M\,,\ 1 \leq j \leq N\,. \tag{1}$$

Let us now consider the single-stage process where each player makes precisely *one* play. The determination of optimal play, defined as that which maximizes return, is straightforward if A is required to move before B and if B can use this information. If A takes the i—th alternative, B chooses $j = j(i)$ so as to maximize b_{ij}. Consequently A chooses i so as to maximize $a_{i,\,j(i)}$. A similar rule determines the choice of j if B is required to move first.

[1] The successors of the algebraic A, B, and C discussed by S. Leacock.

The only interesting case is that where both players are required to move simultaneously, without knowing of the other's choice.

In these circumstances, they can protect themselves by mixing their choices, which is to say they will randomize their choices in a certain fashion. Let us assume then that A makes the i—th choice with probability p_i and B the j—th choice with probability q_j. The vector $p = (p_1, p_2, \ldots, p_M)$ specifies A's probability distribution and the vector $q = (q_1, q_2, \ldots, q_N)$ specifies B's probability distribution.

As in our discussion of the stochastic processes of the previous chapters, we can no longer speak of *the* return, but must agree to consider some average return. The simplest such, as usual, is the expected return. The expected return to A will be

(2) $$E_A(p, q) = \sum_{i=1}^{M} \sum_{j=1}^{N} a_{ij} p_i q_j ,$$

while that for B is

(3) $$E_B(p, q) = \sum_{i=1}^{M} \sum_{j=1}^{N} b_{ij} p_i q_j .$$

The first player would like to choose p so as to maximize E_A, while the second player would like to choose q so as to maximize E_B.

§ 3. The min-max theorem

In order to obtain definitive results, we must assume that the players are in direct opposition, expressed by the relation

(1) $$b_{ij} = -a_{ij} .$$

In this case, the game is called *zero-sum*, and only in this case does a satisfactory general theory exist. We then have

(2) $$E_B(p, q) = -E_A(p, q) ,$$

from which it is clear that any choice of p and q which increases $E_A(p, q)$ decreases $E_B(p, q)$, and vice versa.

It is sufficient then to consider $E_A(p, q)$ in our further discussion. We can, using this expression, define two values of the game,

(3) $$V_A = \underset{q}{\text{Min}}\ \underset{p}{\text{Max}}\ E_A(p, q)$$

$$V_B = \underset{p}{\text{Max}}\ \underset{q}{\text{Min}}\ E_A(p, q) .$$

The first is the expected return to A if B is required to choose q before A chooses p, while the second is the value to A if the situation is reversed.

It is a remarkable fact (the min-max theorem of von Neumann), the basic result in the theory of games, that

$$V_A = V_B. \tag{4}$$

This common value is called the *value* of the game. We shall assume this result without proof here.

The interpretation of this result is that A can announce his probability distribution p in advance, and likewise B can announce q, without either gaining from this advance knowledge. This is neither an intuitive, nor a trivial, result, but it is true.

§ 4. Continuous games

Let us now suppose that in place of choosing one of a discrete set of moves, A must choose from a continuum and similarly B. As a simple example, suppose that A must choose a real number x in the interval $[0, 1]$, and B similarly a real number y in $[0, 1]$. Considering the zero-sum case only, there is now a payoff function $K(x, y)$ which measures the value of this set of moves to A, with $-K(x, y)$ the value to B.

If A chooses a distribution $F(x)$ to govern the frequency with which he chooses x, and B the distribution function $G(y)$, the expected gain to A will be

$$V_A = \int_0^1 \int_0^1 K(x, y)\, dF(x)\, dG(y). \tag{1}$$

The continuous analogue of the min-max theorem is the result:

$$\operatorname*{Max}_F \operatorname*{Min}_G V_A = \operatorname*{Min}_G \operatorname*{Max}_F V_A, \tag{2}$$

where the variation is over the space of functions defined by

$$\text{(a) } dF \geq 0, \int_0^1 dF(x) = 1, \tag{3}$$

$$\text{(b) } dG \geq 0, \int_0^1 dG(y) = 1,$$

provided that $K(x, y)$ is jointly continuous in x, y over the unit square.[2] If $K(x, y)$ is not continuous, (2) need not hold, and $V_A(F, G)$ need not even exist for all F and G.

[2] This theorem is a very fine illustration of the utility of the Stieltjes integral, since the result is not valid if we consider only functions $F(x)$ and $G(y)$ which are integrals, i.e., $dF(x) = \varphi(x)\, dx$, $dG(y) = \psi(y)\, dy$.

§ 5. Finite resources

In many situations involving multi-stage play, the above model is not satisfactory. This is particularly so in multi-stage processes where each player possesses finite resources. Here the choice of plays depends upon the quantity of resources available, and the game terminates when either player has no resources. Consequently we cannot consider the set of N games as consisting of N disjoint plays.

Let us consider a simple example. Suppose that A has a quantity x and B a quantity y. At each stage each player may allocate 1 or 2 units of his resources with the return a_{ij} to A if A makes an allocation of i and B an allocation of j, and a return of $-a_{ij}$ to B, where $i, j = 1, 2$.

Here, for the sake of initial simplicity, the return a_{ij} is in units different from those of x and y, and so cannot be reconverted into resources.

Let us take the process to terminate as soon as either side has no resources and suppose that each plays to maximize his total return. Assume that we may define the function

(1) $f(x, y) =$ expected return from the process to A when A has x and B has y initially, and each employs an optimal policy.

On the first move, A mixes his choices according to the probability distribution $p = (p_1, p_2)$ and B according to the distribution $q = (q_1, q_2)$, where p and q will, in general, be functions of x and y.

An enumeration of possibilities yields the relation

$$f(x, y) = \sum_{i=1}^{2} \sum_{j=1}^{2} p_i q_j [a_{ij} + f(x - i, y - j)] , \tag{2}$$

for an optimal policy, assuming for the moment that the principle of optimality is equally valid for multi-stage games. A proof of this will be given in § 9. Thus the functional equation for $f(x, y)$ is

$$\begin{aligned} f(x, y) &= \operatorname*{Max}_{p} \operatorname*{Min}_{q} \left\{ \sum_{i=1}^{2} \sum_{j=1}^{2} p_i q_j [a_{ij} + f(x - i, y - j)] \right\} \\ &= \operatorname*{Min}_{q} \operatorname*{Max}_{p} \left\{ \sum_{i=1}^{2} \sum_{j=1}^{2} p_i q_j [a_{ij} + f(x - i, y - j)] \right\} \end{aligned} \tag{3}$$

for $x, y > 0$, with the boundary conditions

$$f(x, y) = 0 \text{ if } x \leq 0 \text{ or } y \leq 0 . \tag{4}$$

§ 6. Games of survival

Returning to the game described in § 2, let us take A to have x, B to have y and assume that the returns a_{ij} and b_{ij} are in the same units as x

and y, say dollars, and that $b_{ij} = -a_{ij}$, the zero-sum case. Let us now suppose that the game is continued until one player is ruined, and that each player attempts to ruin the other. A game of this type we call a *game of survival*. It is a generalized "gambler's ruin" problem.

Assuming the existence of the function

(1) $f(x, y)$ = probability that A ruins B when A has x, B has y, and each player employs an optimal policy,

and proceeding as before, we obtain the functional equation

$$f(x, y) = \underset{p}{\text{Max}}\ \underset{q}{\text{Min}} \sum_{i,j} f(x + a_{ij}, y - a_{ij})\, p_i\, q_j = \underset{q}{\text{Min}}\ \underset{p}{\text{Max}} \sum_{i,j} f(x + a_{ij}, y - a_{ij})\, p_i\, q_j, \tag{2}$$

$x, y > 0$, with the boundary conditions

$$f(x, y) = 1,\ x > 0,\ y \leq 0, \quad = 0,\ x \leq 0,\ y > 0. \tag{3}$$

Since the game is zero-sum, the quantity of resources in the game remains constant. Thus the state of the process is specified by x, the quantity possessed by A. Setting $x + y = c$, and $f(x, y) = f(x)$, we obtain the simpler equation

$$f(x) = \underset{p}{\text{Max}}\ \underset{q}{\text{Min}} \sum_{i,j} f(x + a_{ij})\, p_i\, q_j = \underset{q}{\text{Min}}\ \underset{p}{\text{Max}} \sum_{i,j} f(x + a_{ij})\, p_i\, q_j, \tag{4}$$

for $0 < x < c$, with $f(x) = 0$ for $x \leq 0$, $f(x) = 1$, $x \geq c$.

§ 7. Pursuit games

Another interesting class of games are those involving the pursuit of one player by another. In some cases there is a question as to whether one player can catch the other, in other cases where capture is certain, the problem is to determine the choice of paths for one player which minimizes the time of capture and for the other player a path which maximizes the time of capture.

The continuous versions of these problems are quite difficult to formulate rigorously, and as a consequence most of the results obtained in this connection pertain to the discrete version.

Consider the following simple problem. The two players, A and B are situated at the points $k\Delta$, $l\Delta$ respectively on the line, where $\Delta > 0$ and k and l are integers or zero. At each move of the game, each player has the choice of moving one unit to the right or to the left. Moves are made simultaneously with full information as to the positions of each player.

After each move there is pay-off from B to A of an amount $g(d)$ where $d = |k - l| \Delta$, the distance between the players. Furthermore, there is a probability $1 - a(d)$ that the process terminates on that move.

The total pay-off of the multi-stage game is taken to be the expected value of the quantity that B pays A before the process terminates. Once again assume that the function

(1) $f(d)$ = expected pay-off if A and B are initially d units apart and both employ optimal strategies,

exists. Then proceeding as before, we obtain the functional equation

$$(2) \quad f(d) = a(d) \operatorname*{Max}_{p} \operatorname*{Min}_{q} [p_1 q_1 f(d) + p_2 q_2 f(d) + p_1 q_2 f(d-2) + p_2 q_1 f(d+2)] + g(d).$$
$$= a(d) \operatorname*{Min}_{q} \operatorname*{Max}_{p} [\ \ldots] + g(d),$$

where p_1, p_2 are the probabilities of A going to the left or the right respectively on any move, and q_1, q_2 are the corresponding probabilities for B. In general, optimal p_1, p_2, q_1 and q_2 will depend upon d.

§ 8. General formulation

Let us now describe, in some generality, a class of multi-stage games we wish to analyze. At any stage of the game, the states of the two players, A and B, are represented by m-dimensional vectors, x and y, which we can think of as "resources."

In order to avoid for the moment the conceptual difficulties of infinite processes, we shall first consider a finite process. At the beginning of each stage of an N-stage process, A allocates a certain quantity of his resources, a vector u, and B a certain quantity of his resources, a vector v; this will be represented symbolically by the notation $0 \leq u \leq x$, $0 \leq v \leq y$, where the inequalities hold component-wise.

As a result of this allocation, there are two consequences. A receives a payoff of $R(u, v; x, y)$, a scalar function, and B a payoff of $-R(u, v; x, y)$ — a zero-sum process. In addition to these payoffs, there is an alteration in their resources; x is transformed into $T(x, y; u, v)$, and y becomes $T'(x, y; u, v)$. The process now continues in the same fashion for $(N - 1)$ additional stages.

The total return to A of the N-stage process is

$$(1) \quad \begin{aligned} R_N &= R_N(u, u_1, u_2, \ldots, u_{N-1}; v, v_1, \ldots, v_{N-1}; x, y) \\ &= R(u, v) + R(u_1, v_1) + \ldots + R(u_{N-1}, v_{N-1}). \end{aligned}$$

There are several ways we can treat this N-stage process. One extreme regards the N-stage game as a single-stage game of complicated type, requiring a choice of a set of vectors $(u, u_1, \ldots, u_{N-1})$ by A and a set $(v, v_1, \ldots, v_{N-1})$ by B, where the choice of u_K and v_K is dependent upon the choice of $u, u_1, \ldots, u_{K-1}, v, v_1, \ldots, v_{K-1}$. Alternatively, we can employ the functional equation approach. For the case of unbounded processes, or processes involving stochastic interaction, the recurrence technique is, in general, the only feasible one. For the case of finite deterministic processes, this technique is usually simpler analytically and computationally.

We shall assume that $R(u, v; x, y)$ is a continuous function of u and v for all finite values of u and v, x and y, and that similarly $T(x, y, u, v)$, $T'(x, y, u, v)$ are continuous functions of x, y, u, and v for all finite values of the vector variables.

The general case where only boundedness and measurability of the functions are assumed may be handled using the same principles, at the expense of introducing Sup-Inf operators in place of Max-Min. The particular case where x, y, u, v, T, T' assume only finite sets of values is also interesting to consider, and has the advantage of avoiding continuity considerations.

One advantage to considering the N-stage process as a single-stage process, as described above, is that it permits us to define the multi-stage game precisely on the basis of known results for the single-stage game and thus the value of the multi-stage game. Once having defined the game, we can prove that recurrence techniques are applicable.

The value of the N-stage game described above is given by the expression

$$(2)\quad v_N = \operatorname*{Max}_{G} \operatorname*{Min}_{G'} \left[\iint R_N \, dG(u, u_1, u_2, \ldots, u_{N-1}) \, dG'(v, v_1, v_2, \ldots, v_{N-1})\right]$$
$$= \operatorname*{Min}_{G'} \operatorname*{Max}_{G} [\ \ldots\ ],$$

where G and G' are distribution functions over regions of quite complicated form defined by the inequalities

$$(3)\qquad \begin{array}{ll} 0 \leq u \leq x, & 0 \leq v \leq y \\ 0 \leq u_1 \leq T, & 0 \leq v_1 \leq T' \\ \vdots & \\ 0 \leq u_{N-1} \leq T_{N-1}, & 0 \leq v_{N-1} \leq T'_{N-1}. \end{array}$$

The quantities T and T' depend upon x, y, u, and v; T_1, T_1' depend upon x, y, u, v, u_1, v_1, and so on.

§ 9. The principle of optimality and functional equations

Let us now change our notation, replacing x by P and y by P', in order to consider more general situations where x and y are not necessarily vectors whose elements are quantities of resources.

Since v_N depends only upon the initial states, we may define the sequence of functions

(1) $$f_N(P, P') = v_N, \ N = 1, 2, \ldots$$

Assuming for the moment that the principle of optimality is valid for multi-stage games, we obtain the following recurrence relations.[3]

(2) $$f_1(P, P') = \operatorname*{Max}_{G} \operatorname*{Min}_{G'} \Big[\iint\limits_{\substack{0 \leq u \leq P \\ 0 \leq v \leq P'}} R(u, v)\, dG(u)\, dG'(v) \Big]$$

$$= \operatorname*{Min}_{G'} \operatorname*{Max}_{G} [\ \ldots\ ],$$

$$f_{N+1}(P, P') = \operatorname*{Max}_{G} \operatorname*{Min}_{G'} \Big[\iint\limits_{\substack{0 \leq u \leq P \\ 0 \leq v \leq P'}} [R(u, v) + f_N(T, T')]\, dG(u)\, dG'(v) \Big]$$

$$= \operatorname*{Min}_{G'} \operatorname*{Max}_{G} [\ \ldots\ ].$$

That this principle is valid for one-person processes where we are attempting to maximize a return, or minimize a "cost" is clear by contradiction. Since its validity may not be as obvious for game processes, let us present a brief proof for the sake of completeness.

The recurrence relation in (2) provides a sequence, not necessarily unique, of pairs of distribution functions, $\{G_N(u, P, P'), G'_N(v, P, P')\}$ which furnish the sequence $\{f_N(P, P')\}$. In order to show that the function $f_N(P, P')$ is actually the value of the N-stage game, it is sufficient to show that A can guarantee an expected return of $f_N(P, P')$ if he chooses u at the first stage of an N-stage process in accordance with the distribution function $G_N(u, P, P')$, when the states of A and B are described by P and P', respectively, and similarly that B can guarantee an expected loss of not more than $-f_N(P, P')$.

To demonstrate this, consider the one-person N-stage process in which A employs the fixed strategy represented by the sequence of distribution functions, $\{G_k(u, P, P')\}$, $k = 1, 2, \ldots, N$, and B attempts to minimize A's expected N-stage return. It is sufficient to consider this process, since any other policy employed by B yields a larger expected return for A. Let

[3] To simplify the notation, we shall write $R(u, v)$ for $R(u, v; P, P')$.

(2) $w_N(P, P')$ = N-stage expected return to A when A employs the fixed strategy $\{G_k(u, P, P')\}$, B employs a minimizing strategy, and A and B are in the states P and P' initially.

Then we have the recurrence relations

$$(3) \qquad w_1(P, P') = \operatorname*{Inf}_{G'} \int_{0 \le v \le P'} \left[\int_{0 \le u \le P} R(u, v)\, dG(u, P, P') \right] dG'(v),$$

$$w_{N+1}(P, P') = \operatorname*{Inf}_{G'} \int_{0 \le v \le P'} \left[\int_{0 \le u \le P} [R(u, v) + w_N(T, T')] \right.$$

$$\left. dG_{N+1}(u, P, P') \right] dG'(v).$$

upon employing the principle of optimality for the one-person process.

Considering the origin of the function G_1, we see that the minimum in the relation for $w_1(P, P')$ in (3) is attained by the function $G' = G_1'$, not uniquely in general. Hence,

$$(4) \qquad w_1(P, P') = v_1(P, P').$$

Since $w_1 \equiv v_1$, the relation for w_2 yields in the same way the fact that $w_2 \equiv v_2$, and thus, inductively, we see that

$$(5) \qquad w_N(P, P') = v_N(P, P').$$

In precisely the same way we show that if B employs the strategy $G_N'(v, P, P')$, A cannot obtain more than $v_N(P, P')$. Hence $v_N(P, P')$ is the value of the N-stage game.

§ 10. More general process

Before presenting some precise statements concerning the processes we have discussed above, let us consider a sequence of more general processes which may be treated by means of the same techniques we shall employ below.

Consider, to begin with, an infinite process of the type described in § 8 in which we allow the transformations T and T', as well as the return R, to depend upon the stage.

We then consider the functions

(1) $f(P, P'; k)$ = the value to A of the infinite process beginning at the k—th stage when A and B possess P and P' at this stage, and both employ optimal strategies.

This sequence, with the usual proviso relating to existence, satisfies the recurrence relation

$$(2)\quad f(P, P'; k) = \underset{G}{\text{Max}}\ \underset{G'}{\text{Min}} \left[\iint_{\substack{0 \le u \le P \\ 0 \le v \le P'}} [R(u, v, k) + f(T_k, T_k'; k+1)]\, dG(u)\, dG'(v) \right]$$
$$= \underset{G'}{\text{Min}}\ \underset{G}{\text{Max}} [\qquad].$$

Let us now complicate the process to a further degree. We have assumed, in the above formulation, that the interaction between the players was perfectly determined once u and v were chosen. In a variety of processes, a choice of u and v determines a distribution of outcomes, which is to say the interaction is stochastic rather than deterministic. Let $K_k(z, t, t'; u, v)$ denote the distribution function, where z is the value of $R_k(u, v)$, t the value of T_k and t' the value of T_k'.

The functional equation of (2) is replaced by

$$(3)\quad f(P, P'; k) = \underset{G}{\text{Max}}\ \underset{G'}{\text{Min}} \left[\iint_{\substack{0 \le u \le P \\ 0 \le v \le P'}} \left[\int [z + f(t, t'; k+1)]\, dK_k\right] dG(u)\, dG'(v) \right]$$
$$= \underset{G'}{\text{Min}}\ \underset{G}{\text{Max}} [\qquad].$$

Finally, let us consider the case where we are concerned with a nonlinear function of the total return, R, rather than the total return itself. A particularly important situation is that where A wishes to maximize the probability of achieving a return of at least R_o, a specified constant. Another interesting utility function is e^{aR}.

Let us assume that A wishes to maximize the expected value of $\varphi(R)$, where φ is a given function of R. To describe this nonlinear situation, we must introduce an additional state variable, a, the total return obtained by A from the previous stages of the process. Defining the function $f(P, P', a; k)$ essentially as in (1), we obtain the associated functional equation

$$(4)\quad f(P, P', a; k) = \underset{G}{\text{Max}}\ \underset{G'}{\text{Min}} \left[\iint_{\substack{0 \le u \le P \\ 0 \le v \le P'}} \left[f(t, t', a + z; k+1)\, dK_k\right] dG(u)\, dG'(v) \right]$$
$$= \underset{G'}{\text{Min}}\ \underset{G}{\text{Max}} [\qquad].$$

None of these functional equations will be discussed here in connection with the existence and uniqueness of solutions since the basic approach is the same for all cases.

§ 11. A basic lemma

Let us present a simple but extremely useful inequality which exhibits the quasi-linearity of the transformation

$$(1) \qquad L(f) = \underset{G}{\text{Max}}\ \underset{G'}{\text{Min}}\ T(P, P'; f; G, G') = \underset{G'}{\text{Min}}\ \underset{G}{\text{Max}}\ T.$$

It will play the same role in the existence and uniqueness proofs of this chapter that Lemma 1 of Chapter 4 played in that chapter.

LEMMA 1.[4] *Let*

$$(2) \quad L(f) = \underset{G}{\text{Max}}\ \underset{G'}{\text{Min}}\ \Big[\iint_{\substack{u \in S \\ v \in S'}} [R(u, v) + h(P, P'; u, v) f(T, T')]\, dG(u)\, dG'(v)\Big]$$

$$= \underset{G'}{\text{Min}}\ \underset{G}{\text{Max}}\ [\qquad\qquad].$$

$$L_1(F) = \underset{G}{\text{Max}}\ \underset{G'}{\text{Min}}\ \Big[\iint_{\substack{u \in S \\ v \in S'}} [R_1(u, v) + h(P, P'; u, v) F(T, T')]\, dG(u)\, dG'(v)\Big]$$

$$= \underset{G'}{\text{Min}}\ \underset{G}{\text{Max}}\ [\qquad\qquad].$$

Then

$$(3) \qquad |L(f) - L_1(F)| \leq \underset{u \in S}{\text{Max}}\ \underset{v \in S'}{\text{Max}}\ [\,|R(u, v) - R_1(u, v)| + |h(P, P'; u, v)|\,|f(T, T') - F(T, T')|\,].$$

PROOF: Let us write

$$(2) \quad L(f) = \underset{G}{\text{Max}}\ \underset{G'}{\text{Min}}\ T(P, P'; f; G, G') = \underset{G'}{\text{Min}}\ \underset{G}{\text{Max}}\ T(P, P' : f, G, G')$$

$$L_1(F) = \underset{G}{\text{Max}}\ \underset{G'}{\text{Min}}\ T_1(P, P'; F; G, G') = \underset{G'}{\text{Min}}\ \underset{G}{\text{Max}}\ T_1(P, P'; F; G, G').$$

Let (G_1, G_1') be a pair of functions yielding the value $L(f)$, and let (G_2, G_2') be a pair of functions yielding the value $L_1(F)$. Then, by virtue of the saddle-point property, we have the following chain of equalities and inequalities:

$$(5) \quad L(f) = T(P, P'; f; G_1, G_1') \geq T(P, P'; f; G_2, G_1')$$
$$\leq T(P, P'; f; G_1, G_2'),$$
$$L_1(F) = T_1(P, P'; F; G_2, G_2') \geq T_1(P, P'; F; G_1, G_2')$$
$$\leq T_1(P, P'; F; G_2, G_1').$$

[4] It is *assumed* that max-min = min-max for each transformation. A similar result holds for the one-sided max-min operator; see § 18.

Combining these inequalities we have

$$(6) \quad L(f) - L_1(f) \geq T(P, P'; f; G_2, G_1') - T_1(P, P'; F; G_2, G_1') \\ \leq T(P, P'; f; G_1, G_2') - T_1(P, P'; F; G_1, G_2').$$

The inequalities in (6) yield

$$(7) \quad L(f) - L_1(F) \geq \iint_{\substack{u \varepsilon S \\ v \varepsilon S'}} [R(u, v) - R_1(u, v) + h(P, P'; u, v)[f(T, T') \\ - F(T, T')]] \, dG_2(u) \, dG_1'(v) \\ \geq \iint_{\substack{u \varepsilon S \\ v \varepsilon S'}} [R(u, v) - R_1(u, v) + h(P, P'; u, v)[f(T, T') \\ - F(T, T')]] \, dG_1(u) \, dG_2'(v).$$

Using as in Chapter 4 the fact that $a \leq c \leq b$ implies $|c| \leq$ Max $(|a|, |b|)$, we obtain from (7) the further inequality

$$(8) \quad |L(f) - L_1(F)| \leq \text{Max}\,\{[\iint_{\substack{u \varepsilon S \\ v \varepsilon S'}} [\,|R(u, v) - R_1(u, v)| \\ + |h(P, P'; u, v)|\,|f(T, T') - F(T, T')|\,] \, dG_2(u) \, dG_1'(v)], \\ [\iint_{\substack{u \varepsilon S \\ v \varepsilon S'}} [\,|R(u, v) - R_1(u, v)| + |h(P, P'; u, v)|\,|f(T, T') \\ - F(T, T')|\,] \, dG_1(u) \, dG_2'(v)],$$

from which (3) follows immediately.

It is easy to make the modifications required to obtain the analogous result for the case where Max Min is replaced by Sup Inf.

§ 12. Existence and uniqueness

Before stating our results, let us introduce some notation. Let P and P' represent n- and n'-dimensional vectors defined over regions D and D', respectively, each containing the origin in its respective space. For all values of u, v, P and P', the transformed vectors $T(P, P'; u, v)$, $T'(P, P'; u, v)$, are required to lie within these same domains, where u and v are k- and k'-dimensional choice vectors, respectively, constrained to domains S and S', which may or may not depend upon P and P'. Since we shall be dealing with shrinking transformations in the theorem below, there is no loss of generality in taking D and D' to be finite.

In each space, let us introduce the norm, $||P||$, equal to the sum of the absolute values of the components of P,

(1) $$||P|| = \sum_{i=1}^{n} |P_i|,$$

$$||P'|| = \sum_{i=1}^{n'} |P_i'|.$$

Actually, these norms need not be identical, and, in some situations, it might be useful to consider norms molded to the structure of the functional equation, rather than standard norms of the above type.

The functional equation we shall consider in some detail is

(2) $$f(P, P') = \operatorname*{Max}_{G} \operatorname*{Min}_{G'} \left[\int_{u \varepsilon S} \int_{v \varepsilon S'} [R(u, v) + h(P, P'; u, v) f(T, T')] \, dG(u) \, dG'(v) \right] = \operatorname*{Min}_{G'} \operatorname*{Max}_{G} [\qquad],$$

where $T = T(P, P'; u, v)$, $T' = T'(P, P'; u, v)$.

To simplify our notation, let us represent the operator appearing within the outer brackets in equation (2) by $T(P, P'; f; G, G')$. The equation in (2) then assumes the form

(3) $$f(P, P') = \operatorname*{Max}_{G} \operatorname*{Min}_{G'} T(P, P'; f; G, G') = \operatorname*{Min}_{G'} \operatorname*{Max}_{G} T(P, P'; f; G, G').$$

There is a question as to whether this should be referred to as one equation or as a pair of equations. We shall refer to (3) as "an equation."

The result we shall demonstrate is

THEOREM 1. *Consider the equation in* (2) *under the following assumptions:*

(4) (*a*) *The functions* $R(u, v)$, $h(P, P'; u, v)$, $T(P, P'; u, v)$ *and* $T'(P, P'; u, v)$ *are continuous functions of* P *and* P', u *and* v, *in any bounded domain of the variables.*

(b) *The choice domains,* $S(P, P')$, $S'(P, P')$, *vary continuously with* P *and* P'.

(c) T *and* T' *are shrinking transformations,* i.e.,

$$\operatorname*{Max}_{\substack{u \varepsilon S \\ v \varepsilon S'}} (||T|| + ||T'||) \leq k(||P|| + ||P'||),$$

where k *is a fixed constant less than* 1.

(d) *Let* $\sum_{n=1}^{\infty} w(k^n c) < \infty$ *for all* $c > 0$, *where*

$$w(c) = \underset{||P|| + ||P'|| \leq c}{\text{Max}} \left(\underset{\substack{u \varepsilon S \\ v \varepsilon S'}}{\text{Max}} |R(u, v)| \right).$$

(e) $$\underset{u, v, P, P'}{\text{Max}} ||h(u, v, P, P')|| \leq 1.$$

If the above conditions are satisfied, we can assert that there is a unique solution of the equation in (2) *within the class of functions* $f(P, P')$ *which are continuous for all finite* P *and* P' *and vanish when* P *and* P' *are both null vectors.*

This solution may be found by the method of successive approximations,

(5) $$f_0(P, P') = \underset{G}{\text{Max}} \underset{G'}{\text{Min}} \left[\iint_{\substack{u \varepsilon S \\ v \varepsilon S'}} R(u, v)\, dG(u)\, dG'(v) \right]$$

$$= \underset{G'}{\text{Min}} \underset{G}{\text{Max}} [\qquad\qquad\qquad\qquad],$$

$$f_{n+1}(P, P') = \underset{G}{\text{Max}} \underset{G'}{\text{Min}}\, T(P, P'; f_n; G, G'),$$

$$= \underset{G'}{\text{Min}} \underset{G}{\text{Max}}\, T(P, P'; f_n; G, G'),\ n \geq 0.$$

The solution is obtained as the limit $f(P, P') = \lim_{n \to \infty} f_n(P, P')$, *in any bounded region of* (P, P') *space.*

We shall further demonstrate

THEOREM 2. *Under the hypothesis of Theorem* 1, *a set of functions* $(G(u), G'(v))$ *furnished by the functional equation constitute a set of optimal strategies for* A *and* B, *respectively, in the multi-stage game described above.*

§ 13. Proof of results

Let us now proceed to the proofs of these results. Let

(1) $$f_0(P, P') = \underset{G}{\text{Max}} \underset{G'}{\text{Min}} \left[\iint_{\substack{u \varepsilon S \\ v \varepsilon S'}} R(u, v)\, dG(u)\, dG'(v) \right],$$

$$= \underset{G'}{\text{Min}} \underset{G}{\text{Max}} [\qquad\qquad\qquad\qquad],$$

and

(2) $$f_{n+1}(P, P') = \underset{G}{\text{Max}} \underset{G'}{\text{Min}}\, T(P, P'; f_n; G, G') = \underset{G'}{\text{Min}} \underset{G}{\text{Max}}\, T,$$

where T is defined as in (4.2) and (4.4).

By virtue of our assumptions concerning the coefficient functions and the domains S and S', we can assert the existence of the saddlepoint in (1), and the continuity of $f_o(P, P')$. Inductively, then, all the $f_n(P, P')$ exist and are continuous for all finite P and P'.

Let us now show that the sequence $\{f_n\}$ converges uniformly in any finite portion of the (P, P')-regions. Using Lemma 1 we obtain the inequality

$$(3)\quad |f_{n+1}(P, P') - f_n(P, P')| \leq \underset{G}{\text{Max}}\ \underset{G'}{\text{Max}} \left[\iint |f_n(T, T') - f_{n-1}(T, T')|\, dG(u)\, dG'(v)\right], \quad n = 2, 3, \ldots.$$

Define the new sequence

$$(4)\quad u_{n+1}(c) = \underset{||P|| + ||P'|| \leq c}{\text{Max}} |f_{n+1}(P, P') - f_n(P, P')|.$$

Then (3) yields, using the assumption of (4a) of § 3,

$$(5)\quad u_{n+1}(c) \leq u_n(kc),\ n = 2, 3, \ldots,$$

Also, we have

$$(6)\quad |f_2(P, P') - f_1(P, P')|, \leq \underset{G}{\text{Max}}\ \underset{G'}{\text{Max}} \iint |R(u, v)|\, dG(u)\, dG'(v),$$

whence

$$(7)\quad u_2(c) \leq w(c).$$

Using our assumption that $\Sigma\, w(k^n c) < \infty$, we see that the series $\sum_n [f_{n+1}(P, P') - f_n(P, P')]$ converges uniformly in any finite region. Hence $f_n(P, P')$ converges uniformly to a function $f(P, P')$ which satisfies the original functional equation.

This completes the proof of existence. Let us now turn to a proof of uniqueness. Let $F(P, P')$ be another solution which is continuous at $P = 0$, $P' = 0$, and bounded in any finite region. We see that $F(P, P')$ is then actually continuous for all finite P and P', although this fact is not necessary for our proof. It does simplify it a bit since we can replace Sup-Inf by Max-Min.

We then have the two equations

$$(8)\quad F(P, P') = \underset{G}{\text{Max}}\ \underset{G'}{\text{Min}}\ T(P, P'; F; G, G') = \underset{G'}{\text{Min}}\ \underset{G}{\text{Max}}\ T$$

$$f(P, P') = \underset{G}{\text{Max}}\ \underset{G'}{\text{Min}}\ T(P, P'; f; G, G') = \underset{G'}{\text{Min}}\ \underset{G}{\text{Max}}\ T$$

Applying Lemma 1, we see that

$$(9) \qquad |F(P,P') - f(P,P')| \leq \underset{G}{\text{Max}} \underset{G'}{\text{Max}} [\iint_{\substack{u \varepsilon S \\ v \varepsilon S'}} |F(T,T') - f(T,T')| \, dGdG'$$

Let

$$(10) \qquad \Delta(c) = \underset{||P|| + ||P'|| \leq c}{\text{Max}} |F(P,P') - f(P,P')| .$$

The (9) yields the relation

$$(11) \qquad \Delta(c) \leq \Delta(kc),$$

which, upon iteration, yields $\Delta(c) \leq \Delta(k^n c)$, $n = 1, 2, \ldots$. Since F and f are both continuous at $P = 0$, $P' = 0$, and have the common value 0 there, we see that $\Delta(k^n c) \to 0$ as $u \to \infty$. Hence $\Delta(c) \equiv 0$ and $F \equiv f$.

This completes the proof of Theorem 1.

§ 14. Alternate proof of existence

In the study of functional equations of this class, the proof of the existence is "cheap," while the proof of the uniqueness requires varying degrees of effort. As far as the functional equations arising from the calculus of variations are concerned, the opposite is true; there, existence is difficult and uniqueness is simple.

Let us indicate how we may establish the existence of a solution of the Sup-Inf equation in the case where we assume that $R(u, v) \geq 0$ and $h(P, P'; u, v) \geq 0$. It follows inductively that the sequence $\{f_n(P, P')\}$ is monotone increasing and bounded. Hence the sequence converges to a function $f(P, P')$.

To show that this function satisfies the functional equation

$$(1) \qquad f(P,P') = \underset{G}{\text{Sup}} \underset{G'}{\text{Inf}} T(P,P';f;G,G') = \underset{G'}{\text{Inf}} \underset{G}{\text{Sup}} T,$$

we proceed as follows. We have

$$(2) \qquad f(P,P') \geq f_{n+1}(P,P') = \underset{G}{\text{Sup}} \underset{G'}{\text{Inf}} T(P,P';f_n;G,G'),$$

and thus

$$(3) \qquad f(P,P') \geq \underset{G}{\text{Sup}} \underset{G'}{\text{Inf}} T(P,P';f;G,G').$$

Conversely, utilizing the positivity of the operator, we have

$$(4) \qquad f_{n+1}(P,P') \leq \text{Sup Inf } T(P,P';f;G,G'),$$

for all n, and, in consequence,

$$f(P, P') \leq \text{Sup Inf } T(P, P'; f; G, G'). \tag{5}$$

Comparing (3) and (5) we see that we have equality.

§ 15. Successive approximations in general

The sequence of approximations, $\{f_n(P, P')\}$, used to construct the function $f(P, P')$ was precisely that obtained from the finite n-stage processes. This is actually not the best sequence to use if we are interested only in the infinite stage process. As we have pointed out in previous pages, approximation in "policy space", here "strategy space", is in many ways a more natural and more important type of approximation.

To justify this and other types of approximations we require

THEOREM 3. *Under the assumptions of Theorem* 1, *the sequence defined by*

$$f_{n+1}(P, P') = \underset{G}{\text{Max}}\ \underset{G'}{\text{Min}}\ T(P, P'; f_n; G, G'),\ n = 0, 1, \ldots \tag{1}$$

$$= \underset{G'}{\text{Min}}\ \underset{G}{\text{Max}}\ T(P, P'; f_n; G, G')$$

converges to the solution of (5.3) *for any initial function* $f_0(P, P')$ *which is continuous in any finite part of the* (P, P')*-domain, and equal to zero at* $P = 0, P' = 0$.

The proof is precisely the same as that given above.

§ 16. Effectiveness of solution

We have established existence and uniqueness of the functional equation derived above under the assumption that the infinite process possessed a value for each player. The question now arises as to whether the functional equation actually yields sufficient information to allow each player to obtain this value. If so, we say that the solution is *effective*, and theoretically, the functional equation is equivalent to the game.[5]

The solution will be effective under the hypotheses of Theorem 1, which is to say, continuity.

To show effectiveness, under the hypotheses of Theorem 2, we must show that if A uses a distribution function $G(u) = G(u; P, P')$ obtained from a pair (G, G') which yield the min-max, then, regardless of what B may do, A can guarantee himself a return of at least $f(P, P')$.

[5] In many ways, however, this is not true. Once the functional equation has been formulated, and the process discarded, we have restricted ourselves to a certain direction of approach which may not be optimal for the derivation of *all* properties of the process. It is well then to keep in mind that the above functional equation is only one of many possible mathematical descriptions of the process.

Employing this fixed strategy, A's return will be, at worst, determined by the solution of the functional equation

$$(1) \qquad F(P, P') = \operatorname*{Min}_{G'} \Big[\iint_{\substack{u \varepsilon S \\ v \varepsilon S'}} [R(u, v) + h(P, P'; T, T') F(T, T')] \, dG(u) \, dG'(v) \Big].$$

It is easy to show, using the techniques of the preceding chapters, together with the assumptions we have made, that this equation has a unique continuous solution which is zero at $P = 0$, $P' = 0$. Furthermore, the solution of this equation may be obtained as the limit of the sequence defined by

$$(2) \qquad F_0(P, P') = \operatorname*{Min}_{G'} \Big[\iint_{\substack{u \varepsilon S \\ v \varepsilon S'}} R(u, v) \, dG(u) \, dG'(v) \Big],$$

$$F_{n+1}(P, P') = \operatorname*{Min}_{G'} \iint_{\substack{u \varepsilon S \\ v \varepsilon S'}} [R(u, v) + h(P, P'; u, v) F_n(T, T')] \, dG(u) \, dG'(v).$$

It is clear, from the derivation of $G(u)$, that $F \equiv f_1$. Hence, inductively, $F_{n+1} \equiv f_{n+1}$, as defined by (14.1). Thus

$$(3) \qquad F(P, P') = \lim_{n \to \infty} F_n = \lim_{n \to \infty} f_n = f(P, P').$$

This demonstrates the effectiveness of the solutions.

With reference to the remarks made in § 6 of Chapter 4, let us now establish

THEOREM 4. *Let*

$$(1) \qquad \Delta(c) = \operatorname*{Max}_{||P|| + ||P'|| \leq c} \ \operatorname*{Max}_{\substack{u \varepsilon S \\ v \varepsilon S'}} |R(u, v) - R'(u, v)|.$$

Then, under the hypotheses of Theorem 1, *the solutions of*

$$(2) \qquad f(P, P') = \operatorname*{Max}_{G} \operatorname*{Min}_{G'} \iint_{\substack{u \varepsilon S \\ v \varepsilon S'}} [R(u, v) + h(P, P'; u, v) f(T, T')] \, dG dG'$$

$$= \operatorname*{Min}_{G} \operatorname*{Max}_{G'} [\ldots],$$

$$F(P, P') = \operatorname*{Max}_{G} \operatorname*{Min}_{G'} \iint_{\substack{u \varepsilon S \\ v \varepsilon S'}} [R'(u, v) + h(P, P'; u, v) F(T, T')] \, dG dG'$$

$$= \operatorname*{Min}_{G'} \operatorname*{Max}_{G} [\ldots]$$

satisfy the inequality

(3) $$|f(P, P') - F(P, P')| \leq \sum_{n=0}^{\infty} \Delta(k^n c).$$

PROOF. Applying the Lemma of § 3, we see that

(4) $$|f(P, P') - F(P, P')| \leq \operatorname*{Max}_{G} \operatorname*{Max}_{G'} \iint_{\substack{u \varepsilon S \\ v \varepsilon S'}} [\, |R - R'|$$
$$+ |f(T, T') - F(T, T')| \,]\, dG dG'.$$

Iteration of this inequality yields the desired result.

§ 17. Further results

The results obtained in the previous sections depended upon the fact that the total resources of the system were diminished as a consequence of the play at any particular stage of the game. Analytically, we may express this by the statement that the transformation $(P, P') \to (T, T')$ is a shrinking transformation.

Let us now introduce a shrinking transformation in another way by assuming that

(1) $$|h(P, P', u, v)| \leq k < 1,$$

for all admissible P, P', u, and v. Provided that we assume that P and P' now be within bounded domains, with T and T' transformations of these domains into themselves for all u and v, we obtain ready analogues of the preceding theorems under the assumption of (1). We shall leave the formulation and proofs of the results as exercises for the reader.

§ 18. One-sided min-max

Let us now consider the equation

(1) $$f(P, P') = \operatorname*{Min}_{v \varepsilon S'} \operatorname*{Max}_{u \varepsilon S} [R(u, v) + h(P, P'; u, v) f(T, T')],$$

which arises from the allocation process described above if the second player is required to announce his choice of v to the first player before each play.

We can obtain an analogue of the basic lemma of § 10 in the following way. For any function $R(u, v)$ permitting the operations we have

(2) $$\operatorname*{Min}_{v \varepsilon S'} \operatorname*{Max}_{u \varepsilon S} R(u, v) = \operatorname*{Min}_{v \varepsilon S'} \operatorname*{Max}_{u(v) \varepsilon S} R(u, v),$$

where $u(v)$ is now a function which maximizes $R(u, v)$ for fixed v. Let $U(v)$ be this function.

Let V be a value of v which minimizes $R(U(v), v)$. Then we have the inequalities

$$(3) \qquad \begin{aligned} R(U(V), V) &\le R(U(v), v)\,, \\ R(U(V), V) &\ge R(u(V), V)\,, \end{aligned}$$

for any other admissible values of u and v. This saddlepoint property yields the analogue of Lemma 1. Having obtained this lemma, the proofs of existence and uniqueness proceed in a straightforward fashion.

§ 19. Existence and uniqueness for games of survival

We shall prove the following result:

THEOREM 5. *Consider the equation*

$$(1) \qquad \begin{aligned} f(x) &= \operatorname*{Min}_q \operatorname*{Max}_p [p_1 q_1 f(x-1) + p_1 q_2 f(x+a) + \\ &\qquad p_2 q_1 f(x+c) + p_2 q_2 f(x-b)]\,, \\ &= \operatorname*{Max}_p \operatorname*{Min}_q [p_1 q_1 f(x-1) + p_1 q_2 f(x+a) + \\ &\qquad p_2 q_1 f(x+c) + p_2 q_2 f(x-b)]\,, \end{aligned}$$

for $x = 1, 2, 3, \ldots d-1$, *associated with the game matrix*

$$(2) \qquad A = \begin{pmatrix} -1 & a \\ c & -b \end{pmatrix},$$

where a, b, and c are positive integers, $a > 1$, and $f(x)$ satisfies the boundary conditions:

$$(3) \qquad f(x) = 0,\ x \le 0,\ f(x) = 1,\ x \ge d\,.$$

There is a unique function $f(x)$ satisfying the inequalities $0 \le f(x) \le 1$, which satisfies (1) *and* (3).

PROOF. To simplify the notation, let us set $V(f(x))$ as the value of the game whose matrix is

$$(4) \qquad \begin{pmatrix} f(x-1) & f(x+a) \\ f(x+c) & f(x-b) \end{pmatrix}$$

The functional equation in (1) has the form

$$(5) \qquad \begin{aligned} f(x) &= V(f(x))\,, & x &= 1, 2, \ldots d-1 \\ f(x) &= 0\,, & x &\le 0 \\ f(x) &= 1\,, & x &\ge d\,. \end{aligned}$$

Let us define the sequence $\{f_n(x)\}$ as follows.

$$(6) \qquad \begin{aligned} f_0(x) &= 1, x \geq d, \\ &= 0, x \leq d-1, \\ f_{n+1}(x) &= V(f_n(x)), n = 0, 1, 2, \ldots, x = 1, 2, \ldots d-1, \\ f_{n+1}(x) &= 1, x \geq d, \\ &= 0, x \leq 0. \end{aligned}$$

It is clear that $f_1(x) \geq f_0(x)$ for all x, and hence inductively that $f_{n+1}(x) \geq f_n(x)$. It follows from the fact that $0 \leq f_n(x) \leq 1$ for all x and n, that $f_n(x)$ converges as $n \to \infty$ for all x to a function $f(x)$. That $f(x)$ satisfies (5) is easily seen. This completes the proof of existence.

Since $f_0(x)$ is a monotone increasing function of x, each function $f_n(x)$ is monotone increasing, and hence $f(x)$ is monotone increasing. Let us now demonstrate the important result that it is actually *strictly* monotone. Upon this fact our proof of uniqueness depends.

We have

$$(7) \qquad f(1) = V\begin{pmatrix} 0 & f(a) \\ f(c) & 0 \end{pmatrix}.$$

If $f(a)$ and $f(c)$ are positive, we have $f(1) > 0$.

To establish the fact that $f(a)$ and $f(c)$ are positive, let us assume, on the contrary, that $f(x) = 0$, for $x = 0, 1, 2, \ldots, k < d$, but $f(k+1) = 0$. Then

$$(8) \qquad f(k) = V\begin{pmatrix} f(k-1) & f(k+a) \\ f(k+c) & f(k-b) \end{pmatrix} = V\begin{pmatrix} 0 & f(k+a) \\ f(k+c) & 0 \end{pmatrix}.$$

Since $f(k+a) \geq f(k+1) > 0, f(k+c) \geq f(k+1) > 0$, it follows that $f(k) > 0$, which is a contradiction unless $k = 0$. Thus $f(1) > 0$.

We have

$$(9) \qquad f(2) = V\begin{pmatrix} f(1) & f(a+2) \\ f(c+2) & f(2-b) \end{pmatrix}.$$

Since $f(1) > 0, f(a+2) \geq f(a+1), f(c+2) \geq f(c), f(2-b) \geq 0$, we must have $f(2) > f(1)$, unless $f(2-b) = 0$ and the solution is $p_2 = q_2 = 1$. This is clearly impossible since it yields $f(2) = 0 < f(1)$ and we know that $f(2) \geq f(1)$.

We thus prove, inductively, that

$$(10) \qquad 0 = f(0) < f(1) < f(2) < \ldots < f(d) = 1,$$

with strict inequality at every step.

The uniqueness now follows readily. Let us set

$$T(p, q, f) = p_1 q_1 f(x-1) + p_1 q_2 f(x+a) + p_2 q_1 f(x+c) + p_2 q_2 f(x-b). \tag{11}$$

Let f and g be solutions of

$$\begin{aligned} f(x) &= \operatorname*{Min}_q \operatorname*{Max}_p T(p, q, f) = \operatorname*{Max}_p \operatorname*{Min}_q T(p, q, f) \\ g(x) &= \operatorname*{Min}_q \operatorname*{Max}_p T(p, q, g) = \operatorname*{Max}_p \operatorname*{Min}_q T(p, q, g), \end{aligned} \tag{12}$$

for $0 < x < d$, and

$$\begin{aligned} f(x) = g(x) &= 0, \; x \leq 0 \\ &= 1, \; x \geq d, \end{aligned} \tag{13}$$

with the further assumption that $g(x)$ is bounded for $0 < x < d$.

Under the assumption that $f(x) \not\equiv g(x)$, set

$$\Delta = \operatorname*{Max}_x |f(x) - g(x)|, \tag{14}$$

and let y be the largest integer in $[0, d]$ for which the maximum, assumed non-zero, is attained.

If we let $p_i = p_i(y)$, $q_i = q_i(y)$, $\bar{p}_i = \bar{p}_i(y)$, $\bar{q}_i = \bar{q}_i(y)$ be sets of values for which the Min Max = Max Min is assumed, we have

$$\begin{aligned} f(y) &= T(p, q, f) \\ g(y) &= T(\bar{p}, \bar{q}, g), \end{aligned} \tag{15}$$

and, as in Lemma 1,

$$\Delta = |f(y) - g(y)| \leq \operatorname*{Max}_{p, q} [\,|T(p, q, f-g)|\,]. \tag{16}$$

Since, for all p and q,

$$|T(p, q, f-g)| \leq \Delta, \tag{17}$$

we see that (16) is an equality, which means that

$$\begin{aligned} T(p, q, f) &= T(\bar{p}, q, f), \\ T(\bar{p}, \bar{q}, f) &= T(\bar{p}, q, f). \end{aligned} \tag{18}$$

Consider the relation

$$\begin{aligned} f(y) - g(y) = {} & \bar{p}_1 q_1 [f(y-1) - g(y-1)] \\ & + \bar{p}_2 q_1 [f(y+c) - g(y+c)] \\ & + \bar{p}_1 q_2 [f(y+a) - g(y+a)] \\ & + \bar{p}_2 q_2 [f(y-b) - g(y-b)]. \end{aligned} \tag{19}$$

Since $\sum_{i,j} \bar{p}_i q_j = 1$, if any of the brackets in (19) have absolute value less than Δ, the corresponding coefficient $\bar{p}_i q_j$ must be zero. By assumption, y was the largest value for which $|f(y) - g(y)| = \Delta$. Hence $\bar{p}_2 q_1 = 0$, $\bar{p}_1 q_2 = 0$.

Since $\bar{p}_1 + \bar{p}_2 = 1$, both $\bar{p}_1$ and $\bar{p}_2$ cannot be zero, which means $q_1 = 0$ or $q_2 = 0$. Turning to the game matrix

$$\begin{pmatrix} f(y-a) & f(y+a) \\ f(y+c) & f(y-b) \end{pmatrix}, \tag{20}$$

we see that the strict monotonicity of $f(x)$ as a function of x makes it impossible for $q_1 = 0$ or $q_2 = 0$ to be optimal play at $x = y$. This yields a contradiction to $\Delta > 0$ and completes the proof of uniqueness.

We see then that the proof of uniqueness of a strictly increasing solution is relatively easy, with the whole difficulty of the complete uniqueness proof centering about the proof of strict monotonicity.

The method we have employed is quite general and applies to large classes of functional equations. It fails, however, to treat the general case where we assume only that the elements a_{ij} of the game matrix A are real quantities.

§ 20. An approximation

Let us now return to the general equation

$$\begin{aligned} f(x) &= \operatorname*{Max}_p \operatorname*{Min}_q \sum_{i,j} p_i q_j f(x + a_{ij}), \\ &= \operatorname*{Min}_q \operatorname*{Max}_p \sum_{i,j} p_i q_j f(x + a_{ij}), \end{aligned} \tag{1}$$

and assume that x is large compared to a_{ij}.

The reasoning we shall employ below, while quite formal, possesses many features of interest. Assume that we can write

$$f(x + a_{ij}) \cong f(x) + a_{ij} f'(x). \tag{2}$$

Then (1) takes the form

$$\begin{aligned} f(x) &\cong \operatorname*{Max}_p \operatorname*{Min}_q \sum_{i,j} p_i q_j [f(x) + a_{ij} f'(x)] \\ &\cong \operatorname*{Max}_p \operatorname*{Min}_q [f(x) + f'(x) \sum_{i,j} p_i q_j a_{ij}], \end{aligned} \tag{3}$$

which leads to

$$\begin{aligned} 0 &\cong \operatorname*{Max}_p \operatorname*{Min}_q [f'(x) \sum_{i,j} p_i q_j a_{ij}] \\ &\cong \operatorname*{Min}_q \operatorname*{Max}_p [f'(x) \sum_{i,j} p_i q_j a_{ij}]. \end{aligned} \tag{4}$$

Assume now that $f'(x) > 0$. Then we obtain the approximate equations for the unknown distributions p and q,

(5)
$$0 = \underset{p}{\text{Max}}\ \underset{q}{\text{Min}} \sum_{i,j} a_{ij}\, p_i\, q_j = \underset{q}{\text{Min}}\ \underset{p}{\text{Max}} \sum_{i,j} a_{ij}\, p_i\, q_j ,$$

an equation which is independent of $f(x)$!

The meaning of this equation is that for large x, with a large number of plays remaining before the end of the game, the play is approximately the same as that employed in the single-stage game where both players wish merely to maximize the expected return from one play.

In taking a_{ij} small compared to x we are essentially passing over to a continuous version of the process. As we noted in § 18 of Chapter 8 in the discussion of the nonlinear utility function, the optimal policy was independent of the form of the criterion function. Here is another manifestation of this general principle, and we shall encounter a further example in § 22 devoted to a similar approximation for non-zero-sum games.

§ 21. Non-zero-sum games—games of survival

Let us now turn to a discussion of the more general situation where $b_{ij} \neq - a_{ij}$. Here there is no generally acceptable theory for the determination of optimal play in a single-stage process. Consequently, we shall turn immediately to the discussion of a multi-stage process. Let us assume once more that the players are both striving to ruin the other, and that the game continues until this occurs. They are now in direct opposition and we can use a Min-Max formulation.

Since the game is non-zero-sum, the state of the process depends upon the fortunes of both A and B, x and y, respectively. Let us define

(1) $f(x, y)$ = probability that A ruins B when A has x, B has y, and both employ optimal policies.

Then $f(x, y)$, provided that it exists, satisfies the functional equation

(2)
$$f(x, y) = \underset{p}{\text{Max}}\ \underset{q}{\text{Min}} \sum_{i,j} p_i\, q_j\, f(x + a_{ij}, y + b_{ij}) = \underset{q}{\text{Min}}\ \underset{p}{\text{Max}} \sum_{i,j} p_i\, q_j\, f(x + a_{ij}, y + b_{ij}) ,$$

with the boundary conditions

(3)
$$\begin{aligned} f(x, y) &= 1,\ x \geq 0,\ y < 0 \\ &= 0,\ x \leq 0,\ y > 0 \\ &= 1/2,\ x = y = 0 \text{ (by convention)}. \end{aligned}$$

It is easy to establish the following result, using the methods we have employed above.

THEOREM 6. *If* $a_{ij} + b_{ij} < 0$ *for all* i, j, *there is a unique bounded solution to* (2), (3).

§ 22. An approximate solution

Let us assume that we are dealing with a process where a_{ij} and b_{ij} are always negative. Then assuming that x and y are large compared to a_{ij} and b_{ij}, and that we may write

$$f(x + a_{ij}, y + b_{ij}) \cong f(x, y) + a_{ij} f_x + b_{ij} f_y, \tag{1}$$

we obtain the approximate equation

$$\begin{aligned} f(x, y) &\cong \operatorname*{Max}_p \operatorname*{Min}_q \sum_{i,j} p_i q_j [f(x, y) + a_{ij} f_x + b_{ij} f_y] \\ &\cong \operatorname*{Min}_q \operatorname*{Max}_p \sum_{i,j} p_i q_j [f(x, y) + a_{ij} f_x + b_{ij} f_y]. \end{aligned} \tag{2}$$

From this we obtain the approximate equations

$$\begin{aligned} 0 &= \operatorname*{Max}_p \operatorname*{Min}_q [f_x \sum_{i,j} a_{ij} p_i q_j + f_y \sum_{i,j} b_{ij} p_i q_j] \\ &= \operatorname*{Min}_q \operatorname*{Max}_p [f_x \sum_{i,j} a_{ij} p_i q_j + f_y \sum_{i,j} b_{ij} p_i q_j]. \end{aligned} \tag{3}$$

Using the same reasoning employed in § 4 of Chapter 9, we see that these yield

$$\begin{aligned} -f_x/f_y &= \operatorname*{Max}_p \operatorname*{Min}_q [\sum_{i,j} b_{ij} p_i q_j / \sum_{i,j} a_{ij} p_i q_j] \\ &= \operatorname*{Min}_q \operatorname*{Max}_p [\sum_{i,j} b_{ij} p_i q_j / \sum_{i,j} a_{ij} p_i q_j]. \end{aligned} \tag{4}$$

This is a very reasonable criterion. Observe that it makes no difference whether we solve for $f_x = f_y$ or f_y/f_x, since maximizing f_x/f_y is equivalent to minimizing f_y/f_x.

In the next section we shall demonstrate that Max Min in (4) actually equals Min Max.

§ 23. Proof of the extended min-max theorem

In this section we wish to prove

THEOREM 7. *If* $\sum_{i,j} b_{ij} p_i q_j \geq d > 0$ *for all distribution vectors* p *and* q, *then*

$$\operatorname*{Max}_p \operatorname*{Min}_q \frac{\sum_{i,j} a_{ij} p_i q_j}{\sum_{i,j} b_{ij} p_i q_j} = \operatorname*{Min}_q \operatorname*{Max}_p \frac{\sum_{i,j} a_{ij} p_i q_j}{\sum_{i,j} b_{ij} p_i q_j}. \tag{1}$$

PROOF. There is no loss of generality in further assuming that $b_{ij} \leq m < 1$ for all i, j so that $\sum_{i,j} b_{ij} p_i q_j \leq m$ for all relevant p and q. Consider the system of recurrence relations

$$(2) \qquad u_0 = \operatorname*{Max}_p \operatorname*{Min}_q \sum_{i,j} a_{ij} p_i q_j = \operatorname*{Min}_q \operatorname*{Max}_p \sum_{i,j} a_{ij} p_i q_j .$$

$$u_{n+1} = \operatorname*{Max}_p \operatorname*{Min}_q \left[\sum_{i,j} a_{ij} p_i q_j + [1 - \sum_{i,j} b_{ij} p_i q_j] u_n \right]$$

$$= \operatorname*{Min}_q \operatorname*{Max}_p \left[\sum_{i,j} a_{ij} p_i q_j + [1 - \sum_{i,j} b_{ij} p_i q_j] u_n \right].$$

It is easy to show, using the methods discussed above, that the sequence $\{u_n\}$ converges to a value u, satisfying the equation

$$(3) \qquad u = \operatorname*{Max}_p \operatorname*{Min}_q \left[\sum_{i,j} a_{ij} p_i q_j + [1 - \sum_{i,j} b_{ij} p_i q_j] u \right]$$

$$= \operatorname*{Min}_q \operatorname*{Max}_p \left[\sum_{i,j} a_{ij} p_i q_j + [1 - \sum_{i,j} b_{ij} p_i q_j] u \right].$$

The condition $0 < 1 - \sum_{i,j} b_{ij} p_i q_j < 1 - d$ yields geometric convergence of the series $\sum_{n=0}^{\infty} (u_{n+1} - u_n)$.

Since u satisfies (3), it is easy to see that it is given by the expression

$$(4) \qquad u = \operatorname*{Max}_p \operatorname*{Min}_q \sum_{i,j} a_{ij} p_i q_j / \sum_{i,j} b_{ij} p_i q_j$$

$$= \operatorname*{Min}_q \operatorname*{Max}_p \sum_{i,j} a_{ij} p_i q_j / \sum_{i,j} b_{ij} p_i q_j ,$$

which establishes the theorem.

§ 24. A rationale for non-zero sum games

The importance of the above result, combined with the approximation procedure discussed in § 14, is that we now have a possible rationale for the play of non-zero sum games, namely one based upon the criterion function

$$(1) \qquad R(p, q) = \frac{\sum a_{ij} p_i q_j}{\sum b_{ij} p_i q_j}.$$

Whether or not to accept this is a matter of individual taste. It must be realized that this question must always arise in two-person processes, where it is not a priori evident that both individuals are employing the same criterion function, or, what is worse, they may not have commensurable utility scales.

Exercises and research problems for Chapter X

1. Consider the following game. Two players, I and II, match coins according to the following rules:

a. I and II both lose one, if a head-head combination occurs,
b. I gains one, II loses one if a tail-tail combination occurs,
c. I loses one, II gains one if head-tail, tail-head occur.

The first player starts with a quantity m and the second player with a quantity n. Each plays so as to ruin the other. Let $p(m, n; x, y)$ be the probability that I will be ruined before or together with II if I shows heads with probability x and II shows heads with probability y.
Define

$q_1 = xy$ = Probability that I and II both display heads
$q_2 = x(1-y) + y(1-x)$ = Probability that a head-tail combination appears
$q_3 = (1-x)(1-y)$ = Probability that both I and II display tails.

Obtain the recurrence relation

$$p(m, n) = q_1 p(m-1, n-1) + q_2 p(m-1, n+1) + q_3 p(m+1, n+1),$$

for $m, n \geq 1$, with the boundary conditions

$$p(m, 0) = 0, m \geq 1, p(0, n) = 1, \geq 0$$

(R. Bellman and D. Blackwell)

2. Show that for $n \geq 2$ we obtain the finite set of equations

$$p(1, n) = (q_1 + q_2) + q_3 p(2, n-1)$$
$$p(2, n-1) = q_1 p(1, n-2) + q_2 p(1, n) + q_3 p(3, n-2)$$
$$\vdots$$
$$p(n-1, 2) = q_1 p(n-2, 1) + q_2 (n-2, 3) + q_3 p(n, 1)$$
$$p(n, 1) = q_2 p(n-1, 2)$$

3. Show that

$$p(2, 1) = \frac{(q_1 + q_2) q_2}{1 - q_2 q_3}$$

and hence that $\operatorname{Min}_x \operatorname{Max}_y p(m, n, x, y) \neq \operatorname{Max}_y \operatorname{Min}_x p(m, n, x, y)$ in general. (It is interesting to note that $\operatorname{Min}_x \operatorname{Max}_y \cong .4397$, $x' = .43$, $y' = .5$, $\operatorname{Max}_y \operatorname{Min}_x \cong .4304$, $x' = .43$, $y' = 1$, where $x' = 1 - x$, $y' = 1 - y$.

4. It follows from the fundamental min-max theorem for continuous games that $\underset{A}{\text{Min}}\ \underset{B}{\text{Max}}\ K(m, n; A, B) = \underset{B}{\text{Max}}\ \underset{A}{\text{Min}}\ K(m, n; A, B)$, where

$$K(m, n; A, B) = \int_0^1 \int_0^1 p(m, n; x, y)\, dA(x)\, dB(y),$$

and A, B range over the space of monotone functions of uniformly bounded variation equal to 1. Show that the solution for $m = 2, n = 1$ is given by

a. II chooses the value of y', y_0, for which $p(2, 1, 0, y') = p(2, 1, 1, y')$, a pure strategy.

b. I chooses a mixed strategy, using either all heads or all tails in the combination $(a, 1 - a)$, where a is chosen so that $ap(2, 1, 0, y') + (1 - a)\, p(2, 1, 1, y')$ has a maximum at $y' = y_0$.

5. Show that the expected probability of ruin for I is $y_0 \cong .4302$, the unique real root of $y' = (1 - y')^2/(1 - y' + y'^2)$.

6. Prove that as $m, n \to \infty$ along any fixed direction, $m/n = r$, player II can choose y so that uniformly in x we have

$$\lim_{m, n \to \infty} p(m, n) = 1\,.$$

7. Show that the above considerations lead to the following principle: In playing a game of this type, I should try to make the stakes as high as possible, whereas II should try to make them as low as possible.

8. Let

$$f_N(u_1, u_2, \ldots, u_N; v_1, v_2, \ldots, v_N) = \underset{y}{\text{Min}}\ \underset{x}{\text{Max}} \Big[\sum_{i,j=1}^{N,N} a_{ij} x_i y_j + \sum_{i=1}^{N} u_i x_i + \sum_{j=1}^{N} v_j y_j \Big] = \underset{x}{\text{Max}}\ \underset{y}{\text{Min}} [\ \ldots\ ],\quad N = 1, 2, \ldots$$

Derive a recurrence relation for $\{f_N\}$.

9. Consider the game of survival described by the matrix

$$A = \begin{pmatrix} 2 & -1 \\ -2 & 1 \end{pmatrix},$$

where the total fortune of both players is 4 and k describes the fortune of the first player. Show that $f(k)$, the probability of survival of the first player, satisfies the equations

$$\begin{aligned} f(1) &= f(3) + f(2)/(f(2) + f(3)) \\ f(2) &= f(3)/(1 + f(3) - f(1)) \\ f(3) &= (1 - f(2)\, f(1)/(2 - f(2) - f(1)) \end{aligned}$$

(Hausner)

10. Hence show that

$$f(1) = 1 - \sqrt{2}/2, f(2) = 1/2, f(3) = \sqrt{2}/2,$$

and that the corresponding optimal strategies are given by

$$p_1 = \sqrt{2} - 1, p_2 = 1/2, p_3 = 2 - \sqrt{2}$$
$$q_1 = \sqrt{2} - 1, q_2 = 1 - \sqrt{2}/2, q_3 = \sqrt{2} - 1$$

11. Consider the game of survival described by

$$A = \begin{pmatrix} a & -1 \\ -1 & b \end{pmatrix},$$

where a and b are positive integers. Let $v_n(k)$ be the probability that A survives when the fortunes of both players total n and A possesses k of this. Show that

$$v_{n+1}(k+1) = v_n(k) + (1 - v_n(k))\, v_{n+1}(1)$$

12. Show that

$$v_{n+1}(1) = \frac{v_{n+1}(1+a)\, v_{n+1}(1+b)}{v_{n+1}(1+a) + v_{n+1}(1+b)}$$

and hence that

$$v_{n+1}(1) = \frac{\sqrt{v_n(a)\, v_n(b)}}{1 + \sqrt{v_n(a)\, v_n(b)}}$$

13. Show that

$$p_n(1) = \frac{v_n(1+b)}{v_n(1+a) + v_n(1+b)} = \frac{v_n(1)}{v_n(1+a)}$$

$$p_{n+1}(k+1) = p_k(k) = p_{n-k+1}(1)$$

14. Prove Theorem 5 by showing that val $(A - B\lambda)$ is a continuous function of λ which is monotone decreasing as a function of λ. Hence show that there is exactly one solution of the equation val $(A - B\lambda) = 0$ which may be represented in the form given in (23.1). (Karlin).

15. Consider the equation

$$u(p) = L(u(p, q, q')) + a(p, q, q'),$$

and the related equation

$$v(p) = \operatorname*{Max}_{q} \operatorname*{Min}_{q'} [L(v(p, q, q')) + a(p, q, q')]$$
$$= \operatorname*{Min}_{q'} \operatorname*{Max}_{q} [\qquad \ldots \qquad].$$

Under what conditions may we write

$$v(p) = \underset{q}{\text{Max}}\ \underset{q'}{\text{Min}}\ u(p) = \underset{q'}{\text{Min}}\ \underset{q}{\text{Max}}\ u(p)\ ?$$

16. Consider, in particular, the system of equations

$$x_i = \underset{q}{\text{Max}}\ \underset{q'}{\text{Min}}\ [c_i(q, q') + \sum_{j=1}^{n} a_{ij}(q, q')\, x_j]\,,$$

$$= \underset{q'}{\text{Min}}\ \underset{q}{\text{Max}}\ [c_i(q, q') + \sum_{j=1}^{n} a_{ij}(q, q')\, x_j],\ i = 1, 2, \ldots, n.$$

under appropriate conditions concerning the matrix $A(q, q') = (a_{ij}(q, q'))$. (L. Shapley)

17. Suppose that we are given the information that a coin has a fixed but unknown probability p of landing heads and a probability $q = 1 - p$ of landing tails, and that p has a known a priori distribution function $dF(p)$. The coin is to be tossed N times and we are to call heads or tails before each toss with the full knowledge of the results of the previous tosses. What policy maximizes the expected number of correct calls?

18. Suppose that we can toss the coin as many times as we please, at a cost of c per toss, and then are required to furnish a value for p, the probability of heads. If p' is the value decided upon, the cost of deviation from the true value is $g(p - p')$, where g is a known function. What policy minimizes the total expected cost?

19. Returning to problem 17, suppose that an opponent has the choice of choosing $F(p)$ so as to minimize the expected number of correct calls obtained using an optimal policy. Can one characterize the optimal selection of $F(p)$ by the statement that the opponent chooses $F(p)$ in such a way as to minimize the information available after any finite set of tosses? On this hypothesis, determine Min-Max.

20. Generalize these results to cases where there are many different possible outcomes at each stage, e.g. a six-sided die.

21. Player A has resources in quantity x, and Player B resources in quantity y. A divides x up into n parts, $x = \sum_i x_i$, $x_i \geq 0$, and B likewise, $y = \sum_i y_i$, $y_i \geq 0$. The payoff to A is

$$P(x, y) = \sum_{i=1}^{n} c_i \,\text{Max}\,(x_i - y_i, 0)\,,$$

and the negative of this to B.

Write

$$f_n(x, y) = \underset{G}{\text{Max}}\ \underset{G'}{\text{Min}} \left[\int P(x, y)\, dG(x_1, x_2, \ldots, x_n)\, dG'(y_1, y_2, \ldots, y_n)\right]$$
$$= \underset{G'}{\text{Min}}\ \underset{G}{\text{Max}} \left[\quad \ldots \quad\right].$$

Obtain the recurrence relations connecting f_n and f_{n-1}. (Colonel Blotto)

22. Let A be a positive matrix, i.e. $a_{ij} > 0$ for all i, j. Show that A has a unique characteristic root of largest absolute value, which is positive, and that the associated characteristic vector may be taken to be positive. Denote this root by $p(A)$, the Perron root of A.

23. Show that

$$p(A) = \underset{x}{\text{Max}}\ \underset{i}{\text{Min}} \sum_{j=1}^{n} a_{ij} x_j / x_i,$$
$$= \underset{x}{\text{Min}}\ \underset{i}{\text{Max}} \sum_{j=1}^{n} a_{ij} x_j / x_i,$$

where the variation is over the region $x_i \geq 0, \sum_i x_i = 1$.

24. Show that

$$p(A) = \underset{R'}{\text{Max}}\ \underset{i}{\text{Min}} \sum_{j=1}^{n} a_{ij} x_j / x_i,$$
$$= \underset{R'}{\text{Min}}\ \underset{i}{\text{Max}} \sum_{j=1}^{n} a_{ij} x_j / x_i,$$

where R' is defined by $x_i \geq d, \sum_i x_i = 1$, and d may be taken to be

$$d = \underset{i,j}{\text{Min}}\ a_{ij} / \underset{i}{\text{Max}} \left(\sum_j a_{ij}\right).$$

25. Prove that $p(A)$ is the unique solution of

$$\lambda = \underset{R'}{\text{Max}}\ \underset{i}{\text{Min}} \left[\sum_{j=1}^{n} a_{ij} x_j + \lambda(1 - x_i)\right],$$

or of

$$\lambda = \underset{R'}{\text{Min}}\ \underset{i}{\text{Max}} \left[\sum_{j=1}^{n} a_{ij} x_j + \lambda(1 - x_i)\right],$$

where R' is as above.

26. Consider the nonlinear recurrence relation

$$u_{n+1} = \underset{R'}{\text{Min}}\ \underset{i}{\text{Max}} \left[\sum_{j=1}^{n} a_{ij} x_j + u_n(1 - x_i)\right].$$

with u_0 arbitrary. Prove that $p(A) = \lim_{n \to \infty} u_n$.

(*Proc. Amer. Math. Soc.*, 1956).

Bibliography and Comments for Chapter X

§ 1. An excellent introduction to the theory of games is J. D. Williams, *The Compleat Strategyst*, McGraw-Hill, 1955. The classic work in the field is J. Von Neumann and O. Morgenstern, *The Theory of Games and Economic Behavior*, Princeton University Press, 1948. An exposition of the application of the mathematical theory of games to the study of card games is contained in R. Bellman and D. Blackwell, "Red Dog, Blackjack and Poker," *Scientific American*, vol. 184 (1951), pp. 44–47; cf. also, the references given in the comment on § 5 of Chapter VIII. There is also a discussion of several poker games in "The Theory of Games and Economic Behavior" cited above, and quite a few authors since have studied particular games. The interested reader may refer to the *Studies in Game Theory* issued by Princeton University Press for further references.

We have completely avoided in this volume any contact with the theory of sequential analysis of Wald, and the general theory of statistical decision processes and the design of experiments. The interested reader may refer to Wald's book, A. Wald, *Statistical Decision Functions*, J. Wiley and Sons, 1950, to the book by D. Blackwell and A. Girshick, *Theory of Games and Statistical Decisions*, J. Wiley and Sons, 1954, and generally to the current and back numbers of the Annals of Mathematical Statistics for papers in this field. See also H. Robbins, "Some Aspects of the Sequential Design of Experiments," *Bull. Amer. Math. Soc.*, vol. 58 (1952), pp. 527–536.

§ 3. A proof the min-max theorem may be found in the Von Neumann-Morgenstern book cited above, as well as in the Blackwell-Girshick volume where extensions are also discussed.

Although there are theories of non-zero-sum games and N-person games, $N > 2$, available, none of them have the elegance or the finality of the two-person zero-sum theory because of the lack of a corresponding min-max theorem. A large part of Von Neumann-Morgenstern is devoted to these questions, and a fundamental result in the field is contained in J. F. Nash, "Equilibrium Points in N-person Games," *Proc. Nat. Acad. Sci.*, vol. 36 (1950), pp. 48–9.

§ 4. For an abstract discussion of continuous games, see S. Karlin, "The Theory of Infinite Games," *Annals of Math.*, 1951 and for some particular results see M. Dresher and S. Karlin, "Solutions of Convex Games as Fixed Points," Contributions to the Theory of Games II, *Annals of Mathematics Study No.* 28, *Princeton University Press*, 1951.

§ 5. As far as we know, the first treatment of games where both players have finite resources, and, in particular, "games of survival," is contained in R. Bellman and J. P. LaSalle, "On Non-Zero Sum Games and Stochastic Processes," RM–212, The RAND Corporation, 1949, R. Bellman and D. Blackwell, "On a Particular Non-Zero Sum Game," RM–250, The RAND Corporation, 1949.

The name "games of survival" was given in the course of some seminar lectures at RAND.

§ 7. The subject of "pursuit games" has been intensively investigated by R. Isaacs, who resolved a number of special games, and developed a general theory of this class of problems. See R. Isaacs, "Games of Pursuit," P–257,

The Rand Corporation, 1955, R. Isaacs, "The Problem of Aiming and Evasion," P–642, The RAND Corporation, 1955, R. Isaacs, "Differential Games—I, II, III, IV," RM–1391, 1399, 1411, 1486, The RAND Corporation, 1955.

§ 8. The results of this and the following sections, 9–17, follow the paper of R. Bellman, "Functional Equations in the Theory of Dynamic Programming—III, Multi-Stage Games," Rendiconti di Palermo, 1957.

§ 18. The technique utilized here was suggested by W. Fleming. Further results on existence and uniqueness were announced by L. Shapley, "Stochastic Games," *Proc. Nat. Acad. Sci.*, vol. 39 (1953), pp. 1095–1100.

§ 19. The proof presented here is contained in R. Bellman, "Introduction to the Theory of Dynamic Programming," RAND, R–245, 1953, Chapter VI. Since the original papers cited above, a good deal of work has been done on the subject. The deepest results so far obtained are contained in M. Peisakoff, "More on Games of Survival," RM–884, The RAND Corporation, 1952, and J. Milnor and L. Shapley, "On Games of Survival," P–622, The RAND Corporation, 1955.

§ 20. The contents of this section and those of 21, 22, and 24 were presented in R. Bellman, "Decision Making in the Face of Uncertainty–II, "Naval Research Logistics Quarterly, vol. 1 (1954), pp. 323–332.

§ 21. This proof of the extended min-max theorem is due to L. Shapley in the references cited in the comment on § 18. The formulation of the theorem and the original proof are due to Von Neumann.

CHAPTER XI

Markovian Decision Processes

§ 1. Introduction

In this chapter we shall study some decision processes of a different form than those previously encountered, giving rise to a new class of functional equations.

We shall consider discrete processes, which lead us to the study of the difference equation

$$(1) \quad x_i(n+1) = \operatorname*{Max}_q \sum_{j=1}^{N} a_{ij}(q)\, x_j(n),\ x_i(0) = c_i,\ i = 1, 2, \ldots, N,$$

and some continuous processes which generate the equation

$$(2) \qquad dx_i/dt = \operatorname*{Max}_q \sum_{j=1}^{N} a_{ij}(q)\, x_j(t),\ x_i(0) = c_i,\ i = 1, 2, \ldots, N,$$

in the one-person case, and the equation

$$(3) \quad dx_i/dt = \operatorname*{Max}_q \operatorname*{Min}_p \Big[\sum_{j=1}^{N} a_{ij}(p, q)\, x_j(t) \Big],\ x_i(0) = c_i,\ i = 1, 2, \ldots, N,$$

$$= \operatorname*{Min}_p \operatorname*{Max}_q [\ \ldots\ ],$$

in the two-person case.

As we shall see, equations of this type have connections with the classical theory of differential and difference equations. We shall, however, reserve any detailed exploration of this liaison until the second volume.

§ 2. Markovian decision processes

Let us describe, in this section, a decision process which motivates the study of a class of nonlinear difference equations, of which (1.1) is a representative. We shall then consider the limiting form, namely (1.2).

Consider a physical system S which at any of the times $t = 0, \Delta, 2\Delta, \ldots$ must be in one of a set of states which we denote by $S_1, S_2, \ldots, S_N$. Assume that at any time t there is a probability $x_i(t)$ that the system is in the ith state, and that transition probabilities exist governing the changeover from one state to another. It is important to realize that these are very strong assumptions concerning the nature of the system.

Let

(1) $a_{ij} =$ the probability that the system will be in state i at $t + \Delta$ if it is in state j at time t.

The relation between the set of probabilities $\{x_i (t + \Delta)\}$ and the set $\{x_i (t)\}$ is then given by the relations

$$x_i (t + \Delta) = \sum_{j=1}^{N} a_{ij}\, x_j (t),\ i = 1, 2, \ldots, N, \tag{2}$$

for $t = 0, \Delta, 2\,\Delta, \ldots,$. Setting $x_i (n\,\Delta) = y_i (n)$, we may write these equations in the simpler form

$$y_i (n + 1) = \sum_{j=1}^{N} a_{ij}\, y_j (n),\ i = 1, 2, \ldots, N. \tag{3}$$

The asymptotic behavior of the state vector $(y_1, y_2, \ldots, y_N)$ as $t \to \infty$ is determined by the algebraic character of the characterisitic roots of the matrix $A = (a_{ij})$. A process of this type is called a *Markoff process*. There exists an extensive mathematical theory of these processes.

Let us now consider Markovian decision processes. Assume that the transition probabilities, a_{ij}, depend upon a parameter q, which may be a vector, and that at each stage of the process q is to be chosen so as to maximize the probability that the system is in the state S_1. In place of the equations in (3) we obtain the nonlinear system

$$\begin{aligned} y_1 (n + 1) &= \operatorname*{Max}_{q} \sum_{j=1}^{N} a_{1j} (q)\, y_j (n), \\ y_i (n + 1) &= \sum_{j=1}^{N} a_{ij} (q^*)\, y_j (n),\ i = 2, 3, \ldots, N. \end{aligned} \tag{4}$$

where $q^* = q^* (n)$ in the remaining $N - 1$ equations is one of the values of q which maximizes $y_1 (n + 1)$.

Since the a_{ij} are transition probabilities, they are restricted by the conditions

$$a_{ij} \geq 0,\ \sum_{i} a_{ij} = 1,\ j = 1, 2, \ldots, N, \tag{4}$$

for all q.

To obtain more general equations, consider the situation in which we have N different types of items and let $x_i (t)$ represent the quantity of the ith item at time t. These items have the property that a unit quantity of the ith item generates an amount a_{ij} of the jth item over the time interval $[t, t + \Delta)$. Here $a_{ij} > 0$ represents production, and the reverse inequality represents consumption. Once again let a_{ij} depend upon a parameter q and let the purpose of the process be to maximize the quantity of the first item available at any time. In this case we obtain the equation in (4) with no restriction on the magnitude or sign of the a_{ij}.

In the limit as $\Delta \to 0$, we obtain in place of (4) the nonlinear differential system

$$\frac{dx_1}{dt} = \operatorname*{Max}_q \sum_{j=1}^{N} b_{1j}(q)\, x_j(t),\ x_1(0) = c_1, \tag{5}$$

$$\frac{dx_i}{dt} = \sum_{j=1}^{N} b_{ij}(q^*)\, x_j(t),\ x_i(0) = c_i,\ i = 2, 3, \ldots, N.$$

To obtain this system, we set, in the usual fashion

$$a_{ij} = b_{ij}\Delta,\ i \neq j \tag{6}$$

$$a_{ii} = 1 - b_{ii}\Delta,$$

and then let $\Delta \to 0$. Having obtained the equations by means of this formalism, we now define the continuous process by means of the equation in (5). In return for this, we must establish existence and uniqueness of solutions, which is to say we must show that this method of defining a process is actually valid.

§ 3. Notation

Taking account of the foregoing remarks, we shall begin by considering the continuous version first. Introducing vector-matrix notation to simplify our notation,

$$x = \begin{pmatrix} x_1 \\ x_2 \\ \cdot \\ \cdot \\ \cdot \\ x_N \end{pmatrix},\ A(q) = (a_{ij}(q)),\ c = \begin{pmatrix} c_1 \\ c_2 \\ \cdot \\ \cdot \\ \cdot \\ c_N \end{pmatrix}, \tag{1}$$

the system

$$\frac{dx_i}{dt} = \operatorname*{Max}_q \sum_{j=1}^{N} a_{ij}(q)\, x_j,\ x_i(0) = c_i,\ i = 1, 2, \ldots, N, \tag{2}$$

takes the form

$$dx/dt = \operatorname*{Max}_q A(q)\, x,\ x(0) = c. \tag{3}$$

where it is understood that the maximum is taken element by element. By this we mean that the set of q's for each row is distinct from the corresponding set for any other row. Thus,

$$(4) \qquad \begin{aligned} a_{1j}(q) &= a_{1j}(q_{11}, q_{12}, \ldots, q_{1k}), \\ a_{2j}(q) &= a_{2j}(q_{21}, q_{22}, \ldots, q_{2k}), \\ &\vdots \\ a_{Nj}(q) &= a_{Nj}(q_{N1}, q_{N2}, \ldots, q_{Nk}), \end{aligned}$$

so that there is no interaction between the various maximizations. After discussing this case, we shall return to the equations obtained in the preceding sections, where interaction occurs.

It is convenient to employ the notation

$$(5) \qquad \begin{aligned} \|x\| &= \sum_{i=1}^{N} |x_i|, \\ \|A\| &= \sum_{i,j=1}^{N} |a_{ij}| \end{aligned}$$

These fulfill the usual requirements for norms, and in addition we have

$$(10) \qquad \|Ax\| \leq \|A\| \, \|x\|.$$

§ 4. A lemma

As is usual in the theory of differential equations, the first step in establishing existence and uniqueness of a solution consists of converting the differential equation into a suitable integral equation. This enables us to take advantage of the smoothing properties of integration.

Considering the more general equation

$$(1) \qquad dx/dt = \operatorname*{Max}_{q} [A(q, t)\, x + b(q, t)], \; x(0) = c,$$

we obtain the integral equation

$$(2) \qquad x = c + \int_0^t \operatorname*{Max}_{q} [A(q, s)\, x + b(q, s)]\, ds$$

which may be written

$$(3) \qquad x = \operatorname*{Max}_{q} [c + \int_0^t b(q, s)\, ds + \int_0^t A(q, s)\, x ds].$$

Since q is a function of t, pointwise maximization yields global maximization.

It is easy to demonstrate the following result in much the same way as Lemma 1 of Chapter 2 was established.

LEMMA. *Let*

$$T_1(x) = \operatorname*{Max}_q \left[b_1(q, t) + \int_0^t A(q, s)\, x ds\right], \tag{4}$$

$$T_2(y) = \operatorname*{Max}_q \left[b_2(q, t) + \int_0^t A(q, s)\, y ds\right].$$

then

$$\| T_1(x) - T_2(y) \| \leq \operatorname*{Max}_q \left[\| b_1(q, t) - b_2(q, t) \| + \int_0^t \| A(q, s) \| \| x - y \| ds \right. \tag{5}$$

This lemma will be the fulcrum of our existence and uniqueness proof.

§ 5. Existence and uniqueness—I

Let us now consider the question of the existence and uniqueness of solutions of the equation

$$dx/dt = \operatorname*{Max}_q [A(q, t)\, x + b(q, t)],\ x(0) = c. \tag{1}$$

There are a number of cases of particular interest, corresponding to different assumptions that can be made concerning the function $A(q, t)$, $b(q, t)$, and the set of admissible functions $q(t)$. We shall discuss one class of equations and leave the matter there, since the method used will illustrate the procedures that may be employed in other cases.

Our first result is

THEOREM 1. *Assume that q is an element of a set of functions S with the property that*

$$\| A(q, t) \|, \quad \| b(q, t) \| \leq f(t), \tag{2}$$

where $f(t)$ is integrable over any finite interval $0 \leq t \leq T$. Assume further that the maximum of $A(q, t)\, x + b(q, t)$ is attained for $q \,\varepsilon\, S$ for any fixed t and x values.[1]

Then there is a unique solution to (1) *satisfying the equation almost everywhere. This solution may be found as the limit of the successive approximations,*

$$x_0 = c, \tag{3}$$

$$x_{n+1} = c + \operatorname*{Max}_q \left[\int_0^t [A(q, s)\, x_n + b(q, s)]\, ds,\ n = 0, 1, \ldots \right.$$

[1] The purpose of this assumption is to handle simultaneously the case where q assumes only a discrete set of values, in which case the maximum is always attained, and the case where q varies continuously.

PROOF. Let us first show that the x_n are uniformly bounded in the interval $[0, T]$. Specifically, we shall show that

$$||x_n|| \le a \exp\left(\int_0^t f(s)\, ds\right), \tag{3}$$

where

$$a = ||c|| + \int_0^t f(s)\, ds. \tag{4}$$

The inequality certainly holds for $n = 0$. Assume that it is valid for $k = 0, 1, \ldots, n$. Then we have, from (3),

$$\begin{aligned} ||x_{n+1}|| &\le ||c|| + \int_0^t \operatorname*{Max}_q ||b(q, s)||\, ds \\ &\quad + \int_0^t (\operatorname*{Max}_q ||A(q, s)||)\, ||x_n||\, ds \\ &\le a + \int_0^t f(s)\, ||x_n||\, ds \end{aligned} \tag{5}$$

Replacing $||x_n||$ by its bound, we have

$$||x_{n+1}|| \le a + \int_0^t f(s) \left[a \exp\left(\int_0^s f(s_1)\, ds_1\right)\right] ds \tag{6}$$

and thus obtain the same bound for $||x_{n+1}||$.

Let us now establish the convergence of the sequence $\{x_n\}$. Applying the Lemma of § 4 to the two relations

$$\begin{aligned} x_{n+1} &= c + \operatorname*{Max}_q \left[\int_0^t [A(q, s)\, x_n + b(q, s)]\, ds\right], \\ x_n &= c + \operatorname*{Max}_q \left[\int_0^t [A(q, s)\, x_{n-1} + b(q, s)]\, ds\right] \end{aligned} \tag{7}$$

we obtain the inequality

$$\begin{aligned} ||x_{n+1} - x_n|| &\le \operatorname*{Max}_q \int_0^t ||A(q, s)||\, ||x_n - x_{n-1}||\, ds \\ &\le \int_0^t f(s)\, ||x_n - x_{n-1}||\, ds, \quad n = 1, 2, \ldots \end{aligned} \tag{8}$$

Iterating this relation, starting with the inequality for $||x_1 - x_0||$, we obtain the inequality

$$||x_{n+1} - x_n|| \le (||c|| + 1) \left(\int_0^t f(s)\, ds\right)^{n+1} / (n+1)! \tag{9}$$

which establishes the uniform convergence of the sequence $\{x_n\}$ in the interval $[0, T]$ to a function $x(t)$. This function is continuous for $0 \leq t \leq T$, satisfies the integral equation

$$x(t) = c + \int_0^t \operatorname*{Max}_q [A(q, s)\, x + b(q, s)]\, ds, \tag{10}$$

and hence the differential equation almost everywhere.

Finally, let us demonstrate uniqueness. Let $y(t)$ be another solution of the equation, existing in some interval $[0, S]$. Then in this interval $y(t)$ satisfies the equation in (10). Applying the lemma of § 4, we derive the inequality

$$\begin{aligned} ||x(t) - y(t)|| &\leq \operatorname*{Max}_q \int_0^t ||A(q, s)||\,||x - y||\, ds \\ &\leq \int_0^t f(s)\, ||x - y||\, ds \end{aligned} \tag{11}$$

This inequality has the form

$$u(t) \leq \int_0^t f(s)\, u(s)\, ds, \tag{12}$$

where $f(s), u(s) \geq 0$.
Hence, for an arbitrarily small positive constant a, we have

$$u(t) \leq a + \int_0^t f(s)\, u(s)\, ds. \tag{13}$$

Dividing through, this yields

$$\frac{f(t)\, u(t)}{a + \int_0^t f(s)\, u(s)\, ds} \leq f(t). \tag{14}$$

Integrating between 0 and s, we have

$$a + \int_0^t f(s)\, u(s)\, ds \leq a\, e^{\int_0^t f(s)\, ds} \tag{15}$$

Combining this with (13), we obtain the inequality

$$u(t) \leq a\, e^{\int_0^t f(s)\, ds} \tag{16}$$

Since a is an arbitrary positive constant, we see that $u(t) = 0$.

An alternative proof proceeds as follows. It is clear that a constant b exists such that $||x - y|| \leq b$ in $[0, s]$. Hence

$$u(t) \leq b \int_0^t f(s)\, ds. \tag{17}$$

Use this inequality on the right side of (12), obtaining

$$u(t) \leq b \int_0^t \left(f(s) \int_0^s f(s_1)\, ds_1\right) ds = b/2! \left(\int_0^t f(s)\, ds\right)^2 \tag{18}$$

Continuing in this way, we have for each $n = 1, 2, \ldots$, the inequality

$$u(t) \leq \frac{b^n}{(n+1)!} \left(\int_0^t f(s)\, ds\right)^{n+1} \tag{19}$$

Letting $n \to \infty$, we see again that $u(t) \equiv 0$.

§ 6. Existence and uniqueness—II

Let us now consider the equation of (2.5). In general, equations of this general type need not have unique solutions, due to multiplicity of maximizing q-values. Consider, for example, the equation

$$\begin{aligned} dx_1/dt &= \operatorname*{Max}_q \left[1 - q^2 (1 - q)^2\right] + x_2, \quad x_1(0) = 0 \\ dx_2/dt &= q^* x_2 \qquad\qquad\qquad\qquad\quad , \; x_2(0) = 1 \end{aligned} \tag{1}$$

Since $q^* = 0$ or 1, we obtain infinitely many sets of solutions, of which the following are representative

$$\begin{aligned} x_1 &= 2t, \quad x_1 = t + (e^t - 1) \\ x_2 &= 1, \quad\; x_2 = e^t. \end{aligned} \tag{2}$$

We can, however, obtain uniqueness theorems if we restrict ourselves to solutions obtained in the following way. First solve the equations

$$\begin{aligned} dx_2/dt &= b_2(q, t) + \sum_{j=1}^{N} a_{2j}(q)\, x_j, \quad x_2(0) = c_2, \\ &\vdots \\ dx_N/dt &= b_N(q, t) + \sum_{j=1}^{N} a_{Nj}(q)\, x_j, \quad x_N(0) = c_N, \end{aligned} \tag{3}$$

for the quantities $x_2, x_3, \ldots, x_N$ in terms of function x_1, regarding q for the moment as some unknown function.

Each x_k, $k = 2, 3, \ldots, N$, will have the form

$$x_k = u_k(q, t) + \int_0^t v_k(q, t, s)\, x_1(s)\, ds \tag{4}$$

Substituting these expressions into the equation

$$(5) \qquad dx_1/dt = \operatorname*{Max}_q \, [b_1 (q, t) + \sum_{j=1}^{N} a_{1j} (q) \, x_j], \quad x_1 (0) = c_1 ,$$

we obtain an equation of the form

$$(6) \qquad dx_1/dt = \operatorname*{Max}_q \, [b (q, t) + a_{11} (q) \, x_1 + \int_0^t v (q, t, s) \, x_1 (s) \, ds] .$$

This equation we write in the form

$$(7) \qquad x_1 = c_1 + \operatorname{Max} [\int_0^t b (q, s) \, ds + \int a_{11} (q) \, x_1 \, ds + \int_0^t [\int_0^{t_1} v (q, t, s) \, x_1 (s) \, ds] \, dt_1]$$

Using the methods employed in § 5, it is easy to establish the existence of a unique solution of this equation under the hypotheses of Theorem 1.

§ 7. Existence and uniqueness—III

It is possible, using the same technique of successive approximation and inequalities, to establish existence and uniqueness theorems for more general systems of differential equations of the form

$$(1) \qquad dx/dt = \operatorname*{Max}_q f (x, q, t), \; x (0) = c .$$

Since these results are more within the province of differential equations than pertinent to the theory of decision processes, we shall leave it for the ambitious reader to frame his own analogues of the classical existence and uniqueness theorems.

§ 8. The Riccati equation

Although we do not wish to penetrate too deeply here into the study of this class of nonlinear differential equations, the following result seems particularly worthy of notice.

The change of variable

$$(1) \qquad v = u'/u$$

converts the general second order linear differential equation

$$(2) \qquad u'' + p (t) \, u' + q (t) \, u = 0 ,$$

into the first order non-linear equation

$$(3) \qquad v' + v^2 + p (t) \, v + q (t) = 0 .$$

This equation is called a Riccati equation. It is clear from the foregoing that the general solution of (3) is equivalent to the general solution of (2), and hence, in general, cannot be obtained explicitly in terms of quadratures.

Let us now show that (3) can be interpreted to be an equation of the general class exhibited above. We begin with the observation that

$$-v^2 = \operatorname*{Min}_{w} (w^2 - 2wv) . \tag{4}$$

Hence (3) may be written

$$v' = \operatorname*{Min}_{w} [w^2 - 2wv - p(t)v - q(t)] , \tag{5}$$

where w now varies over all functions of t.

For fixed w, let $V(w, t)$ represent the solution of

$$V' = w^2 - 2wV - p(t)V - q(t) , \tag{6}$$

fixed by the condition $V(0) = v(0) = c$. This solution has the explicit representation

$$V = ce^{-\int_0^t (p(s)+2w)\,ds} + \int_0^t (w^2 - q(s))\, e^{-\int_s^t (p(s_1)+2w)\,ds_1}\, ds. \tag{7}$$

obtained in the usual way by means of an integrating factors.

Let us now show that

$$v = \operatorname*{Min}_{w} V(w, t) . \tag{8}$$

For an arbitrary function $w = w(t)$, we have

$$v' \leq w^2 - 2wv - p(t)v - q(t) , \tag{9}$$

which shows that $v \leq V(w, t)$. Hence $v \leq \operatorname*{Min}_{w} V(w, t)$. On the other hand $v = V(w^*, t)$ for the minimizing value w^*, which is actually $v(t)$. Hence the equality in (8) holds.

We thus have an explicit representation for the solution of the Riccati equation in terms of quadratures and a minimization.

§ 9. Approximation in policy space

As we have discussed in the preceding chapters, there are two types of successive approximations in the theory of dynamic programming, one based upon approximation to the functions which satisfy the functional equation, and the other based upon approximation to the policies which yield these solutions. We have applied the traditional method above in § 3. Let us now discuss the second method.

Consider the scalar equation

$$du/dt = \operatorname*{Max}_q \, [b(q, t) + a(q, t)\, u],\ u(0) = c, \tag{1}$$

where we impose the restrictions $|a(q, t)|, |b(q, t)| \leq f(t), \int_0^{t_0} f(t)\, dt < \infty$.

We begin by guessing an initial policy function $q_0 = q_0(t)$, and determining u_0 by means of the equation

$$du_0/dt = b(q_0, t) + a(q_0, t)\, u_0,\ u_0(0) = c. \tag{2}$$

Next determine q_1 by the condition that it maximize the function $b(q, t) + a(q, t)\, u_0$, and compute u_1 as the solution of

$$du_1/dt = b(q_1, t) + a(q_1, t)\, u_1,\ u_1(0) = c. \tag{3}$$

Continuing is this way we determine a sequence of functions $\{u_n\}$ and a sequence of policies $\{q_n\}$. It remains to show that this sequence $\{u_n\}$ actually converges.

We have

$$\begin{aligned} du_1/dt &= b(q_1, t) + a(q_1, t)\, u_1,\ u_1(0) = c, \\ du_0/dt &= b(q_0, t) + a(q_0, t)\, u_0 \\ &\leq b(q_1, t) + a(q_1, t)\, u_0,\ u_0(0) = c, \end{aligned} \tag{4}$$

referring to the definition of q_1.

The solution of

$$dv/dt = g(t)\, v + h(t),\ v(0) = c, \tag{5}$$

has the form

$$v = c e^{\int_0^t g(s)\, ds} + \int_0^t h(s)\, e^{\int_s^t g(t_1)\, dt_1}\, ds. \tag{6}$$

which we may write as $L(h)$, an operator on the function h.

Since $e^{\int_s^t g(s) ds} \geq 0$, it follows that

$$L(h_1) \geq L(h_2) \tag{7}$$

if $h_1(t) \geq h_2(t)$ for $t \geq 0$. Hence

$$u_0 \leq u_1, \text{ for } 0 \leq t \leq T \tag{8}$$

Proceeding in the same fashion, we see inductively that $u_n \leq u_{n+1}$ for

$n = 0, 1, 2, \ldots$ Since each member of the sequence $\{u_n\}$ is uniformly bounded by $(c + \int_0^T f(s)\, ds)\, e^{\int_0^T f(s)\, ds}$, it follows that the sequence $\{u_n(t)\}$ converges to a function $u(t)$. This limit function satisfies the integral equation

$$(9) \qquad u(t) = c + \int_0^t \operatorname*{Max}_q \,[b(q, s) + a(q, s)\, u]\, ds ,$$

and hence the differential equation almost everywhere. We see then that approximation in policy space leads to convergent sequences in the one-dimensional case.

Let us turn now to the corresponding question for systems of the form

$$(10) \qquad dx/dt = \operatorname*{Max}_q \,[b(q, t) + A(q, t)\, x],\ x(0) = c$$

Using the same procedure as above, it is easy to see that the problem reduces to determining conditions upon the matrix $A(q, t)$ which will ensure that $f(t) \geq 0$ for $t \geq 0$ ensures that $y \geq 0$ for $t \geq 0$, where y is the solution of

$$(11) \qquad dy/dt = A(q, t)\, y + f(t),\ y(0) = 0$$

Since the solution of (11) is given by

$$(12) \qquad w = \int_0^t Y(t)\, Y^{-1}(s)\, f(s)\, ds ,$$

where $Y(t)$ is the matrix solution of

$$(13) \qquad dY/dt = A(q, t)\, Y,\ Y(0) = I ,$$

we see that a necessary and sufficient condition is

$$(14) \qquad Y(t)\, Y^{-1}(s) \geq 0 \text{ for } t \geq s \geq 0 ,$$

and uniformly for $q \in S$.

Since this is a difficult condition to verify, we shall content ourselves with the remark that $a_{ij}(q, t) \geq 0,\ i \neq j$, is a sufficient condition.

If then the condition $a_{ij}(q, t) \geq 0,\ i \neq j$, is satisfied for $t \geq 0$ and all $q \in S$, we have the desired convergence in policy space.

§ 10. Discrete versions

In this section we wish to ascertain the asymptotic behavior of the sequence $\{x_i(n)\}$, $i = 1, 2, \ldots, N$, determined by the recurrence relations

$$(1) \qquad x_i(n+1) = \operatorname*{Max}_q \sum_{j=1}^N a_{ij}(q)\, x_j(n),\ i = 1, 2, \ldots, N,\ n \geq 0$$

under certain assumptions concerning the initial values $c_i = x_i(0)$ and the coefficient matrix $A(q)$.

We shall begin by considering the homogeneous equation

$$\lambda y_i = \operatorname*{Max}_q \sum_{j=1}^{N} a_{ij}(q)\, y_j,\ i = 1, 2, \ldots, N, \tag{2}$$

where we impose the following conditions

(3) (a) $q = (q_1, q_2, \ldots, q_N)$ runs over a set S with the property that the maximum is attained in (1) for any set of parameters $(y_1, y_2, \ldots, y_N)$.

(b) $0 < a_{ij}(q) \leq m < \infty$ for $q \in S$ and $i, j = 1, 2, \ldots, N$.

(c) for any q, let $\varphi(q)$ denote the characteristic root of $A(q) = (a_{ij}(q))$ of largest absolute value, the Perron root. It is assumed that $\varphi(q)$ assumes its maximum for $q \in S$.

Let us now prove

THEOREM 2. *Under these assumptions, there exists a unique positive constant λ with the property that the homogeneous system in* (2) *has a positive solution $y_i > 0$, $i = 1, 2, \ldots, N$. This solution is unique up to a multiplicative constant, and*

$$\lambda = \operatorname*{Max}_{q \in S} \varphi(q) \tag{4}$$

PROOF. We shall begin by establishing the existence of a positive λ and a positive solution $\{y_i\}$. The simplest, though least elementary, method employs the Brouwer fixedpoint theorem. Consider the region defined by

$$y_i \geq 0,\ \sum_{i=1}^{N} y_i = 1 \tag{5}$$

The normalized transformation

$$y_i' = [\operatorname*{Max}_q \sum_{j=1}^{N} a_{ij}(q)\, y_j] \,/\, \sum_{i=1}^{N} \operatorname*{Max}_q [\sum_{j=1}^{N} a_{ij}(q)\, y_j], \tag{6}$$

is a continuous mapping of this region into itself. It follows that there exists a fixed point $\{y_i\}$, constituting the required positive solution since $a_{ij}(q) > 0$. The parameter λ is the denominator in (6).

To show that this solution is unique up to multiplicative constant, let $[\mu, z]$ be another solution of (2) with $\mu > 0$ and z a positive vector. Let $\{q\}$ be a set of values for which the maximum is attained in (2) and $\{\bar{q}\}$ a similar set associated with z. We have

(7) $$\lambda y_i = \sum_j a_{ij}(q)\, y_j \geq \sum a_{ij}(\bar{q})\, y_j,\ i = 1, 2, \ldots, N,$$

$$\mu z_i = \sum_j a_{ij}(\bar{q})\, z_j$$

Assume, without loss of generality that $\lambda < \mu$, and thus that y and z are non-proportional vectors. If y and z are proportional, then $y = z$. Let ε be a positive constant chosen so that one, at least, of the components $y_i - \varepsilon z_i$ is zero, one at least is positive, and the others are non-negative. If i is an index for which $y_i - \varepsilon z_i$ is zero, we have

(8) $$0 = \mu (y_i - \varepsilon z_i) > \lambda y_i - \varepsilon \mu z_i \geq \sum_{j=1}^{N} a_{ij}(\bar{q})\, (y_j - \varepsilon z_j) > 0,$$

since $a_{ij}(\bar{q}) > 0$, a contradiction. Hence y and z are proportional, which means that $\lambda = \mu$.

To show that $\lambda = \operatorname{Max}_q \varphi(q)$, we proceed as follows. Let $\mu = \operatorname{Max}_q \varphi(q)$. It is clear that λ, as the characteristic root of some $A(q)$, satisfies the inequality $\lambda \leq \mu$. Assume for the moment that $\lambda < \mu$. Let $z = (z_1, z_2, \ldots, z_N)$ be a positive characteristic vector associated with μ and $\bar{q}$ a set of q-values which yield $\mu = \varphi(\bar{q})$. Then we have

(9) $$\mu z_i = \sum_{j=1}^{N} a_{ij}(\bar{q})\, z_j \leq \operatorname{Max}_q \sum_{j=1}^{N} a_{ij}(q)\, z_j$$

Since each y_i is positive, we can find a positive constant m such that

(10) $$z_i \leq m y_i,\ i = 1, 2, \ldots, N.$$

Then (9) yields

(11) $$\mu z_i \leq \operatorname{Max}_q \Big(\sum_{j=1}^{N} a_{ij}(q)\, y_j \Big)\, m = m \lambda y_i$$

Thus, instead of (10) we obtain the result $z_i \leq m y_i\, \lambda/\mu$. Iterating this, we obtain $z_i \leq m y_i (\lambda/\mu)^k$ for arbitrary k. Since $\lambda/\mu < 1$, by assumption, this yields $z_i = 0$, a contradiction. Hence $\lambda = \mu$.

§ 11. The recurrence relation

Returning to the recurrence relation of (10.1), let us prove

THEOREM 3. *If, in addition to the conditions of* (10.3), *we assume that there is a unique q for which the maximum value of q is assumed, and that $c_i \geq 0$, then*

(1) $$x_i(n) \sim a y_i \lambda^n,$$

as $n \to \infty$, where $a = a(c_1, c_2, \ldots, c_N)$.

PROOF. Let us take $c_i > 0$, without loss of generality. There are then two positive constants k and K such that $ky_i \leq c_i \leq Ky_i$ for $i = 1, 2. \ldots, N$. Let us show inductively that

$$ky_i \lambda^n \leq x_i(n) \leq Ky_i \lambda^n \tag{2}$$

Assume that we have the result for n, then

$$\begin{aligned} x_i(n+1) &\leq K \lambda^n \operatorname*{Max}_q \sum_{j=1}^{N} a_{ij}(q) y_j = K \lambda^{n+1} y_i \\ &\geq k \lambda^n \operatorname*{Max}_q \sum_{j=1}^{N} a_{ij}(q) y_j = k \lambda^{n+1} y_i \end{aligned} \tag{3}$$

To establish the asymptotic behavior we show that for sufficiently large n, the set of q's which furnish the maximum in (10.1) is precisely the set which yields $\operatorname*{Max}_q \varphi(q)$.

Assume the contrary. This means that infinitely often in the recurrence relation of (10.1) we will employ a set $\{\bar{q}\}$ which is not identical with the set, $\{q\}$, which furnishes the maximum of $\varphi(q)$.

We then have

$$\begin{aligned} x_i(n+1) &= \sum_{j=1}^{N} a_{ij}(\bar{q}) x_j(n), \quad i = 1, 2, \ldots, N, \\ &\leq \left(\sum_{j=1}^{N} a_{ij}(\bar{q}) y_j \right) K \lambda^n \end{aligned} \tag{4}$$

For some index i we must have

$$\sum_{j=1}^{N} a_{ij}(\bar{q}) y_j < \lambda y_i, \tag{5}$$

with strict inequality. For if $\sum_{j=1}^{N} a_{ij}(\bar{q}) y_j \geq \lambda y_i$ for all i, the characteristic root of $A(\bar{q}) = (a_{ij}(\bar{q}))$ of largest absolute value would at least equal $\lambda = \operatorname*{Max}_q \varphi(q)$, contradicting the assumption concerning the uniqueness of the maximum of $\varphi(q)$.

Hence for some component, say the first, we have

$$x_1(n+1) \leq \theta K \lambda^{n+1} y_1, \quad 0 < \theta < 1 \tag{6}$$

Since $a_{ij}(q^*) > 0$ for all i, j, where q^* is the value of q for which $\lambda = \varphi(q^*)$, we see that, for $i = 1, 2, \ldots, N$,

(7) $$x_i(n+2) \leq K\lambda^{n+1}\left[\sum_{j=2}^{N} a_{ij}(q^*)\, y_j + \theta\, a_{i1}(q^*)\, y_1\right],$$
$$\leq \theta_1 K \lambda^{n+2} y_i,$$

where $\theta_1 < 1$.

Consequently, if a set of q's distinct from q^* are used R times, we obtain

(8) $$x_i(n) \leq \theta_1^R K \lambda^n y_i,$$

for n large. Since $0 < \theta_1 < 1$, we eventually contradict the lower bound for $x_i(n)$ if R is large. Hence a set of q's distinct from q^* can only be used a bounded number of times, with the bound determined by k and K.

§ 12. Min-max

The same method we employed to demonstrate Theorem 1 establishes the following result

THEOREM 4. *Consider the equation*

(1) $$dx/dt = \operatorname*{Max}_p \operatorname*{Min}_q [A(p, q, t)\, x + b(p, q, t)]$$
$$= \operatorname*{Min}_q \operatorname*{Max}_p [\ \ldots\ ],\ x(0) = c,$$

where we assume that

(2) (a) *For fixed values of x and t, the max-min in* (1) *is equal to the min-max, where p and q range over some set of admissible vectors S.*

(b) $\operatorname*{Max}_S \|A(p, q, t)\|$, $\operatorname*{Max}_S \|b(p, q, t)\| \leq f(t)$ *for* $t \geq 0$, *where*

$$\int_0^T f(t)\, dt < \infty.$$

Then there is a unique solution to (1) *in* $0 \leq t \leq T$ *which satisfies the equation almost everywhere, and may be found as the limit of the following sequence*

(3) (a) $x_0 = c$,

(b) $$x_{n+1} = c + \int_0^t \operatorname*{Max}_p \operatorname*{Min}_q [A(p, q, s)\, x_n + b(p, q, s)]\, ds$$
$$= c + \int_0^t \operatorname*{Min}_q \operatorname*{Max}_p [A(p, q, s)\, x_n + b(p, q, s)]\, ds$$

§ 13. Generalization of a von Neumann result

In the chapter devoted to multi-stage games, we established the result that

$$\text{(1)} \qquad \operatorname*{Max}_p \operatorname*{Min}_q \frac{(Ap, q)}{(Bp, q)} = \operatorname*{Min}_q \operatorname*{Max}_p \frac{(Ap, q)}{(Bp, q)},$$

where A and B are matrices, and p and q are probability vectors, provided that $(Bp, q) \geq d > 0$ for all p and q.

Let us now obtain the following generalization

THEOREM 5. *Consider the scalar equation*

$$\text{(2)} \qquad du/dt = \operatorname*{Max}_p \operatorname*{Min}_q [(Ap, q) - (Bp, q)\, u],\ u(0) = c,$$
$$= \operatorname*{Min}_q \operatorname*{Max}_p [(Ap, q) - (Bp, q)\, u].$$

If $(Bp, q) \geq d > 0$ *for all probability vectors* p *and* q, *we have*

$$\text{(3)} \qquad \lim_{t \to \infty} u(t) = \operatorname*{Max}_p \operatorname*{Min}_q \frac{(Ap, q)}{(Bp, q)}$$
$$= \operatorname*{Min}_q \operatorname*{Max}_p \frac{(Ap, q)}{(Bp, q)}$$

PROOF. The classical min-max theorem of Von Neumann guarantees the equality of max-min and min-max of $(Ap, q) - (Bp, q)\, u$ for each u. The other conditions of Theorem 4 are satisfied and ensure the existence and uniqueness of $u(t)$.

To obtain the asymptotic behavior, consider first the scalar equation

$$\text{(4)} \qquad du/dt = a - bu,\ u(0) = c,$$

where a and b are constants and where $b > 0$. It is easy to see that the solution is bounded as $t \to \infty$, and we can show that $\lim_{t \to \infty} u(t) = a/b$ by means of the following simple argument. Whenever $du/dt = 0$, we must have $u = a/b$. Hence $u(t)$ can have at most one turning point for $t \geq 0$, and thus is ultimately monotone. Since $u(t)$ is bounded, it approaches a finite limit which must be a/b.

Consider the nonlinear equation

$$\text{(5)} \qquad du/dt = \operatorname*{Max}_p [a(p) - b(p)\, u],\ u(0) = c,$$

where $b(p) \geq b > 0$ for all p, $|a(p)| \leq M$ for all p, and $a(p)$ and $b(p)$ are such that the maximum is assumed. At any turning point of u, we must have

$$\text{(6)} \qquad u = \operatorname*{Max}_p a(p) / b(p).$$

Consequently, $u(t)$ must be ultimately monotone and approach the finite limit given in (6).

We see that precisely the same argument works for the equation of (2). At a turning point we must have

$$u = \operatorname*{Max}_{p} \operatorname*{Min}_{q} \frac{(Ap, q)}{(Bp, q)} = \operatorname*{Min}_{q} \operatorname*{Max}_{p} \frac{(Ap, q)}{(Bp, q)} \tag{7}$$

Exercises and Research Problems for Chapter XI

1. A merchant has n identical items and a length of time t to dispose of these. Goods may be sold at times 0, 1, 2, . . ., t, and the probability of selling an item depends upon its price. Let $\varphi(z)$ be the probability that an item of price z is sold when displayed at a particular time.

Define $f_n(t)$ to be the maximum expected return from n items over a maximum sale period of t. Assuming independence of sales, obtain the recurrence relation

$$f_n(t) = \operatorname*{Max}_{z \geq 0} \left[\sum_{k=0}^{n} \binom{n}{k} \varphi(z)^k (1 - \varphi(z))^{n-k} [f_{n-k}(t-1) + kz] \right],$$

with $f_n(0) = 0$. (Darling)

2. Assume that the items are on sale continuously, and that $\varphi(z)\, dt$ represents the probability that an item of price z will be sold in a time-interval $(t, t + dt)$. Show that the limiting form of the above recurrence relation is

$$f_N'(t) = \operatorname*{Max}_{z \geq 0} [-N\varphi(z) f_N(t) + N\varphi(z) f_{N-1}(t) + N\varphi(z)]$$

$$f_N(0) = 0,$$

$$N \geq 1, f_0(t) = 0.$$

3. Consider the case $N = 1$ in Problem 4. Show that if we solve the equation

$$F'(t) = -\varphi(z) F(t) + \varphi(z), F(0) = 0,$$

obtaining

$$F(t) = F(t, z) = \int_0^t e^{-\int_s^t \varphi(z)\, dt} {}_1\varphi(z)\, ds,$$

then $f_1(t) = \operatorname*{Max}_{z \geq 0} F(t, z)$.

4. Show that the equation

$$f_1'(t) = \operatorname*{Max}_{z \geq 0} [-\varphi(z) f_1(t) + z\varphi(z)], f_1(0) = 0$$

is equivalent to the two equations

$$f_1'(t) = -\varphi(z) f_1(t) + z\varphi(z), f_1(0) = 0$$
$$0 = -\varphi'(z) f_1(t) + \varphi(z) + z\varphi'(z).$$

5. Consider in detail the particular cases where

a. $\varphi(z) = b\, e^{-bz}$

b. $\varphi(z) = (k - z)/k, \ 0 \leq z \leq k$
$\quad = 0, \qquad z \geq k$

6. Obtain the solution of the equations in Exercise 2 for general N.

7. Consider the similar situation in which we have the same item in two price ranges. How do we set the prices?

8. Consider the process in Problem 1 in which we reduce the price per item for multiple orders. How should this be done to maximize expected profit?

9. Establish existence and uniqueness theorems for integral equations of the form

$$u(t) = \operatorname*{Max}_q [a(q, t) + \int_0^1 K(q, t, s)\, u(s)\, ds],$$

under appropriate assumptions.

10. Obtain results corresponding to those in § 8 for the equations

$$u' = u^k + p(t)\, u + q(t),$$

for $k > 1$ and $0 < k < 1$.

11. Consider the general case where

$$u' = g(u, t),$$

and g is either strictly convex in u for all t, or strictly concave.

12. Consider the Riccati equation

$$\frac{du}{dt} = u^2 + a(t), \ u(0) = c,$$

and the sequence of successive approximations defined by

$$\frac{du_0}{dt} = 2u_0 v_0 - v_0^2 + a(t), \ \ u_0(0) = c,$$

$$\frac{du_{n+1}}{dt} = 2u_{n+1} u_n - u_n^2 + a(t), \ u_{n+1}(0) = c,$$

where $v_0(t)$ is an arbitrary continuous function.

Show that this method is equivalent to a certain approximation in policy space, and that $u_0 \leq u_1 \leq \ldots \leq u_n \leq \ldots$, in a common interval of definition.

13. Similarly, consider the sequence defined by

$$\frac{du_{n+1}}{dt} = \varphi(u_n, t) + (u_{n+1} - u_n) \frac{\partial \varphi}{\partial u_n}, \; u_{n+1}(0) = c,$$

in connection with the equation $du/dt = \varphi(u, t)$, $u(0) = c$.

14. What is the connection between this method of successive approximations and Newton's method for solving equations?

15. What is the connection between the approximation schemes outlined above and the concept of approximation in policy space?

Bibliography and Comments for Chapter XI

§ 1. A discussion of these processes, and formulation of the corresponding functional equations is contained in R. Bellman, "Functional Equations in the Theory of Dynamic Programming—IV, Multi-Stage Decision Processes of Continuous Type," Rand Paper p. 705, June 1955.

§ 2. For a treatment of Markoff processes, we refer to the book W. Feller, *Probability Theory*, John Wiley and Sons, 1948.

§ 5. Existence and uniqueness theorems for this type of nonlinear differential equation were given in R. Bellman, "Functional Equations in the Theory of Dynamic Programming—II, Nonlinear Differential Equations," *Proc. Nat. Acad. Sci.*, vol. 41 (1955), pp. 482–5.

The consequence of the inequality in (5.13), stated in (5.16), is a fundamental inequality in the study of boundedness and stability of the solutions of linear and nonlinear differential equations. It was first used in this way in R. Bellman, "The Stability of Solutions of Linear Differential Equations," *Duke Math. Jour.*, vol. 10 (1943), pp. 643–7, and other applications may be found in R. Bellman, *Stability Theory of Differential Equations*, McGraw-Hill, 1952.

§ 8. For an application of this device to other classes of functional equations see R. Bellman, "Functional Equations in the Theory of Dynamic Programming—V, Positivity and Linearity," *Proc. Nat. Acad. Sci.*, vol. 41

(1955), pp. 743–6. and also R. Bellman, "On the Explicit Solutions of Some Algebraic Equations," *Math. Mag.*, 1956.

§ 10. The contents of this section appeared originally in R. Bellman, "On a Class of Quasi-Linear Equations," *Can. Jour. of Math.*, 1956.

§ 13. This extension was announced in the paper cited above in the comment on § 8.

INDEX OF APPLICATIONS

NAME AND SUBJECT INDEX

A CATALOG OF SELECTED

DOVER BOOKS

IN SCIENCE AND MATHEMATICS

A CATALOG OF SELECTED

DOVER BOOKS

IN SCIENCE AND MATHEMATICS

Astronomy

BURNHAM'S CELESTIAL HANDBOOK, Robert Burnham, Jr. Thorough guide to the stars beyond our solar system. Exhaustive treatment. Alphabetical by constellation: Andromeda to Cetus in Vol. 1; Chamaeleon to Orion in Vol. 2; and Pavo to Vulpecula in Vol. 3. Hundreds of illustrations. Index in Vol. 3. 2,000pp. 6⅛ x 9¼.
23567-X, 23568-8, 23673-0 Three-vol. set

THE EXTRATERRESTRIAL LIFE DEBATE, 1750–1900, Michael J. Crowe. First detailed, scholarly study in English of the many ideas that developed from 1750 to 1900 regarding the existence of intelligent extraterrestrial life. Examines ideas of Kant, Herschel, Voltaire, Percival Lowell, many other scientists and thinkers. 16 illustrations. 704pp. 5⅜ x 8½. 40675-X

A HISTORY OF ASTRONOMY, A. Pannekoek. Well-balanced, carefully reasoned study covers such topics as Ptolemaic theory, work of Copernicus, Kepler, Newton, Eddington's work on stars, much more. Illustrated. References. 521pp. 5⅜ x 8½.
65994-1

AMATEUR ASTRONOMER'S HANDBOOK, J. B. Sidgwick. Timeless, comprehensive coverage of telescopes, mirrors, lenses, mountings, telescope drives, micrometers, spectroscopes, more. 189 illustrations. 576pp. 5⅝ x 8¼. (Available in U.S. only.)
24034-7

STARS AND RELATIVITY, Ya. B. Zel'dovich and I. D. Novikov. Vol. 1 of *Relativistic Astrophysics* by famed Russian scientists. General relativity, properties of matter under astrophysical conditions, stars, and stellar systems. Deep physical insights, clear presentation. 1971 edition. References. 544pp. 5⅝ x 8¼. 69424-0

Chemistry

CHEMICAL MAGIC, Leonard A. Ford. Second Edition, Revised by E. Winston Grundmeier. Over 100 unusual stunts demonstrating cold fire, dust explosions, much more. Text explains scientific principles and stresses safety precautions. 128pp. 5⅜ x 8½. 67628-5

THE DEVELOPMENT OF MODERN CHEMISTRY, Aaron J. Ihde. Authoritative history of chemistry from ancient Greek theory to 20th-century innovation. Covers major chemists and their discoveries. 209 illustrations. 14 tables. Bibliographies. Indices. Appendices. 851pp. 5⅜ x 8½. 64235-6

CATALYSIS IN CHEMISTRY AND ENZYMOLOGY, William P. Jencks. Exceptionally clear coverage of mechanisms for catalysis, forces in aqueous solution, carbonyl- and acyl-group reactions, practical kinetics, more. 864pp. 5⅜ x 8½.
65460-5

THE HISTORICAL BACKGROUND OF CHEMISTRY, Henry M. Leicester. Evolution of ideas, not individual biography. Concentrates on formulation of a coherent set of chemical laws. 260pp. 5⅜ x 8½. 61053-5

A SHORT HISTORY OF CHEMISTRY, J. R. Partington. Classic exposition explores origins of chemistry, alchemy, early medical chemistry, nature of atmosphere, theory of valency, laws and structure of atomic theory, much more. 428pp. 5⅜ x 8½. (Available in U.S. only.) 65977-1

GENERAL CHEMISTRY, Linus Pauling. Revised 3rd edition of classic first-year text by Nobel laureate. Atomic and molecular structure, quantum mechanics, statistical mechanics, thermodynamics correlated with descriptive chemistry. Problems. 992pp. 5⅜ x 8½. 65622-5

Engineering

DE RE METALLICA, Georgius Agricola. The famous Hoover translation of greatest treatise on technological chemistry, engineering, geology, mining of early modern times (1556). All 289 original woodcuts. 638pp. 6¾ x 11. 60006-8

FUNDAMENTALS OF ASTRODYNAMICS, Roger Bate et al. Modern approach developed by U.S. Air Force Academy. Designed as a first course. Problems, exercises. Numerous illustrations. 455pp. 5⅜ x 8½. 60061-0

DYNAMICS OF FLUIDS IN POROUS MEDIA, Jacob Bear. For advanced students of ground water hydrology, soil mechanics and physics, drainage and irrigation engineering and more. 335 illustrations. Exercises, with answers. 784pp. 6⅛ x 9¼. 65675-6

ANALYTICAL MECHANICS OF GEARS, Earle Buckingham. Indispensable reference for modern gear manufacture covers conjugate gear-tooth action, gear-tooth profiles of various gears, many other topics. 263 figures. 102 tables. 546pp. 5⅜ x 8½. 65712-4

MECHANICS, J. P. Den Hartog. A classic introductory text or refresher. Hundreds of applications and design problems illuminate fundamentals of trusses, loaded beams and cables, etc. 334 answered problems. 462pp. 5⅜ x 8½. 60754-2

MECHANICAL VIBRATIONS, J. P. Den Hartog. Classic textbook offers lucid explanations and illustrative models, applying theories of vibrations to a variety of practical industrial engineering problems. Numerous figures. 233 problems, solutions. Appendix. Index. Preface. 436pp. 5⅜ x 8½. 64785-4

STRENGTH OF MATERIALS, J. P. Den Hartog. Full, clear treatment of basic material (tension, torsion, bending, etc.) plus advanced material on engineering methods, applications. 350 answered problems. 323pp. 5⅜ x 8½. 60755-0

A HISTORY OF MECHANICS, René Dugas. Monumental study of mechanical principles from antiquity to quantum mechanics. Contributions of ancient Greeks, Galileo, Leonardo, Kepler, Lagrange, many others. 671pp. 5⅜ x 8½. 65632-2

METAL FATIGUE, N. E. Frost, K. J. Marsh, and L. P. Pook. Definitive, clearly written, and well-illustrated volume addresses all aspects of the subject, from the historical development of understanding metal fatigue to vital concepts of the cyclic stress that causes a crack to grow. Includes 7 appendixes. 544pp. 5⅜ x 8½. 40927-9

STATISTICAL MECHANICS: Principles and Applications, Terrell L. Hill. Standard text covers fundamentals of statistical mechanics, applications to fluctuation theory, imperfect gases, distribution functions, more. 448pp. 5⅜ x 8½. 65390-0

THE VARIATIONAL PRINCIPLES OF MECHANICS, Cornelius Lanczos. Graduate level coverage of calculus of variations, equations of motion, relativistic mechanics, more. First inexpensive paperbound edition of classic treatise. Index. Bibliography. 418pp. 5⅜ x 8½. 65067-7

THE VARIOUS AND INGENIOUS MACHINES OF AGOSTINO RAMELLI: A Classic Sixteenth-Century Illustrated Treatise on Technology, Agostino Ramelli. One of the most widely known and copied works on machinery in the 16th century. 194 detailed plates of water pumps, grain mills, cranes, more. 608pp. 9 x 12. 28180-9

ORDINARY DIFFERENTIAL EQUATIONS AND STABILITY THEORY: An Introduction, David A. Sánchez. Brief, modern treatment. Linear equation, stability theory for autonomous and nonautonomous systems, etc. 164pp. 5⅜ x 8¼. 63828-6

ROTARY WING AERODYNAMICS, W. Z. Stepniewski. Clear, concise text covers aerodynamic phenomena of the rotor and offers guidelines for helicopter performance evaluation. Originally prepared for NASA. 537 figures. 640pp. 6⅛ x 9¼. 64647-5

INTRODUCTION TO SPACE DYNAMICS, William Tyrrell Thomson. Comprehensive, classic introduction to space-flight engineering for advanced undergraduate and graduate students. Includes vector algebra, kinematics, transformation of coordinates. Bibliography. Index. 352pp. 5⅜ x 8½. 65113-4

HISTORY OF STRENGTH OF MATERIALS, Stephen P. Timoshenko. Excellent historical survey of the strength of materials with many references to the theories of elasticity and structure. 245 figures. 452pp. 5⅜ x 8½. 61187-6

ANALYTICAL FRACTURE MECHANICS, David J. Unger. Self-contained text supplements standard fracture mechanics texts by focusing on analytical methods for determining crack-tip stress and strain fields. 336pp. 6⅛ x 9¼. 41737-9

Mathematics

HANDBOOK OF MATHEMATICAL FUNCTIONS WITH FORMULAS, GRAPHS, AND MATHEMATICAL TABLES, edited by Milton Abramowitz and Irene A. Stegun. Vast compendium: 29 sets of tables, some to as high as 20 places. 1,046pp. 8 x 10½. 61272-4

THE FUNCTIONS OF MATHEMATICAL PHYSICS, Harry Hochstadt. Comprehensive treatment of orthogonal polynomials, hypergeometric functions, Hill's equation, much more. Bibliography. Index. 322pp. 5⅜ x 8½. 65214-9

THREE PEARLS OF NUMBER THEORY, A. Y. Khinchin. Three compelling puzzles require proof of a basic law governing the world of numbers. Challenges concern van der Waerden's theorem, the Landau-Schnirelmann hypothesis and Mann's theorem, and a solution to Waring's problem. Solutions included. 64pp. 5⅜ x 8½. 40026-3

CALCULUS REFRESHER FOR TECHNICAL PEOPLE, A. Albert Klaf. Covers important aspects of integral and differential calculus via 756 questions. 566 problems, most answered. 431pp. 5⅜ x 8½. 20370-0

THE PHILOSOPHY OF MATHEMATICS: An Introductory Essay, Stephan Körner. Surveys the views of Plato, Aristotle, Leibniz & Kant concerning propositions and theories of applied and pure mathematics. Introduction. Two appendices. Index. 198pp. 5⅜ x 8½. 25048-2

INTRODUCTORY REAL ANALYSIS, A.N. Kolmogorov, S. V. Fomin. Translated by Richard A. Silverman. Self-contained, evenly paced introduction to real and functional analysis. Some 350 problems. 403pp. 5⅜ x 8½. 61226-0

APPLIED ANALYSIS, Cornelius Lanczos. Classic work on analysis and design of finite processes for approximating solution of analytical problems. Algebraic equations, matrices, harmonic analysis, quadrature methods, much more. 559pp. 5⅜ x 8½. 65656-X

AN INTRODUCTION TO ALGEBRAIC STRUCTURES, Joseph Landin. Superb self-contained text covers "abstract algebra": sets and numbers, theory of groups, theory of rings, much more. Numerous well-chosen examples, exercises. 247pp. 5⅜ x 8½. 65940-2

SPECIAL FUNCTIONS, N. N. Lebedev. Translated by Richard Silverman. Famous Russian work treating more important special functions, with applications to specific problems of physics and engineering. 38 figures. 308pp. 5⅜ x 8½. 60624-4

QUALITATIVE THEORY OF DIFFERENTIAL EQUATIONS, V. V. Nemytskii and V.V. Stepanov. Classic graduate-level text by two prominent Soviet mathematicians covers classical differential equations as well as topological dynamics and ergodic theory. Bibliographies. 523pp. 5⅜ x 8½. 65954-2

NUMBER THEORY AND ITS HISTORY, Oystein Ore. Unusually clear, accessible introduction covers counting, properties of numbers, prime numbers, much more. Bibliography. 380pp. 5⅜ x 8½. 65620-9

THEORY OF MATRICES, Sam Perlis. Outstanding text covering rank, nonsingularity and inverses in connection with the development of canonical matrices under the relation of equivalence, and without the intervention of determinants. Includes exercises. 237pp. 5⅜ x 8½. 66810-X

POPULAR LECTURES ON MATHEMATICAL LOGIC, Hao Wang. Noted logician's lucid treatment of historical developments, set theory, model theory, recursion theory and constructivism, proof theory, more. 3 appendixes. Bibliography. 1981 edition. ix + 283pp. 5⅜ x 8½. 67632-3

CALCULUS OF VARIATIONS, Robert Weinstock. Basic introduction covering isoperimetric problems, theory of elasticity, quantum mechanics, electrostatics, etc. Exercises throughout. 326pp. 5⅜ x 8½. 63069-2

THE CONTINUUM: A Critical Examination of the Foundation of Analysis, Hermann Weyl. Classic of 20th-century foundational research deals with the conceptual problem posed by the continuum. 156pp. 5⅜ x 8½. 67982-9

CHALLENGING MATHEMATICAL PROBLEMS WITH ELEMENTARY SOLUTIONS, A. M. Yaglom and I. M. Yaglom. Over 170 challenging problems on probability theory, combinatorial analysis, points and lines, topology, convex polygons, many other topics. Solutions. Total of 445pp. 5⅜ x 8½. Two-vol. set.
Vol. I: 65536-9 Vol. II: 65537-7

A SURVEY OF NUMERICAL MATHEMATICS, David M. Young and Robert Todd Gregory. Broad self-contained coverage of computer-oriented numerical algorithms for solving various types of mathematical problems in linear algebra, ordinary and partial, differential equations, much more. Exercises. Total of 1,248pp. 5⅜ x 8½. Two volumes. Vol. I: 65691-8 Vol. II: 65692-6

INTRODUCTION TO PARTIAL DIFFERENTIAL EQUATIONS WITH APPLICATIONS, E. C. Zachmanoglou and Dale W. Thoe. Essentials of partial differential equations applied to common problems in engineering and the physical sciences. Problems and answers. 416pp. 5⅜ x 8½. 65251-3

THE THEORY OF GROUPS, Hans J. Zassenhaus. Well-written graduate-level text acquaints reader with group-theoretic methods and demonstrates their usefulness in mathematics. Axioms, the calculus of complexes, homomorphic mapping, *p*-group theory, more. Many proofs shorter and more transparent than older ones. 276pp. 5⅜ x 8½. 40922-8

DISTRIBUTION THEORY AND TRANSFORM ANALYSIS: An Introduction to Generalized Functions, with Applications, A. H. Zemanian. Provides basics of distribution theory, describes generalized Fourier and Laplace transformations. Numerous problems. 384pp. 5⅜ x 8½. 65479-6

Math–Decision Theory, Statistics, Probability

ELEMENTARY DECISION THEORY, Herman Chernoff and Lincoln E. Moses. Clear introduction to statistics and statistical theory covers data processing, probability and random variables, testing hypotheses, much more. Exercises. 364pp. 5⅜ x 8½. 65218-1

Math–Geometry and Topology

ELEMENTARY CONCEPTS OF TOPOLOGY, Paul Alexandroff. Elegant, intuitive approach to topology from set-theoretic topology to Betti groups; how concepts of topology are useful in math and physics. 25 figures. 57pp. 5⅜ x 8½. 60747-X

COMBINATORIAL TOPOLOGY, P. S. Alexandrov. Clearly written, well-organized, three-part text begins by dealing with certain classic problems without using the formal techniques of homology theory and advances to the central concept, the Betti groups. Numerous detailed examples. 654pp. 5⅜ x 8½. 40179-0

EXPERIMENTS IN TOPOLOGY, Stephen Barr. Classic, lively explanation of one of the byways of mathematics. Klein bottles, Moebius strips, projective planes, map coloring, problem of the Koenigsberg bridges, much more, described with clarity and wit. 43 figures. 210pp. 5⅜ x 8½. 25933-1

CONFORMAL MAPPING ON RIEMANN SURFACES, Harvey Cohn. Lucid, insightful book presents ideal coverage of subject. 334 exercises make book perfect for self-study. 55 figures. 352pp. 5⅝ x 8¼. 64025-6

THE GEOMETRY OF RENÉ DESCARTES, René Descartes. The great work founded analytical geometry. Original French text, Descartes's own diagrams, together with definitive Smith-Latham translation. 244pp. 5⅜ x 8½. 60068-8

THE THIRTEEN BOOKS OF EUCLID'S ELEMENTS, translated with introduction and commentary by Sir Thomas L. Heath. Definitive edition. Textual and linguistic notes, mathematical analysis. 2,500 years of critical commentary. Unabridged. 1,414pp. 5⅜ x 8½. Three-vol. set.
Vol. I: 60088-2 Vol. II: 60089-0 Vol. III: 60090-4

GEOMETRY OF COMPLEX NUMBERS, Hans Schwerdtfeger. Illuminating, widely praised book on analytic geometry of circles, the Moebius transformation, and two-dimensional non-Euclidean geometries. 200pp. 5⅝ x 8¼. 63830-8

DIFFERENTIAL GEOMETRY, Heinrich W. Guggenheimer. Local differential geometry as an application of advanced calculus and linear algebra. Curvature, transformation groups, surfaces, more. Exercises. 62 figures. 378pp. 5⅜ x 8½. 63433-7

CURVATURE AND HOMOLOGY: Enlarged Edition, Samuel I. Goldberg. Revised edition examines topology of differentiable manifolds; curvature, homology of Riemannian manifolds; compact Lie groups; complex manifolds; curvature, homology of Kaehler manifolds. New Preface. Four new appendixes. 416pp. 5⅜ x 8½. 40207-X

TOPOLOGY, John G. Hocking and Gail S. Young. Superb one-year course in classical topology. Topological spaces and functions, point-set topology, much more. Examples and problems. Bibliography. Index. 384pp. 5⅝ x 8¼. 65676-4

LECTURES ON CLASSICAL DIFFERENTIAL GEOMETRY, Second Edition, Dirk J. Struik. Excellent brief introduction covers curves, theory of surfaces, fundamental equations, geometry on a surface, conformal mapping, other topics. Problems. 240pp. 5⅜ x 8½. 65609-8

Math–History of

A SHORT ACCOUNT OF THE HISTORY OF MATHEMATICS, W. W. Rouse Ball. One of clearest, most authoritative surveys from the Egyptians and Phoenicians through 19th-century figures such as Grassman, Galois, Riemann. Fourth edition. 522pp. 5⅜ x 8½. 20630-0

THE HISTORY OF THE CALCULUS AND ITS CONCEPTUAL DEVELOPMENT, Carl B. Boyer. Origins in antiquity, medieval contributions, work of Newton, Leibniz, rigorous formulation. Treatment is verbal. 346pp. 5⅜ x 8½. 60509-4

THE HISTORICAL ROOTS OF ELEMENTARY MATHEMATICS, Lucas N. H. Bunt, Phillip S. Jones, and Jack D. Bedient. Fundamental underpinnings of modern arithmetic, algebra, geometry and number systems derived from ancient civilizations. 320pp. 5⅜ x 8½. 25563-8

A HISTORY OF MATHEMATICAL NOTATIONS, Florian Cajori. This classic study notes the first appearance of a mathematical symbol and its origin, the competition it encountered, its spread among writers in different countries, its rise to popularity, its eventual decline or ultimate survival. Original 1929 two-volume edition presented here in one volume. xxviii+820pp. 5⅜ x 8½. 67766-4

GAMES, GODS & GAMBLING: A History of Probability and Statistical Ideas, F. N. David. Episodes from the lives of Galileo, Fermat, Pascal, and others illustrate this fascinating account of the roots of mathematics. Features thought-provoking references to classics, archaeology, biography, poetry. 1962 edition. 304pp. 5⅜ x 8½. (Available in U.S. only.) 40023-9

OF MEN AND NUMBERS: The Story of the Great Mathematicians, Jane Muir. Fascinating accounts of the lives and accomplishments of history's greatest mathematical minds–Pythagoras, Descartes, Euler, Pascal, Cantor, many more. Anecdotal, illuminating. 30 diagrams. Bibliography. 256pp. 5⅜ x 8½. 28973-7

HISTORY OF MATHEMATICS, David E. Smith. Nontechnical survey from ancient Greece and Orient to late 19th century; evolution of arithmetic, geometry, trigonometry, calculating devices, algebra, the calculus. 362 illustrations. 1,355pp. 5⅜ x 8½. Two-vol. set. Vol. I: 20429-4 Vol. II: 20430-8

A CONCISE HISTORY OF MATHEMATICS, Dirk J. Struik. The best brief history of mathematics. Stresses origins and covers every major figure from ancient Near East to 19th century. 41 illustrations. 195pp. 5⅜ x 8½. 60255-9

PHYSICAL PRINCIPLES OF THE QUANTUM THEORY, Werner Heisenberg. Nobel Laureate discusses quantum theory, uncertainty, wave mechanics, work of Dirac, Schroedinger, Compton, Wilson, Einstein, etc. 184pp. 5⅜ x 8½. 60113-7

ATOMIC SPECTRA AND ATOMIC STRUCTURE, Gerhard Herzberg. One of best introductions; especially for specialist in other fields. Treatment is physical rather than mathematical. 80 illustrations. 257pp. 5⅜ x 8½. 60115-3

AN INTRODUCTION TO STATISTICAL THERMODYNAMICS, Terrell L. Hill. Excellent basic text offers wide-ranging coverage of quantum statistical mechanics, systems of interacting molecules, quantum statistics, more. 523pp. 5⅜ x 8½. 65242-4

THEORETICAL PHYSICS, Georg Joos, with Ira M. Freeman. Classic overview covers essential math, mechanics, electromagnetic theory, thermodynamics, quantum mechanics, nuclear physics, other topics. First paperback edition. xxiii + 885pp. 5⅜ x 8½. 65227-0

PROBLEMS AND SOLUTIONS IN QUANTUM CHEMISTRY AND PHYSICS, Charles S. Johnson, Jr. and Lee G. Pedersen. Unusually varied problems, detailed solutions in coverage of quantum mechanics, wave mechanics, angular momentum, molecular spectroscopy, more. 280 problems plus 139 supplementary exercises. 430pp. 6½ x 9¼. 65236-X

THEORETICAL SOLID STATE PHYSICS, Vol. 1: Perfect Lattices in Equilibrium; Vol. II: Non-Equilibrium and Disorder, William Jones and Norman H. March. Monumental reference work covers fundamental theory of equilibrium properties of perfect crystalline solids, non-equilibrium properties, defects and disordered systems. Appendices. Problems. Preface. Diagrams. Index. Bibliography. Total of 1,301pp. 5⅜ x 8½. Two volumes. Vol. I: 65015-4 Vol. II: 65016-2

A TREATISE ON ELECTRICITY AND MAGNETISM, James Clerk Maxwell. Important foundation work of modern physics. Brings to final form Maxwell's theory of electromagnetism and rigorously derives his general equations of field theory. 1,084pp. 5⅜ x 8½. Two-vol. set. Vol. I: 60636-8 Vol. II: 60637-6

OPTICKS, Sir Isaac Newton. Newton's own experiments with spectroscopy, colors, lenses, reflection, refraction, etc., in language the layman can follow. Foreword by Albert Einstein. 532pp. 5⅜ x 8½. 60205-2

THEORY OF ELECTROMAGNETIC WAVE PROPAGATION, Charles Herach Papas. Graduate-level study discusses the Maxwell field equations, radiation from wire antennas, the Doppler effect and more. xiii + 244pp. 5⅜ x 8½. 65678-5

INTRODUCTION TO QUANTUM MECHANICS With Applications to Chemistry, Linus Pauling & E. Bright Wilson, Jr. Classic undergraduate text by Nobel Prize winner applies quantum mechanics to chemical and physical problems. Numerous tables and figures enhance the text. Chapter bibliographies. Appendices. Index. 468pp. 5⅜ x 8½. 64871-0

METHODS OF THERMODYNAMICS, Howard Reiss. Outstanding text focuses on physical technique of thermodynamics, typical problem areas of understanding, and significance and use of thermodynamic potential. 1965 edition. 238pp. 5⅜ x 8½. 69445-3

TENSOR ANALYSIS FOR PHYSICISTS, J. A. Schouten. Concise exposition of the mathematical basis of tensor analysis, integrated with well-chosen physical examples of the theory. Exercises. Index. Bibliography. 289pp. 5⅜ x 8½. 65582-2

RELATIVITY IN ILLUSTRATIONS, Jacob T. Schwartz. Clear nontechnical treatment makes relativity more accessible than ever before. Over 60 drawings illustrate concepts more clearly than text alone. Only high school geometry needed. Bibliography. 128pp. 6⅛ x 9¼. 25965-X

THE ELECTROMAGNETIC FIELD, Albert Shadowitz. Comprehensive undergraduate text covers basics of electric and magnetic fields, builds up to electromagnetic theory. Also related topics, including relativity. Over 900 problems. 768pp. 5⅝ x 8¼. 65660-8

GREAT EXPERIMENTS IN PHYSICS: Firsthand Accounts from Galileo to Einstein, edited by Morris H. Shamos. 25 crucial discoveries: Newton's laws of motion, Chadwick's study of the neutron, Hertz on electromagnetic waves, more. Original accounts clearly annotated. 370pp. 5⅜ x 8½. 25346-5

RELATIVITY, THERMODYNAMICS AND COSMOLOGY, Richard C. Tolman. Landmark study extends thermodynamics to special, general relativity; also applications of relativistic mechanics, thermodynamics to cosmological models. 501pp. 5⅜ x 8½. 65383-8

LIGHT SCATTERING BY SMALL PARTICLES, H. C. van de Hulst. Comprehensive treatment including full range of useful approximation methods for researchers in chemistry, meteorology and astronomy. 44 illustrations. 470pp. 5⅜ x 8½. 64228-3

STATISTICAL PHYSICS, Gregory H. Wannier. Classic text combines thermodynamics, statistical mechanics and kinetic theory in one unified presentation of thermal physics. Problems with solutions. Bibliography. 532pp. 5⅜ x 8½. 65401-X